W9-AUD-922

Climate Change

An Encyclopedia of Science and History

Climate Change

An Encyclopedia of Science and History

VOLUME 2: E–G

Brian C. Black, General Editor

David M. Hassenzahl, Jennie C. Stephens, Gary Weisel, and Nancy Gift, Associate Editors

Santa Barbara, California • Denver, Colorado • Oxford, England

Library of Congress Cataloging-in-Publication Data

Climate change : an encyclopedia of science and history / Brian C. Black, general editor ; David M. Hassenzahl, Jennie C. Stephens, Gary Weisel, and Nancy Gift, associate editors.

volumes cm

Includes bibliographical references and index.

ISBN 978-1-59884-761-1 (hardcover : alk. paper) — ISBN 978-1-59884-762-8 (ebook)

1. Climatic changes—Research—Encyclopedias. 2. Climatic changes—History—Encyclopedias. I. Black, Brian, 1966– editor of compilation.

QC903.C48 2013

363.738'7403—dc23 2012034673

ISBN: 978-1-59884-761-1

EISBN: 978-1-59884-762-8

17 16 15 14 13 1 2 3 4 5

This book is also available on the World Wide Web as an eBook.
Visit www.abc-clio.com for details.

ABC-CLIO, LLC
130 Cremona Drive, P.O. Box 1911
Santa Barbara, California 93116-1911

This book is printed on acid-free paper ♾
Manufactured in the United States of America

ENDSHEETS

Front recto: Colorado's Arapahoe Glacier as it appeared in 1898 (top) and again in 2003. (Glacier Photograph Collection/National Snow and Ice Data Center)

Front verso: Alaska's Columbia Glacier in 1980 (top) and again in 2005. (Glacier Photograph Collection/National Snow and Ice Data Center)

Back recto: Alaska's McCall Glacier in 1958 (top) and again in 2003. (Glacier Photograph Collection/National Snow and Ice Data Center)

Back verso: Peru's Qori Kalis Glacier in 1978 (top) and again in 2004. (Glacier Photograph Collection/National Snow and Ice Data Center)

Contents

List of Entries

VOLUME 2

VOLUME 4

E

Early Industrialization

Although energy is always an integral part of human life, there have been clear separations between eras, depending on humans' uses and applications of power systems. Of these various shifts, likely the most significant is the Industrial Revolution, which marked a fundamental shift in the way that humans did many things. Economic historians argue that it was profound because it remade methods of trade—goods moved between human communities with more regularity, increasing a whole set of diverse connections. Outcomes of these trade patterns, though, whether the Silk Road, the Atlantic System, the Panama Canal, or the Internet, are indicators of a radical change in the human condition—particularly in the human use and application of energy.

The intellectual shifts at the heart of industrialization preceded the high-energy era that created problematic emissions; however, the incorporation of early machines was a necessary precursor of burning fossil fuels. Predating larger-scale manufacturing that emphasized burning of fossil fuels, many undertakings in earlier phases of industrial development were much more passive in their energy input.

Reorganizing Human Life

Humans began the management and harvest of Earth's energy with their existence as hunter-gatherers. As their ability to manage and manipulate the surrounding natural systems matured, humans passed through what historians and archaeologists refer to as the Agricultural Revolution. This shift in human life occurred at different moments throughout the globe. Adapting to climatic variations, humans in different regions took control of the natural cycles of energy—primarily of the sun and photosynthesis—and learned to condition their behavior, resulting in a relatively consistent supply of food. Once food management had allowed humans to become more sedentary, they only had to make a slight adaptation to their living patterns in order to exploit and develop practices that we refer to as early industry.

Some renewable energy technologies, including water and wind power, were closely related to agricultural undertakings. These power sources have been used in milling for centuries. For instance, mills to grind grain into flour were powered by waterwheels since at least the first century. The Domesday Book survey of 1086 counted 5,624 mills in the south and east of England. Similar technology could be found throughout Europe and elsewhere and were used for milling or other tasks, including pressing oil or even making wire. Most often, each of these industrial establishments was an entirely local, limited endeavor. A few exceptions also grew well beyond the typical village center. For instance, the Romans built a mill with 16 wheels and an output of over 40 horsepower near Arles in France. In each case, though, the energy was harvested and applied to a specific activity; it did not necessarily alter the way most humans lived their daily lives.

The organization and adaptation that historians refer to as the Industrial Revolution, though, came much later when technical innovations grew to form dominant patterns in human life. First, however, these energy sources were utilized at areas where human and capital concentration made it more possible. Some of the earliest milling technology arrived in England through use in religious communities, including monasteries.

Monasteries at this time were self-sufficient religious communities, producing their own food and other goods. Often they were referred to as estates, and they seem to have resembled diversified plantations. One of these enterprises had monks turning wool into various forms of cloth. The name of this process was fulling. This process was revolutionized when the Cistercians at Quarr Abbey set up a mill that would full the wool by using waterpower. Although this was not the first fulling mill in England, historians credit it with initiating the enterprise on the Isle of Wight, which became world renowned for its kerseys, a coarse cloth made on a narrow loom.

Previously, the cloth would be placed in a trough filled with the fulling liquor, and then it would be walked on with bare feet to complete the process. With access to a waterwheel, the monks created a series of large wooden mallets that would pound on the fabric while it was in the liquid, making the cleaning process much more rigorous and even. Perfecting these methods allowed merchants to prepare for important shifts that were arriving in European history.

The wars of the Renaissance and Reformation eras proved to be a great boon for merchants and manufacturers supplying armed forces. Many of these new industries and systems of transportation would ultimately be put to peacetime uses as well. However, by most modern measures, the manufacturing taking place from the 1300s to the 1500s was of a very limited scale. Between 1500 and 1750, changes in manufacturing continued but would not accelerate remarkably until after 1750.

During this early era of manufacturing, most enterprises garnered energy from passive means, including from rivers and the wind. Each source of power proved extremely limited in energy and reliability. Of course, this meant that manufacturing also could not be reliable and could only expand to a limited level.

The manufacturing that did develop was most often based on technologies that European merchants brought from other regions, particularly from Asia. For instance, Europeans perfected the art of making porcelain imitations of Chinese crafts. And from India, Europeans imported methods for manufacturing silk and textiles. While perfecting these technologies, European business leaders also linked specialized craft production into larger-scale systems that also placed small batch into the class of manufacturing.

The basis for this system of manufacturing was improved energy resources. Ultimately the outcome was a large-scale shift in economic and social patterns in Europe that culminated with the formation of an entirely new organization to society. These living patterns ultimately led up to and fostered the Industrial Revolution.

New Opportunities and Confidence in the Human Mind

By approximately 1200, the economics of Western Europe had absorbed many of the technological and cultural details that Islam and the Orient had to offer. Ideas, techniques, and even actual inventions traveled back to Europe by way of trade networks. Cultural appropriation, of course, is an intrinsic part of human society. What really distinguished European society was what it did with this intellectual raw material. By building on ideas drawn from great societies of the world, European society surged ahead of every other region during the three centuries that followed.

New ways for developing ideas and technology into profitable opportunities defined the era after feudalism and led to the rise of guilds and craftsmanship that also stimulated invention. With these economic developments behind them, cities grew rapidly in medieval Europe and fueled new economic growth and the revival of trade. Most of these towns and cities owed their urban status to the presence of military garrisons during the Middle Ages. They were not very large and were distinguished from the surrounding countryside largely by the possession of protective walls. In the medieval era,

Following the work of Copernicus, Galileo, and Kepler, British scientist Sir Isaac Newton supported the theory that the heavens and the earth were part of one cohesive solar system with the sun at its center. (Perry-Castaneda Library)

though, trade networks made towns capable of growing at a rate unknown in previous centuries.

Intellectual changes initiated economic changes, and European society became increasingly willing to consider new ideas. Ultimately economic development contributed to an increasing desire for new technologies. In addition, the printing press provided an outlet for new ideas that quickly spurred innovations in thought and, eventually, new technologies as well. At the ground level, though, it was this revolution in thought that was essential to ushering in a starkly different era in evolution, such as the Industrial Revolution.

As a bridge between these eras of such different approaches to new technology and ideas, no one is more crucial than Sir Isaac Newton. In addition to his invention of the infinitesimal calculus and a new theory of light and color, Newton transformed the structure of physical science with his three laws of motion and the law of universal gravitation. As the keystone of the scientific revolution of the 17th century, Newton's work combined the contributions of Copernicus, Kepler, Galileo, Descartes, and others into a new and powerful synthesis of understanding. This new way of approaching knowledge formed a new foundation for human thought and technology.

For Newton, his intellectual quest is said to have begun in 1666 when he observed an apple falling from a tree in his garden at Woolsthorpe. From this moment, he derived the concept of universal gravitation. Newton's writings combined with others to stimulate further intellectual questioning. Growth and development became principles understood in whole new ways in European society. This same willingness to change and adapt ideas spurred incredible new developments in the design of new machines and technology. Unlike previous eras, though, the machines being developed after the late 1600s possessed an entirely new scale and scope.

Urbanization and Manufacturing

Technological innovations carry with them social and cultural implications of great importance. For instance, industry brought new importance for people to settle and live in clustered communities. Even limited energy development, such as the waterwheel, spurred urbanization in human history. Changes in manufacturing from 1300 to 1650 brought with it major alterations to the economic organizations of European society as well as the availability of goods and services. In addition, though, patterns such as urbanization helped to foster other factors that helped a singular innovation move into the realm of industrial development.

For instance, the growth of urban areas brought profound changes in banking and in the technology that supported manufacturing. A class of big businessmen arose and in connection with it an urban working class, often referred to as the proletariat. For this new urban society, new types of legal institutions and property tenure had to be devised. A mercantile law, or law merchant, class grew up to settle cases arising from trade disputes. Property holding was set free from the complex network of relationships and obligations that had burdened it, and it became possible for city dwellers to hold property outright. This liberation and flexibility of capital was critical to later economic developments.

One of the most distinctive characteristics of urban life was new freedom that had not been seen in the feudal countryside. Towns grew and flourished; trade, banking, and manufacturing became established on a new scale; and more and more persons achieved the legal status of free men. To accommodate these changes, vast tracts of land, which had been uninhabitable forest or swamp, were cleaned, drained, and subjected to cultivation. A new order and urgency came to the landscape of production that ushered in the scale and scope of industry.

Rapidly these early industries made flexibility a valuable commodity and increased the potential of undertakings that did not rely on geographical features such as wind, tidal flow, and river power. For instance, early industries began to quickly impact Europe's supply of wood. During this early period of industry, Western Europe's forests largely disappeared, as they served as the raw material for shipbuilding and metallurgy. This shortage led English ironmasters, however, to utilize a new source of energy that would greatly multiply the scale and scope of industrial potential. The English use of coal and more specifically of coke revolutionized the scale and scope of the manufacturing that followed throughout the world.

Although Western Europe had abundant supplies of ordinary coal, it had proven useless for smelting ore. Its chemical impurities, such as phosphorus, prohibited its generation of strong iron. For this reason, smelting was fired with charcoal, which was made from wood.

Western Europe's lack of wood made it lag behind other regions during these decades. However, in approximately 1709, Abraham Darby discovered that he could purify coal by partly burning it. The resulting coke could then be used as a smelting fuel for making iron. Darby released this knowledge for public use in 1750. This process proved to be a launching point for the reliance on fossil fuels that would power the Industrial Revolution.

The Intellectual Underpinnings of the Machine

The period introduced above, which lasted from 1500 and 1750, can best be described as one containing great technological developments but no genuine revolution of industrial expansion. In an era in which scientific and technological innovations were frowned upon and when energies and monetary support were focused on exploring the globe, it is relatively remarkable that any developments occurred at all. Simply, society of the Reformation was not conducive to new technological developments. The pressure to conform in this era slowed technological change and kept the implications of energy development fairly focused.

Social changes did occur, however, that bore a significant impact on later uses of technology. Industry began to move outside of cities. The nation-states that began to develop slowly became somewhat supportive of select technologies. In particular, technologies and machines that might be used in battle included designing fortifications, casting cannons, and improving naval fighting ships.

But more important to most members of society, during the 18th century a series of inventions transformed the manufacture of cotton in England and gave rise to a new mode or production that became known as the factory system. Based on a series of related innovations, the new factory-based society that took shape made machines part of nearly every worker's life.

During these years, other branches of industry stimulated comparable advances, and all these together, mutually reinforcing one another, made possible an entire era grown, at least partly, on the back of technological gains. The age would be organized around the substitution of machines for human skill and effort. Heat made from inanimate objects took over for animals and human muscle. Furthermore, this shift enhanced the amount—the scale and scope—of the work that could be undertaken.

After 1750, of course, the steam engine and related developments generated a bona fide industrial revolution. As Joel Mokyr has written, "if European technology had stopped dead in its tracks—as Islam's had done around 1200, China's by 1450, and Japan's by 1600—a global equilibrium would have settled in that would have left the status quo intact." Instead, in the next two centuries human life changed more than it had in its previous 7,000 years. At the root of this change lay machines and an entrepreneurial society committed to applying new technologies to everyday life, each one relying on new flexible and expandable sources of energy.

Case Studies of Early Industry

Waterpower in the Early Republic

Throughout much of the 1700s, the American colonies had defined themselves as the suppliers of raw materials to the industry of Europe. By the late 18th century, efforts abounded to keep the raw materials for our own profit by creating the nation's own industrial infrastructure. One of the first examples of such planning arrived in the 1790s with Alexander Hamilton's effort to develop Paterson, New Jersey. Fearing that it would lose its technological edge, England passed laws forbidding the export of machinery or the emigration of those who could operate it.

Despite these laws, one of the world's first brain drains occurred when laborers in the British textile industry secretly immigrated to the United States. Samuel Slater, who was born in England, became involved in the textile industry at 14 years of age when he was apprenticed to Jedediah Strutt, a partner of Richard Arkwright and the owner of one of the first cotton mills in Belper. Slater spent 8 years with Strutt before rising to oversee Strutt's mill. In this management position, Slater gained a comprehensive understanding of Arkwright's machines.

Believing that the textile industry in England had reached its peak, Slater posed as a farm laborer in order to secretly immigrate to America in 1789. While others with textile manufacturing experience had emigrated from England before him, Slater was the first who knew how to build as well as operate textile machines. Slater, with funding from Providence investors and assistance from skilled local artisans, built the first successful water-powered textile mill in Pawtucket in 1793. Slater's Mill was staffed primarily with children aged 7–12 years of age and women. The laborers worked with machines to spin yarn, which local weavers then turned into cloth. Slater added housing in order to attract poorer families to work in mills. His plan concentrated the workforce within easy walking distance to the mills. Slater also established company stores and paid the workers in credit that could only be used at the company stores. He also established churches and schools for his workers nearby. Slater had created a template for early industrial development in the United States.

The millwrights and textile workers who trained under Slater contributed to the rapid proliferation of textile mills throughout New England in the early 19th century. The Rhode Island System of small rural spinning mills set the tone for early industrialization in the United States. By 1800 Slater's Mill employed more than 100 workers. A decade later 61 cotton mills turning more than 31,000 spindles were operating in the United States, with Rhode Island and the Philadelphia region the main manufacturing centers. The textile industry was established, although factory operations were limited to carding and spinning. By the time other firms entered the industry, Slater's organizational methods had become the model for his successors in the Blackstone River Valley. Based on Slater's model, new models also quickly emerged on other American rivers.

The Merrimack River possessed the raw power to surpass the Passaic and the Blackstone as an industrial center. Located just outside of Boston, the

Slater Mill in Pawtucket, Rhode Island, in 1927. Slater Mill was the first cotton mill in the United States. (Library of Congress)

Merrimack became the next center of American industry when the businessman Francis Cabot Lowell used Slater's idea but exploded the scale in order to create an entire industrial community entirely organized around turning the power of the river into textile manufacture. The workable power loom and the integrated factory, in which all textile production steps take place under one roof, made the mill town of Lowell the model for future American industry.

The city's brick mills and canal network were, however, signs of a new human domination of nature in America. Urban Lowell contrasted starkly with the farms and villages in which the vast majority of Americans lived and worked in the early 19th century. Farming represented humans' efforts to work with and accommodate natural patterns; Lowell followed more of a bulldozer approach—Mill owners prospered by regimenting that world. For instance, they imposed a regularity on the workday that was radically different from the normal agricultural routine that followed seasons and sunlight. Mills ran an average of 12 hours per day, 6 days per week, for more than 300 days per year. Mill owners resisted seasonal rhythms in order to set their own schedule: operating the mills longer in summer or extending the winter workday with whale oil lamps.

The power behind the factory began with the river. Simply damming the existing waterway did not create enough power to run the mills. Lowell's industrial life was sustained by naturally falling water. At Pawtucket Falls, just above the Merrimack's junction with the Concord, the river drops more than 30 feet in less than a mile—a continuous surge of kinetic energy from which the mills harnessed over 10,000 horsepower. Without the falls, Lowell's success would have been impossible. In addition, however, Lowell relied on the construction of canals to better position the Merrimack's water. To increase efficiency, mill owners dammed it, even ponding water overnight for use the next day. Anticipating seasonal dry spells, planners turned the river's watershed into a giant millpond. They were aggressive in purchasing water rights in New Hampshire, storing water in lakes in the spring and releasing it into the Merrimack in the summer and autumn.

The rise of Lowell in the second quarter of the 19th century prompted the rhetoric of poets and politicians who homed in to make it a national model for development. Massachusetts governor Edward Everett wrote that the city's tremendous growth "seems more the work of enchantment than the regular process of human agency." The poet John Greenleaf Whittier described Lowell as "a city springing up . . . like the enchanted palaces of the Arabian Tales, as it were in a single night—stretching far and wide its chaos of brick masonry. . . . [The observer] feels himself . . . thrust forward into a new century." The city became an obligatory stop for Europeans touring the United States.

Although each of these examples relied on the power systems devised during the Middle Ages, American industrialists had made important new changes in harnessing waterpower. Before human labor could go to work in the mills of Paterson, Rhode Island, or Lowell, the water's power needed to be harnessed. The tool for managing this natural resource was the waterwheel or turbine. Until the second half of the 19th century, waterpower was the major mechanical power source in the United States.

Lowell also marked an important moment in the large-scale manipulation of a river for industrial use. In this case, water was channeled out of the river at a certain height in a power canal. This canal led to a point from which the water would fall to a lower level. During its fall, it fills the buckets in a waterwheel, its weight driving the wheel around. The turbine was later substituted for the waterwheel. The first turbines were designed by Uriah Boyden and adapted by James B. Francis to power Lowell's mills. In this system, the water entered the wheel at its center and was directed outward by stationary vanes to turn another set of moving vanes. By 1858, Lowell employed 56 Boyden turbines, each rated at 35 to 650 horsepower. In both the waterwheel and turbine systems, the power was transferred by wooden or metal gears and leather belts to the mill's main power shaft or drive pulley.

Making Iron in the Early Republic

Just as iron manufacture marked one of Europe's early industries, settlers also brought the undertaking to the New World. Iron plantations were one of the first inland industries introduced to North America. Americans began to expand iron making in the early 1700s. Many of these bloomeries, fineries, and furnaces were soon built west and north of Philadelphia. In each case, the industry was powered by wood, a renewable biofuel.

Often, British immigrants established these furnaces with the know-how they brought with them from industry abroad. Many of the blast furnaces were on plantations, which were largely self-sufficient communities with large landholdings to supply fuel, ore, and flux for running the furnaces. The process revolved around heating raw ore in order to create a more pure pig or bar iron. Usually, charcoal provided the best fuel. Therefore, other sites on the plantation would burn the felled lumber to convert it into charcoal.

The wood was hauled to the coaling areas and made into charcoal during the spring, summer, and autumn by skilled colliers. This was done by slowly charring it in pits, a careful process carried out to expel the tar, moisture, and other substances from the wood without consuming the wood itself. Once the process was complete, the charcoal was raked out, cooled, and taken by wagon to the furnace, where it was stored in the charcoal house.

Historians Patrick M. Malone and Robert B. Gordon explain that "An acre of woodland in sustained production on a 20-year rotation in the Middle Atlantic region yielded between 500–1200 bushels of charcoal. The largest annual consumption of charcoal among the New Jersey ironworks listed in 1850 was 200,000 bushels a year; so between 167 and 385 acres would have been cut each year, and between 3300 and 7700 would have sufficed for sustained operation." This ethic, of course, helped to make iron manufacturing a temporary mining industry.

The early iron industry stimulated the development of related industries, including small rail lines to move raw materials around a confined site. Some of these would combine with additional technological advances to make such furnaces obsolete by 1850. The primary reason, though, was the climactic shift in energy sources.

Energy Sources Fuel the Industrial Transition

What historians of technology refer to as the "great transition" is not necessarily the emergence of the Industrial Revolution in the mid-1700s. In order to reach that revolution, a great transition was necessary in intellectual thought and in the availability of energy resources. Biomass fuels such as wood and charcoal had been in use for centuries, but they did not necessarily support an entirely new infrastructural system of machines. Coal, on the other hand, emerged as a prime mover during the 1600s and did exactly that.

After England experienced serious shortages of wood in the 1500s, domestic coal extraction became the obvious alternative. Most of the existing coalfields in England were opened between 1540 and 1640. By 1650, the annual coal output exceeded 2 million tons. It would rise to 10 million tons by the end of the 1700s. Mining technology, of course, needed to be quickly developed to provide the fuel to power this new era. In the new energy resource of coal, industrialists found potential power that far exceeded any sources then in use. Thus, new industrial capabilities became possible. Primary among these was the steam engine.

The basic idea of the steam engine grew from the exploration of some of the revolutionary intellects of this new era in human history. Scientific minds were becoming more and more free to openly explore innovations that might significantly alter human life. For instance, the idea of the piston, which was the basis of the engine, only came about after the realization of the existence of Earth's atmosphere. Although other societies had thought about the concept of an atmosphere and pressure holding things to Earth, it was Europeans who began to contemplate the possibilities of replicating this effect in miniature.

In the mid-1600s English engineers began contemplating a machine that utilized condensation in order to create a repeating vacuum to yield a source of power. The first model of such a device is attributed to Denis Papin, who in 1691 created a prototype piston that was moved within a cylinder using steam. This device remained unreliable for use, though, because the temperature could not be controlled.

In 1712 Thomas Newcomen used atmospheric pressure in a machine that he alternatively heated and cooled in order to create the condensation pressure necessary to generate force. Additionally, Newcomen's engine was fairly simple to replicate by English craftsmen. Employed to pump out wells and for other suction purposes, the Newcomen engine spread to Belgium, France, Germany, Spain, Hungary, and Sweden by 1730. Although it lacked efficiency and could not generate large-scale power, the Newcomen engine was a vision of the future. It marked the first economically viable machine to transfer thermal energy into kinetic energy. This concept, powered by a variety of energy sources, was the flexible prime mover that would lead the Industrial Revolution.

The need for energy sources and the trade networks forming in the Atlantic provided another portion of the raw material to spread industry. Linked by ships, European powers sought necessary resources in other regions. Soon this led the mercantalist nations to establish colonies. In North America, settlement grew from agriculture; however, as the United States developed, it emphasized industries—using technologies perfected in Europe and new ones that blazed important new paths. The key connecting each undertaking was that energy was the necessary raw material to develop the young nation.

Conclusion: Tapping New Energy Resources

Between 1500 and 1750 there were great technological developments but no genuine revolution. These developments, however, formed a critical foundation—cultural, social, and economic—for what would happen later. Most important, cultural acceptance of innovation grew. In an era in which scientific and technological innovation were often frowned upon and when energies and monetary support were focused on exploring the globe, it is relatively remarkable that any developments occurred at all.

In this earlier era in industry, human skill and effort remained essential; however, machines were entering enterprises. Of course, this shift enhanced the amount—the scale and scope—of the work that could be undertaken. After 1750, the steam engine and related developments generated a bona fide industrial revolution. At the root of this change lay machines and an entrepreneurial society committed to applying new technologies to everyday life.

Brian C. Black

Further Reading

Beaudreau, Bernard. *Energy and the Rise and Fall of Political Economy.* Westport, CT: Greenwood, 1999.

Britnell, R. H. *The Commercialization of English Society, 1000–1500.* New York: Cambridge University Press, 1993.

Daumas, Maurice, ed. *A History of Technology and Invention,* Vol. 3, *The Expansion of Mechanization, 1450–1725.* New York: Crown, 1969.

Hunt, Edwin, and James M. Murray. *A History of Business in Medieval Europe, 1200–1550.* New York: Cambridge University Press, 1999.

Landes, David. *The Unbound Prometheus: Technological Change and Industrial Development in Europe.* New York: Cambridge University Press, 1969.

McNeill, William H. *Gunpowder Empires.* New York: American Historical Association, 1990.

Mokyr, Joel. *Twenty-Five Centuries of Technological Change.* New York: Harwood Academic, 1990.

Mokyr, Joel, ed. *The Economics of the Industrial Revolution.* Totowa, NJ: Rowman and Allanheld, 1985.

Mumford, Lewis. *Technics and Civilization.* New York: Harcourt, 1963.

Singer, Charles, et al., eds. *A History of Technology,* Vol. 2, *The Mediterranean Civilizations and the Middle Ages.* London: Oxford University Press, 1956.

Smil, Vaclav. *Energy in World History.* Boulder, CO: Westview, 1994.

Stearns, Peter N. *The Industrial Revolution in World History.* Boulder, CO: Westview, 1998.

Earth First! and Ecoradicalism

Serious environmental concerns fueled grassroots movements, such as that in Love Canal, to demand governmental or industrial action. Social movements such as the one initiated by Lois Gibbs were often linked together under the phrase "Not in my Backyard," or NIMBY. At the root of this movement was a change in middle-class Americans' expectations of their own safety and health. This represented just one faction of environmentalism, though.

Mainstream environmental nongovernmental organizations helped to clear the way for more extreme organizations that held ideas unpalatable to most middle-class Americans. Many critics claim that these extremists give environmentalism a bad name. The activities of environmental extremists, argue critics, make many Americans less likely to support environmental ideals.

One reason for this criticism is that in addition to holding extreme philosophical stands, many of these organizations also go about their activities in a much more confrontational manner. The best known of these extreme environmental organizations was Earth First!, which was led by Dave Foreman. These activists argued that protests and writing letters was not sufficient.

Earth Firsters sought out more active methods of action, which became known as ecoradicalism or ecoterrorism.

Earth First! is an international movement composed of small bioregionally organized groups of supporters. The goal of the movement is for each group to learn about the ecosystem in their region and identify the most immediate and serious threats to that ecosystem. The movement follows many of the ideas of Deep Ecology, a branch of ecological philosophy that views humans as an integral part of the environment rather than a superior entity. Intellectually, Earth Firsters and other extreme environmentalists also introduced ideas more extreme than those broadly entertained by environmentalists of previous eras. In the past, an individual such as Henry David Thoreau or John Muir presented an extreme philosophical stance, and interested Americans steered their minds in that general direction. Now, however, fairly large groups of Americans were willing to entertain concepts such as the need to focus less on human needs and more on those of other portions of nature.

One of the best known of these philosophical stances is referred to as deep ecology. Subscribers to deep ecology argue that nature contains its own purposes, energy, and matter and its own self-validating ethics and aesthetics. Calling themselves defenders of wilderness, such thinkers included Arne Ness, David Rothenberg, William Duvall, and George Sessions. For these deep ecologists, even preservation was based on science that grew from values that remained committed to the control of nature by industrial society. Deep ecologists urged environmentalists to turn their backs on society and adopt a radical new position for humans within nature. All decisions begin by asking what benefits the entire natural system. This holistic perspective prioritizes wildness over any utilitarian view of the environment. Earth First! recognizes an intrinsic value in all aspects of the environment and in all living things.

Unlike many mainstream environmental groups, today Earth First! considers itself a movement rather than an organization. Because of this, Earth First! does not have members but instead has participants who are drawn together by the belief that the life of Earth comes before the comforts, wants, and needs of humans. Additionally, within Earth First! the only leaders who exist are those participants who are achieving their goals.

Earth First! believes in using all types of tactics in order to achieve it goals. Thus, its participants can be found working to achieve their goals through educating the public, working through the legal system, and participating in creative civil disobedience. This creative civil disobedience has historically consisted of not only protests and demonstrations but also a variety of monkey-wrenching activities. Earth Firsters state that they are the

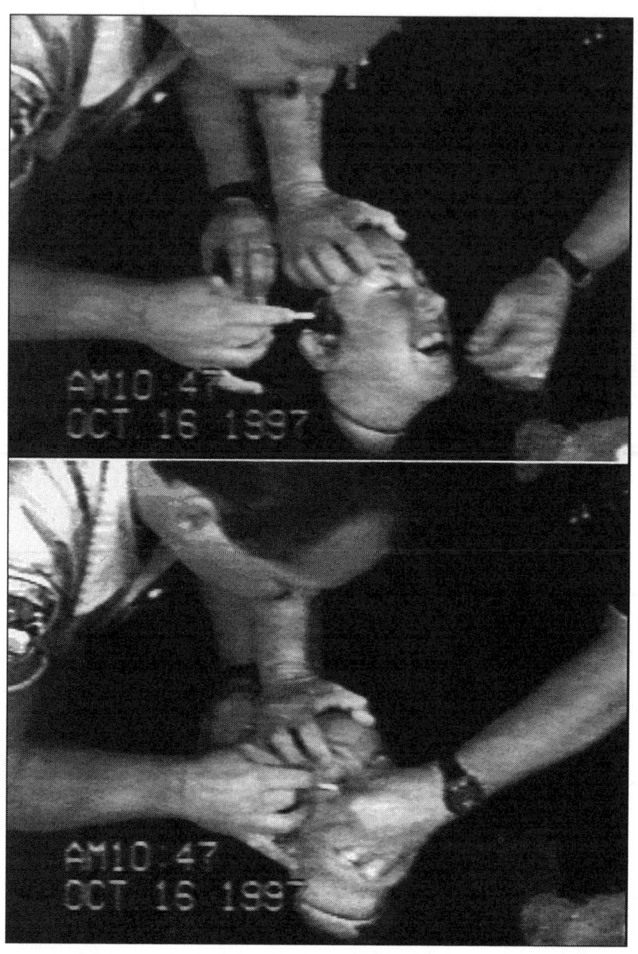

Still photographs from a videotape of Eureka, California police officers swabbing the eyes of a protester with pepper spray. The action occurred when environmentalists from Earth First! and the Headwaters Forest Defenders demonstrated against logging in the office of Congressman Frank Riggs on October 16, 1997. Although the tactics of the group Earth First! are considered by many to be extreme, the police officers' actions were widely believed to be an abuse of police power and an unreasonable response to civil disobedience. (AP/Wide World Photos)

most committed and uncompromising of environmental activists and as such may appear radical compared to other environmentalists. Despite this, Earth First! instructs its participants to weigh all options before taking actions that will get them arrested. Earth First! understands that although being arrested may draw more media attention to the issue at hand, once arrested, the Earth

Firster is no longer free and thus will not be able to continue taking actions to help inform others about the issue.

Earth First! was not so much created as named in 1980 when a group of activists were traveling across the southwestern desert from northern Sonora, Mexico, to New Mexico. Exacerbated by mainstream environmental groups and their willingness to make compromises concerning wilderness, the activists developed the idea of a movement to link together individuals working to help preserve ecologies across the United States. Specifically, the movement was meant to address the selling-out of the mainstream environmental groups. The event that drew particular ire from this group of activists was the U.S. Forest Service's Roadless Areas and Review Evaluation (RARE II) in 1979. The activists wanted to be a part of something that "put the Earth First." Drawing from the ideas of Rachel Carson, Aldo Leopold, and Edward Abbey, this group of activists moved forward with their desire for ecological preserves across the United States.

Carrying the banner of antiestablishment direct action, Earth First! claimed the slogan "no compromise in defense of Mother Earth." Initially, the organization adopted many of the tactics of the American Civil Rights Movement. Earth Firsters were inspired by the writing of Abbey, who in his 1975 book *The Monkey Wrench Gang* wrote of a band of activists wreaking havoc on development efforts in the American West. Earth Firsters swiftly intensified their actions to include ecotage (environmental terrorism), which became standard fare, particularly in the American West. During the first years, Earth First! took the initiative to propose biocentric-focused wilderness proposals—beyond the ideas being advocated by mainstream environmentalists—along with more creative actions such as putting a depiction of a crack onto the Glen Canyon Dam. At this time the movement also published *Earth First! The Radical Environmental Journal,* a periodical created to convey Earth Firsters' perspectives on environmental issues. Over time Earth First! participants became involved in more protests, particularly against logging in the Northwest. It was during one of these protests in Oregon that activist Mike Jakubal devised and carried out the first tree sit. The event lasted for less than a day, but it started a new form of struggle against the destruction of wildlife habitats and wilderness.

After the creation of tree sitting, Earth First! participants became mainly associated with actions to prevent logging and the construction of dams. This change attracted a new type of activist to Earth First!, bringing the movement closer to the ideas championed by anarchists and the counterculture. This change in direction made many of the original Earth Firsters uncomfortable. Thus, by 1990 many of the original members and most of the group of activists who created the movement had severed ties from Earth First! It was at this time when Earth First! changed from an organization with a leadership to a movement greatly influenced by anarchist philosophy. Another change to Earth First! occurred in 1992 when some participants pushed to have recognition as a mainstream movement. The Earth First! activists who did not want to give up their criminal acts started an offshoot focused more on monkey wrenching than Earth First! This group became known as the Earth Liberation Front.

By 1992, the Earth First! movement appeared to be made up of small groups and individuals who attend or set up protests and educational campaigns, along with supporting civil disobedience in the form of tree sitting, road blockades, and occasionally ecotage. However, more recently the movement, led by Judi Bari, has begun to renounce ecotage and work with small logging businesses in order to defeat large-scale corporate logging, particularly in northern California.

Of course, perspectives such as deep ecology made it impossible for subscribers to also support middle-class American ideas. Therefore, quite inadvertently, the radical end of environmentalism functioned to make the mainstream movement seem more reasonable.

Opponents of Earth First! have always focused on the radical element of the movement, pointing out the terrorism-like activities in which some of the members partake. These actions are responsible for economic losses to industries such as logging and mining, mainly due to protests that stop the extraction of resources. Proponents of Earth First! counter their critics with claims that the value of the environment they protected is greater than the economic losses taken by the industries. In addition, supporters of Earth First! also point to their more recent change to work with small local businesses as a means to show a move away from their more radical periods.

Over time the Earth First! movement has changed. The members' desire to put Earth above humans' comforts has remained, but the actions taken to support this effort have been adjusted to fit the people and issues involved.

Brian C. Black

Further Reading

Andrews, R. N. L. *Managing the Environment, Managing Ourselves.* New Haven, CT: Yale University Press, 1999.

Cronon, William, ed. *Uncommon Ground: Rethinking the Human Place in Nature.* New York: Norton, 1996.

Davis, J., ed. *The Earth First! Reader: Ten Years of Radical Environmentalism.* Salt Lake City, UT: Peregrine Smith Books, 1991.

Foreman, D. *Confessions of an Eco-Warrior.* New York: Three River, 1993.

Foreman, D. *Ecodefense: A Field Guide to Monkeywrenching.* 3rd ed. Chico, CA: Abbzugg, 1994.

Gottleib, R. *Forcing the Spring: The Transformation of the American Environmental Movement.* Washington, DC: Island Press, 1993.

Karliner, Joshua. "A Brief History of Greenwash." CorpWatch, March 22, 2001, http://www.corpwatch.org/article.php?id=243.

Kirk, Andrew. *Counterculture Green.* Lawrence: University Press of Kansas, 2007.

Lee, M. *Earth First! Environmental Apocalypse.* Syracuse, NY: Syracuse University Press, 1995.

Nabhan, G. P. *Coming Home to Eat: The Pleasures and Politics of Local Foods.* New York: Norton, 2002.

Naess, A. "The Shallow and the Deep, Long-Range Ecology Movement." *Inquiry* 16 (1973): 95–100.

Nash, Roderick. *Wilderness and the American Mind.* 4th ed. New Haven, CT: Yale University Press, 2001.

Opie, J. *Nature's Nation.* New York: Harcourt Brace, 1998.

Price, J. *Flight Maps.* New York: Basic Books, 2000.

Rothman, H. K. *The Greening of a Nation.* New York: Harcourt, 1998.

Rothman, H. K. *Saving the Planet: The American Response to the Environment in the 20th Century.* Chicago: Ivan R. Dee, 2000.

Scarce, R. *Eco-Warriors: Understanding the Radical Environmental Movement.* Chicago: Nobel Press, 2006.

Wall, D. *Earth First! And the Anti-Roads Movement: Radical Environmentalism and Comparative Social Movements.* London: Routledge, 2001.

Eco-Art: Loss of Habitat, Species Migration, and Extinction

We live in an age when human actions are responsible for an unprecedented degree of habitat destruction, whether directly or indirectly through changing climates. Many scientists have described the resulting devastation to Earth's wildlife as the sixth mass extinction, the only one so far caused by humans. As species disappear, concerned response sounds from all areas of the cultural spectrum. Over the last several decades, visual artists have been confronting these issues. Through imagination, critical thinking, and creativity, they investigate our place within the world and the seemingly unlimited potential of human impact on our larger environment, both for better and for worse.

The collaborative team of Helen and Newton Harrison, widely considered pioneers of the eco-art movement, have been making work that encourages environmental awareness since the early 1970s. In their recent piece *The Mountain in the Greenhouse* (2001), the Harrisons worked with Viennese biologist Georg Grabherr. Through video animation, the artists developed a visual image to illustrate Grabherr's discovery of Alpine plant migration up-mountain in response to warming temperatures. The video opens with a blue screen on which lines of white text appear one after another:

> This is a little drama entitled the Mountain in the Greenhouse.
> The theme is the disruption living systems will undergo
> As the perturbations of global warming
> Reverberate through the European high grounds.
> It is a drama being enacted
> In fast time if you happen to be a glacier.

The video then cuts to an image of a pristine snow-capped mountain with animated clouds gently rolling past in the background and a thermometer below the frame. One by one, square images of various flora, identified by their scientific names, appear on the screen and are correspondingly populated on the mountain face. The translucent frame of a greenhouse slides into place, enclosing the mountain. As the red bar of the thermometer rises, the plants move up the no longer snow-capped mountain until they eventually disappear altogether in the now dry and barren terrain. Interspersed between the images are black screens with white text, poetry based on conversations between the artists and the scientist detailing the dramatic narrative of this single mountain and the severity of the detrimental effects of climate change. In the final image of the video as the camera pans out and we view the scene from outside the greenhouse walls, we are left with the image of the arid mountain, now completely depopulated, behind the lines of the greenhouse architecture that have become a cell.

When the piece was shown in the 2007 exhibition *Weather Report: Art & Climate Change* at the Boulder Museum of Contemporary Art, curator Lucy Lippard described it as "one of those pieces where all of a sudden I could really grasp the whole concept." For most people, the discoveries of scientific research can seem rather peripheral to daily life. Artists, through the creation of tangible imagery, put form to abstract ideas and help propel

Maya Lin's *The Listening Cone*. (Bruce Damonte Photography / San Francisco Arts Commission.)

critical scientific discourse into a broader arena. Lippard puts it simply: "an emblematic image can really make a difference."

The work of another husband and wife team, Louisa Conrad and Lucas Farrell, who began their artistic collaboration in 2006, has focused on northern climates and the effects that climate is having on fragile environments in these regions. During an artist residency in 2009 in Skagastrond, Iceland, the couple began compiling *A People's Guide to Icelandic Butterflies.* The project is in association with 350.org, "an international campaign dedicated to building a movement to unite the world around solutions to the climate crisis."

Conrad and Farrell chose the butterfly as a focus in part due to the creature's sensitivity to subtle changes in its environment, which serves as an indicator of climate change. The artists note that Iceland has never offered a climate hospitable to butterflies, but with rising temperatures in northern regions, Iceland could soon become home to its first recorded butterfly populations. The artists write that "*A People's Guide to Icelandic Butterflies* seeks to capture and collect and present those imaginings. The *People's Guide* hopes to incite the creation of 350 Iceland-bound butterflies, thereby sparking discussion of current and impending ecological dynamics associated with climate change in the arctic." The call for contributions in the form of drawings submitted over the Internet at 350.org opened to the public in October 2010 and garnered responses from artists and schoolchildren throughout the world.

Another important contribution to this artistic movement is Maya Lin's *The Listening Cone* (2009). Lin, known for her design for the Vietnam Veterans Memorial in Washington, D.C., conceived of the sculptural piece as a memorial that "will focus attention on species and places that have gone extinct or will most likely disappear within our lifetime." The work takes the form of a large cone, like a simple megaphone, lying on its side. It is crafted from bronze and lined with reclaimed redwood. There is a screen at the narrow back end of the cone that is viewable from the larger opening. Video, accompanied by audio and text, is back-projected on the screen. The audio and images present wildlife species that are in decline, and the text offers simple information on these creatures and the condition of the planet's biodiversity.

Lin, discussing the larger project "*What Is Missing?*"—of which *The Listening Cone* is one component—states that "*Missing* is not just about specific extinct or endangered species. It's about absence, and it's about a more fundamental level of not knowing what we're losing, and that we need to link species loss to habitat loss and really focus as much on the habitat." Making these connections clearly and poignantly, as artists are able to do, is important to understanding and combating climate change.

Many other artists have worked within these topics and will continue to do so. This small sample of artists focusing on habitat loss, species migration, and extinction in relation to climate change provides a sense of the variety of this work. Imagination and creativity are central to human existence and also central to the development of solutions to social problems. The artist Joseph Beuys went as far as to claim that art is the only force capable of human evolution. Moreover, if the crisis of climate change is ever to capture the public consciousness in a significant way, it must be presented through powerful and visceral imagery. The movement to find solutions to this environmental crisis needs art. We need artists who will respond with skill and imagination so that we may *see* what climate change looks like.

Aaron Bos-Wahl

Further Reading

Conrad, Louisa. "A People's Guide to Icelandic Butterflies." October 24, 2010. http://www.350.org/en/icelandicbutterflies.

Dederer, Claire. "ART: Looking for Inspiration in the Melting Ice." *New York Times,* September 23, 2007.

Harrison Studio. *The Mountain in the Greenhouse—Video.* 2011. http://www.theharrisonstudio.net /?page_id=454.

Maya Lin Studio. "Memorials: What Is Missing?" 2010. http://www.mayalin.com/.

Lin, Maya. *What Is Missing?* 2009. Youtube.com/watch?v=LO3Qmgv1mLM&feature=related.

Trice, Emilie. "To the Thawing Wind." 2010. http://www.emilietrice.com/exhibitions/wind/wind.html.

What is Missing? 2010. http://www.whatismissing.net/.

Eco-Art and Ocean Acidification

In Earth's natural carbon cycle, a large portion of atmospheric carbon is absorbed by the oceans, where it reacts chemically with water to form carbonic acid, hydrogen ions, and carbonate that is metabolized to form the shells and skeletons of marine organisms. Over the last two centuries human activities have disrupted the equilibrium of this cycle, causing the oceans to become approximately 30 percent more acidic. The effects of ocean acidification are devastating for marine biodiversity, disrupting the food chain and corroding coral reefs, which, like terrestrial forests, provide a habitat for many other organisms.

The process of ocean acidification is difficult to visualize, and as a consequence it is a challenging subject for artists to represent. Approaches have ranged from documentary photography to direct action aimed at raising awareness of the problem, such as the work of John Quigley, who organized more than 100 boats to spell out "S.O.S. Acid Ocean" in the bay of Homer, Alaska, in 2009. In addition to Quigley's work, there are several artists dealing with the issue in various ways. Three mixed-media projects work to reconfigure our visual and sensory grasp of the problem of ocean acidification and imagine its future consequences.

In the Caribbean waters between Cancun and Isla Mujeres, Mexico, the British artist James de Caires Taylor has submerged on the ocean's floor a crowd of more than 400 human figures cast from a cross section of Mexican society. The vast underwater sculptural installation titled *The Silent Evolution* (2009–2010), one of four works that compose the Museo Subacuático de Arte (MUSA), is made of a special concrete compound that is meant to encourage colonization by coral. Like the shrinking coral reefs typically frequented by Caribbean tourists, the sculptures can be viewed through glass-bottom boats and underwater photography and by snorkeling or scuba diving. The work intervenes in the problem of coral reef destruction by a double means: first by providing a hospitable habitat for the formation of an artificial reef and second by drawing tourists and their destructive habits away from the remaining natural reefs—a disconcerting recognition of the fact that tourism is both a cause of environmental destruction and a motivation for its conservation. Whereas artists once strove to create in the manner of nature, Taylor seeks to restore nature's own dwindling power to sustain and regenerate itself in the face of human-induced environmental change. At the same time, the work helps to render the effects of ocean acidification intelligible by drawing a powerful visual analogy between the corrosive effects of acid rain on Mexico's archaeological sites and the effects of rising ocean acidity on coral reefs.

This underwater world is mirrored by the *Hyperbolic Crochet Coral Reef* (*HCCR*), initiated in 2005 by artists and sisters Margaret and Christine Wertheim, directors of the Los Angeles–based Institute for Figuring. The uncanny visual resemblance between live coral and the colorful crocheted pieces that compose this artificial reef, with their endless variations of curling lacelike patterns, is based in a precise mathematical identity. The project originated with the work of Dr. Daina Taimina, a mathematician who developed the use of crochet to create visual models of hyperbolic geometry, which describes the structure of corals, kelps, and other organisms. By sharing the techniques of hyperbolic crochet in a series of workshops, the Wertheims have spawned an international collaboration between hundreds of craftspeople, who have contributed to a series of exhibitions and created a series of independent satellite reefs in schools and communities worldwide. Without the need for a snorkel, the *HCCR* makes visible the extraordinary beauty of the myriad species of coral. The work also depicts the impact of acidification and rising ocean temperatures through pieces such as the *Bleached Reef,* crocheted in yarns of varying shades of white. Coral polyps live in symbiotic relationship with microscopic algae, which supply the polyps with crucial nutrients through photosynthesis and also give them their extraordinary colors. Under conditions of environmental stress, corals eject their symbiotic partners and become translucent, revealing their white calcium skeletons. Because they cannot survive long in this state, bleaching events, such as that which occurred in the Great Barrier Reef in 2005, serve as striking indicators of the degradation of marine environments. As a site of visual and tactile embodiment of these issues, the *HCCR* generates an accessible space for thinking through the human relationships and responsibilities across geographic and species boundaries.

The questions of how the environment will be affected by human activities and how in turn humans will adjust to a radically altered environment are projected into the future in Marina Zurkow's *Slurb* (2009), an animated video that imagines a world in which jellyfish thrive and people struggle to retain the ways of life that have constituted our species' identity. Drawing inspiration from scientific studies on the effects of climate change on ocean temperature and acidity and from fictional works such as J. G. Ballard's novel *Drowned World* (1962), Zurkow immerses the viewer in an otherworldly version of our own pressing reality. What is perhaps most evocative about the video is the manner in which its haunting music, a hypnotic chant combined with the endless crashing of waves and rainfall, blends with the visual movement to burn into the mind the rhythms of a world in a state of radical transition. It is as though through art and science we can begin to perceive by other means the inaudible sounds of our tread upon the reefs.

Dehlia Hannah

El Niño–Southern Oscillation

El Niño–Southern Oscillation (ENSO) is a fluctuation of the coupled ocean-atmosphere interactions across the Indo-Pacific ocean characterized by unusual warming (El Niño) or cooling (La Niña) of the ocean and the associated atmospheric changes (i.e., Southern Oscillation) in trade winds, precipitation, and circulation patterns.

It is generally accepted that the El Niño events last about 18 months and have a period of 2 to 7 years alternated with La Niña events. Understanding and forecasting ENSO events is critical, for it affects the weather globally and has major socioeconomic impacts at regional and global scales.

What's in the Name?

It is believed that the term "El Niño" (after the Christ child in Spanish) was originally used by the local fishermen communities in Peru in the 19th century to describe a warm current of water along the coast of South America during Christmas season. The term was later expanded by the scientific community in reference to an extensive warming across the east-central region of the Pacific Ocean.

Climate change is another ongoing issue in Micronesia, as it is throughout most Pacific Islands. Although Micronesia's partly mountainous terrain leaves it less vulnerable to rising sea levels (caused by global warming) than other Pacific Islands that are almost entirely at low elevations, Micronesia's low-lying areas are still under threat. Additionally, frequent El Niño cycles, which some scientists believe to be a by-product of climate change, cause violent storms throughout the islands and have resulted in freshwater shortages. In 1998 a climate change monitoring station was installed in Micronesia in collaboration with the Australian government, part of a chain of stations throughout the Pacific Islands.

What Happens during an ENSO Event?

In normal conditions, the average sea-surface temperature (SST) of the equatorial Pacific Ocean is cooler on the east side (South American coast) than the west side (Asia). This temperature asymmetry is associated with an atmospheric pressure asymmetry between the eastern Pacific (higher pressure, cooler temperatures) and the western Pacific (lower pressure, warmer temperatures). The interaction between the ocean and atmosphere systems promotes the westward movement of surface air (trade winds) and water off the South American coast. The water that accumulates on the western Pacific raises the height of the sea surface slightly (about half a meter) and results in a deeper thermocline on the western Pacific compared to the eastern side of the ocean. In addition to the increase of sea-surface height, the winds also drive the upwelling of cold water below the thermocline along the coasts of Peru and Ecuador, which in turn reinforces the east-west temperature/pressure asymmetry. This upwelling of cold and nutrient-rich water is critical for sustaining the primary productivity of the ocean along the South American coast, because it serves as a source of food for the plankton and fish populations (anchovies in particular).

The warm phase of an ENSO event (El Niño) is characterized by sustained SST anomalies above 2°C (4°F) or more between the eastern and central

Pacific Ocean (see figure 1a) paired with lower pressure over the eastern equatorial Pacific and higher pressure over the western Pacific than usual. The trade winds are weakened, which promotes an eastward movement of warm water. The eastward movement of water deepens the thermocline along the coasts of Peru and Ecuador. As a result, the upwelling of cold water is reduced, and the SST increases along the South American coast. This reduction on the upwelling of cold water limits the amount of nutrients and food available across the food chain and has had dramatic impacts on the local fishery communities in the past.

In contrast, the cool phase of ENSO (La Niña) is characterized by cooler than normal SSTs on the tropical eastern Pacific (see figure 1b). The trade winds are stronger than usual and push surface air and water westward due to the abnormally high atmospheric pressure on the eastern Pacific. During La Niña, the thermocline on the South American coast is shallower than normal, thus reinforcing the upwelling of cool nutrient-rich water from the deeper levels of the ocean.

"I have lived in El Nido most of my life and am a living testament to how the years have changed it. Before, we had a defined dry and wet season. Typhoons never reached our area. We always had fish and squid in abundance and boasted of a water system that reached all families—a year-round freshwater supply for rice fields and households.

The El Niño phenomenon that rocked the country in 1998 gave us our first experience of coral bleaching and its costly aftermath. We were very hard-hit because the fish yield has significantly decreased since then—and a lot of people have livelihoods that depend on the bounty of the sea. My brother used to fish in front of the town and as a family, we caught squid. Nowadays? You're lucky if you can come up with five or ten kilos.

Today, typhoons are common—even in the Calamianes islands up north. We bear the brunt of the heavy flooding and the soil erosion that comes with it. The coconut trees that once dotted our coastlines are no more and floods now reach the town. Freshwater is scarce now—it does not reach everyone. We are still trying to find a good water source. On top of all this, more and people are migrating to El Nido, further straining the resources.

Leonor Corral, mayor of El Nido, Phillippines, April 2010

Children collect water in buckets in order to clean out mud and debris left in their homes by flooding in Aguas Verdes, in northern Peru, Monday, March 9, 1998. Massive flooding, caused by heavy rains attributed to a weather phenomenon El Niño, forced thousands of Peruvians from their homes. These events are expected to come more frequently. (Scott Dalton/AP/Wide World Photos)

Major ENSO events and Associated Impacts

ENSO has major impacts in the tropics but also affects the weather worldwide due to changes in global precipitation and temperature patterns. The impacts of ENSO are quite complex, and although the duration and magnitude of ENSO events varies, scientists have observed strong consistencies among these events.

During the warm ENSO phase, the eastern equatorial Pacific experiences warmer conditions and unusually higher precipitation due to the rise and cooling of warm air masses. As a result, floods and landslides are more likely in Ecuador, Peru, and southern California (particularly between February and March). On the other hand, Indonesia, Malaysia, and northern Australia as well as southeastern Africa and northern Brazil experience drier conditions during the boreal winter, thus increasing the likelihood of forest fires and crop failures. Southern Mexico and the Caribbean also experience drier and warmer temperatures during the summers following El Niño events. Other

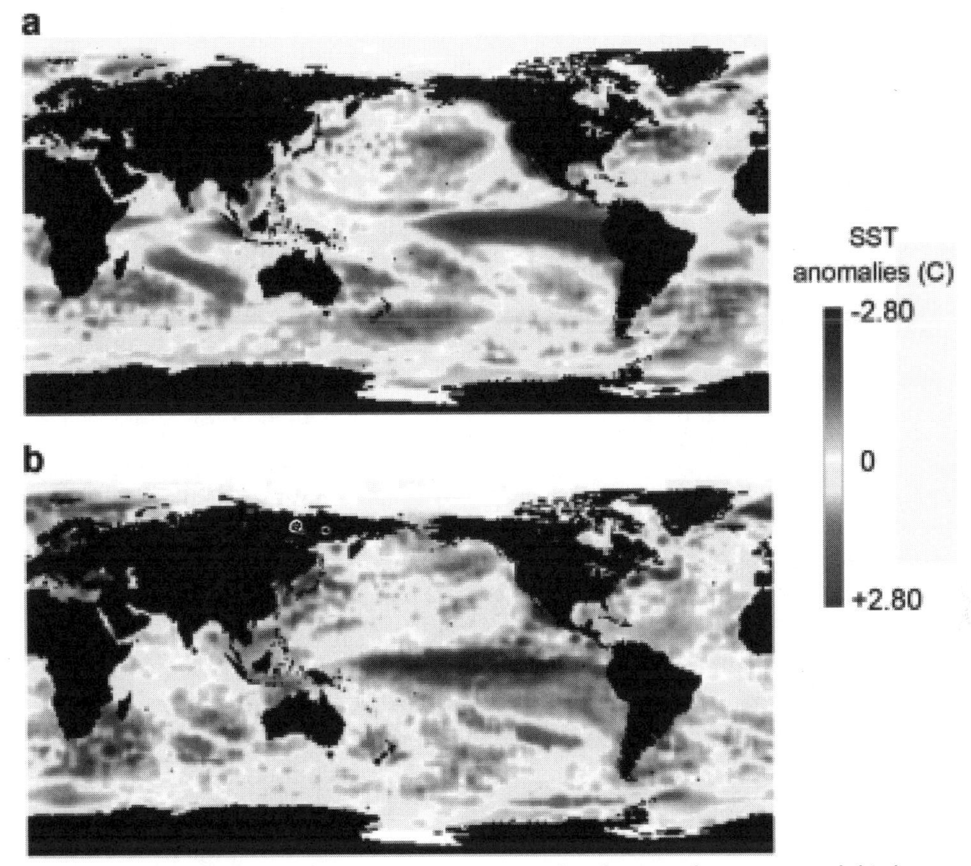

Figure 1 SST during (a) the ENSO warm episode of December 1997 and (b) the ENSO cool episode of December 1988
Base period for anomalies calculation: 1982–2009.
Source: NOAA Optimum Interpolation Sea Surface Temperature Analysis OI.v2 (Reynolds et al. 2002). Data provided by the Physical Sciences Division of the Earth System Research Laboratory, National Oceanic and Atmospheric Administration, Boulder, Colorado.

effects associated with El Niño are milder winters west of Canada and the northwestern United States as well as wetter than usual winters over the southeastern United States (Texas to Florida). At a local level, the warming of the ocean waters reduces the productivity of the ocean over South American coasts, which has had devastating effects on the fishery industry of Peru.

La Niña is associated with cooler and drier conditions in the eastern equatorial Pacific and wetter boreal winters in northern Australia as well as in Brazil, Indonesia, and Malaysia. Central East Africa (Liberia to Benin) is cooler, as is Southeast Africa, which is also wetter between December and

Climate-History Connection

The Little Ice Age in Southeast Asia

In the 17th century a group of interrelated factors—demographic, agricultural, political, economic, and climatic—resulted in a general global crisis. Parts of the Northern Hemisphere underwent a significant warming period during the late Middle Ages that then led to a period of cooling during the early modern era in what scholars call the Little Ice Age. Prior to the last 30 years, debates about the Little Ice Age (which lasted from approximately the 16th century to the mid-19th century) centered on Europe. Although scholars do not agree on the exact time period, current research indicates that climate change significantly impacted parts of the Southern Hemisphere as well, including Southeast Asia.

Climate change can broadly impact people's ability to produce crops and maintain healthy communities. The cooling period that led to the Little Ice Age resulted in a reduction of rainfall. This critical change in climate was disastrous for human communities, contributing to high food prices, disease, and high mortality. Unfavorable weather conditions also led to health epidemics, famine, harvest failures, and economic depression for people living in the region.

The climate in Southeast Asia was impacted by a variety of climate change patterns including volcanic activity, solar activity, increased El Niño–Southern Oscillation (ENSO) cycles, and decreased monsoon rainfall. Researchers have recently uncovered evidence that the droughts in Java during the 17th century and El Niño activity were connected. Volcanic activity also impacted the climate as volcanic ash exploded into the atmosphere, covering Earth and blocking normal solar radiation, which led to a cooling of Earth's temperature. Unlike the amount of climate data available for Europe, data from Southeast Asia is sparse, with less written documentation available. Therefore, our current understanding of the climate during the early modern period comes from piecing together evidence from various data sources.

Communities in India, Indonesia, and the Philippines depended on the cycle of wet and dry monsoons. Overflowing rivers provided resources for rice planting. A disruption in the growing season, such as the rice shortages in various parts of Indonesia (Makassar in 1624, Bali in 1633, and throughout other cities from 1673 to 1676), took a terrible toll on humans living in the area. Detailed records and tree-ring data show that the Chinese population dropped 40 percent between 1585 and 1645 due to severe drought conditions that caused the starvation of thousands and sparked a rebellion that marked the end of the Ming dynasty.

Indian records from the 17th century indicate that infrequent monsoons seriously depleted water supplies. Areas of India, Siam, Burma, and Thailand suffered harvest failures and famine in the late 1630s and 1640s due to drought. Analysis of tree-ring samples from Java's teak forests provides rainfall data for the period 1514–1929. The period 1598–1679 showed the lowest levels of rainfall. As a result, several regions experienced drought during specific periods: 1605–1616, 1633–1638, and 1643–1671, the latter being the most critical period with the least amount of rainfall.

Climate changes also contributed to increased disease in areas. Drought leads to crop failure, malnourishment, and a lack of clean water. Between 1613 and 1659, several regions experienced epidemics that significantly reduced the population, including Maluku (1613) and Java (1625–1626) in Indonesia, Kedah in Malaysia (1614), and Siam in Thailand (1659). Smallpox was the most common epidemic during this time. The tree-ring samples from Java showed that 1664 and 1665 were the driest years in a 400-year period. During those same years, most of Indonesia was greatly impacted by the worst epidemic of the early modern period.

The available data show correlations between drought and population decline, as evidenced by records from the Philippines and Indonesia indicating significant decreases in population during the driest periods and increases in population during the more favorable climate conditions of the 1680s and 1690s. The impact of the Little Ice Age and climate change in Southeast Asia in the centuries preceding and following the early modern period is not well understood. Scholars continue to uncover and interpret climate change data for Southeast Asia not only to understand the impact of climate during the early modern period but also to understand the long-term and complex effects of climate change in today's world.

Kim Kennedy White

Further Reading

Boomgaard, Peter. *Southeast Asia: An Environmental History.* Santa Barbara, CA: ABC-CLIO, 2007.

Chaujar, Ravinder Kumar. "Climate Change and Its Impact on the Himalayan Glaciers: A Case Study on the Chorabari Glacier, Garhwal Himalaya, India." *Current Science* 96(5) (2009): 703–708.

Lamb, Hubert H. *Climate, History and the Modern World.* London: Routledge, 1995.

Lieberman, Victor. *Strange Parallels: Southeast Asia in Global Context, c. 800–1830,* Vol. 1, *Integration on the Mainland.* Cambridge: Cambridge University Press, 2003.

Parker, Geoffrey, and Lesley M. Smith, eds. *The General Crisis of the Seventeenth Century.* London: Routledge, 1997.

Reid, Anthony. "The Seventeenth-Century Crisis in Southeast Asia." *Modern Asian Studies* 24(4) (1990): 639–659.

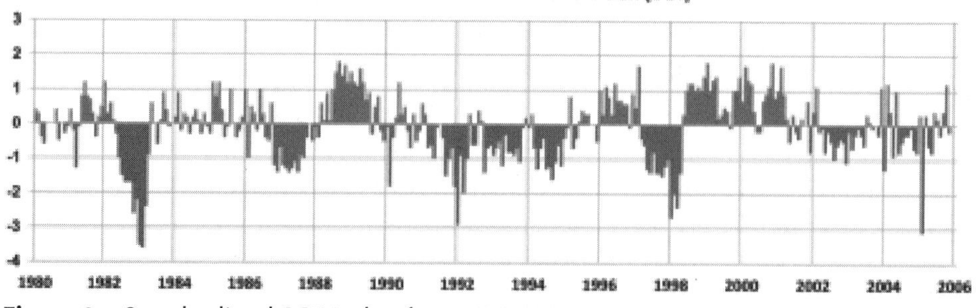

Figure 2 Standardized SOI index from 1980–2005
Base period for anomalies calculation: 1981–2010.
Source: Data provided by the Climate Prediction Center of the National Weather Service, National Oceanic and Atmospheric Administration, College Park, Maryland, USA, http://www.cpc.ncep.noaa.gov/data/indices/soi.

February. During La Niña, the winters are also cooler over western Canada and the U.S. Northwest, while the U.S. Southeast and northern Mexico are likely to experience drier and milder winters. In contrast, the Caribbean tends to be cooler in the summer months (June–August).

ENSO also has important implications for hurricane activity. During El Niño, the frequency of hurricanes is likely to increase over the eastern Pacific and decrease over the Atlantic, the Gulf of Mexico, and the Caribbean Sea. La Niña conditions, however, are associated with higher hurricane activity over the Atlantic, the continental United States, and the Caribbean Sea.

Measuring ENSO

Several indicators and indices are being used to measure the strength of the ENSO events based on atmospheric and ocean components:

Sea-Surface Temperature

El Niño events can be identified by measuring the SST along different established regions of the equatorial Pacific using tropical atmosphere ocean (TAO) moored buoys from NOAA and/or satellite imagery. These ocean regions are numbered from 1 to 4 from east to west. The Oceanic Niño Index (ONI) is an SST-based index calculated over the El Niño 3.4 region (5°N–5°S, 120–170°W) as a three-month running mean of SST anomalies based on both satellite and buoy/ships measurements. An El Niño event is defined as five consecutive months with +0.5 or above anomaly (–0.5 for La

Niña events). Anomalies in the range of ±1.5 or more characterize strong ENSO events.

Atmospheric Pressure

The Southern Oscillation Index (SOI) is one of the best-known descriptors of ENSO (figure 2). Its calculation is based on the fluctuations in normalized air pressure difference between Tahiti and Darwin (Australia). Consistent negative SOI values below –1, such as those observed from March 1997, indicate El Niño conditions, while prolonged positive values above 1 (e.g., 1989 and 1999) indicate stronger trade winds (La Niña conditions).

Elia Machado

Further Reading

"El Niño—Southern Oscillation (ENSO)." National Oceanic and Atmospheric Administration, http://www.cpc.ncep.noaa.gov/products/precip/CWlink/MJO/enso.shtml.

Howard, A. B., and J. E. Oliver, eds. *The Global Climate System: Patterns, Processes, and Teleconnections.* Cambridge: Cambridge University Press, 2006.

Reynolds, R. W., N. A. Rayner, T. M. Smith, D. C. Stokes, and W. Wang. "An Improved In Situ and Satellite SST Analysis for Climate." *Journal of Climate* 15 (2002): 1609–1625.

Trenberth, K. E., and D. P. Stepaniak. "Indices of El Niño Evolution." *Journal of Climate* 14 (2001): 1697–1701.

"Southern Oscillation Index Data." National Oceanic and Atmospheric Administration, http://www.cpc.ncep.noaa.gov/data/indices/soi.

"Tropical Atmosphere Ocean Project." National Oceanic and Atmospheric Administration, http://www.pmel.noaa.gov/tao/proj_over/map_array.html.

Emergence of Ecology

While conservation was taking active form during the early 1900s, an intellectual revolution was changing the scientific ideas of humans' relationship to the natural world. Ultimately these ideas would converge; however, during this era, conservation remained largely focused on managed development, and the science of ecology remained limited to the biology community. In fact, the public at large heard very little about new ideas of the ecological complexity of the world around humans. This would begin to change in the 1930s. The intellectual shift, however, had roots in the late 1800s.

Centered around the midwestern United States, a group of scientists grew in the first decades of the 20th century and eventually led to the development of the field of ecology. Henry Chandler Cowles, a plant ecologist, helped to lead this group of scientists. When he began graduate studies in 1895, his faculty at the University of Chicago introduced him to the ideas of a Danish scientist named Eugenius Warming. Cowles supplemented his study of botany with the science of physiography. This combination helped him to better appreciate the importance of landforms as a factor in shaping the patterns of plant life. He incorporated these combined approaches to form the theory of dynamic vegetational succession, which he first expressed in his 1898 PhD dissertation, "The Ecological Relations of the Vegetation on the Sand Dunes of Lake Michigan."

In his dissertation, Cowles used the southern shore of Lake Michigan, which is an area of beaches, sand dunes, bogs, and woods, to demonstrate that the natural succession of plant forms over time could be traced in physical space as one moved inland from the open lake beach across ancient shorelines through the shifting dunes to the interior forest. Along this route, scrubby beach grass would give way to flowers and more substantial woody plants, cottonwoods and pines would be seen yielding to oaks and hickories, and one would finally encounter the climax forest of beeches and maples. If left alone, Cowles argued, nature had a systematic structure of growth and development all of its own. Of course, this made it clear that humans were a disturbing agent in the natural world. Cowles's thesis had an immediate and far-reaching impact. Published serially in 1899 in the *Botanical Gazette,* "Ecological Relations" became one of the most influential works in American plant science and quickly established Cowles's reputation as a pioneering American ecologist.

Cowles went on to apply the theory of ecology that he had developed in the Indiana Dunes to the entire range of plant communities found throughout the Midwest. He demonstrated that the natural processes of succession and climax were not confined to the isolated dunes. This demonstrated that plant life in any setting had a great deal in common. Therefore, the patterns of change in plant communities could be more effectively tied to climatic or regional variables.

Cowles and his theories attracted students from all over the world. As these former students became active scientists, the influence of Cowles's ideas grew larger. One study of scientific influences by Douglas Sprugel in

1980 concluded that of the 77 recognized American scientists dominant in the field of ecology from 1900 to the early 1950s, no fewer than 46 were students of Cowles or were directly influenced by professional mentors who had been students of Cowles.

These new concepts possessed an intrinsically new way of viewing nature outside of human existence. Although it would not immediately impact American life, the intellectual principles clearly constructed a worldview that relegated humans to simply being one part of a larger natural story.

The International Phytogeographic Excursion of 1913

The United States acted as an incubator for some of these new scientific understandings that eventually moved abroad. One catalyst for the spread of these ideas occurred from July 1913 to September 1913, when American ecologists led by Cowles hosted the International Phytogeographic Excursion in America. The excursion was a scientific tour of significant natural environments in the United States by a visiting party of the leading European botanical experts of the time.

Members of the party traveled between cities by rail and made tours of local environments and plant communities. The route of the excursion was east to west, beginning in New York City on July 27. English botanist and ecologist Arthur Tansley reported on the International Phytogeographic Excursion and wrote that "Certainly no member of the international party will ever forget the overwhelming impressions we received of American landscapes and vegetation, designed truly on the grand scale." In particular, he praised the work of American ecologists when he wrote that "In the vast field of ecology America has secured a commanding position and from the energy and spirit with which the subject is being pursued by very numerous workers and in its most varied aspects, there can be little doubt that her present preeminence in this branch of biology—one of the most promising of all modern developments—will be maintained."

Ultimately ecology spread the concept of ecosystems throughout the biological sciences. This term "ecosystem" is credited to Tansley, who in the 1940s argued that nature occurred in self-sufficient (except for solar energy) ecological systems. He would go on to add that such systems could overlap and interrelate. The existence of such systems, of course, began to suggest that the human agent existed as the interloper in any system. Although the spread of ecological understanding among scientists was a significant change in humans' relationship with nature, the new ideas of ecology now needed to find their way to the general public.

Clements's Idea of Plant Succession

Cowles was not alone in recognizing the significance of dynamic succession for the study of plant communities. Independently, Frederic E. Clements of the University of Nebraska and the University of Minnesota developed the principles of ecology and based them on his studies of the grasslands and sandhills of Nebraska. Simultaneously, Cowles's former student Victor E. Shelford extended ecological theory into the realm of animal communities.

Clements began his ecological work shortly after Cowles published his influential work. In his own work, Clements argued that vegetation must be understood as a complex organism. Beginning in 1913, Clements and his wife established a lab on Pikes Peak in Colorado, where they began conducting systematic studies of plant succession (or self-replacement) in the surrounding mountains. Clements documented the environmental influences (temperature, sunlight, evaporation, etc.) on specific plants so that any shift could be explained. Their work demonstrated the complex interrelationship that mountain plants have with surrounding insects and animals. Ultimately his work established verifiable natural patterns for succession within plant species.

In Shelford's work, he applied Cowles's theories of ecology to the animal world. Specifically, Shelford analyzed the impact of ecological variables on the life histories and habits of tiger beetles. His later studies applied these ideas to species including fishes, moths, antelope, lemmings, owls, and termites. Ultimately Clements and Shelford combined their ideas of plant and animal ecology in *Bio-Ecology* (1939). Their essential argument was that neither plants nor animals exist in a vacuum; instead, they must be understood within a complex set of factors and variables. This was a crucial step toward creating the ecosystem concept that would organize ecological thought. Ultimately these ecological ideas would form the foundation of all of environmental thought.

Aldo Leopold and the Land Ethic

The ideal of wilderness received scientific definition through the growing science of ecology and the related development of the concept of ecosystems. One important figure who carried forward the ideas of Clements, Cowles, and others was Aldo Leopold. Eventually Leopold would be one of the earliest voices to bring these new scientific principles to the public.

After completing a degree in forestry at Yale in 1909 (from the School of Forestry begun by Gifford Pinchot), Leopold worked for the U.S. Forest

The influential writings of 20th-century environmentalist Aldo Leopold changed the way that many Americans viewed nature. An employee of the U.S. Forest Service, Leopold was instrumental in the creation of Gila National Forest, the nation's first protected wilderness area. (Library of Congress)

Service for 19 years. Primarily he worked in the Southwest (New Mexico and Arizona) until he was transferred in 1924 to the Forest Products Lab in Madison, Wisconsin. It was while working in the Gila National Forest that Leopold came to a new understanding about the role of the U.S. Forest Service in managing nature. In *A Sand County Almanac* (1949), he writes:

> We were eating lunch on a high rimrock, at the foot of which a turbulent river elbowed its way. We saw what we thought was a doe fording the

> torrent, her breast awash in white water. When she climbed the bank toward us and shook out her tail, we realized our error: it was a wolf. A half-dozen others, evidently grown pups, sprang from the willows and all joined in a welcoming melee of wagging tails and playful maulings. What was literally a pile of wolves writhed and tumbled in the center of an open flat at the foot of our rimrock.
>
> In those days we had never heard of passing up a chance to kill a wolf. In a second we were pumping lead into the pack, but with more excitement than accuracy; how to aim a steep downhill shot is always confusing. When our rifles were empty, the old wolf was down, and a pup was dragging a leg into impassable side-rocks.
>
> We reached the old wolf in time to watch a fierce green fire dying in her eyes. I realized then, and have known ever since, that there was something new to me in those eyes—something known only to her and to the mountain. I was young then, and full of trigger-itch; I thought that because fewer wolves meant more deer, that no wolves would mean hunters' paradise. But after seeing the green fire die, I sensed that neither the wolf nor the mountain agreed with such a view.
>
> Since then I have lived to see state after state extirpate its wolves. I have watched the face of many a newly wolfless mountain, and seen the south-facing slopes wrinkle with a maze of new deer trails. I have seen every edible bush and seedling browsed, first to anaemic desuetude, and then to death. I have seen every edible tree defoliated to the height of a saddlehorn. Such a mountain looks as if someone had given God a new pruning shears, and forbidden Him all other exercise. In the end the starved bones of the hoped-for deer herd, dead of its own too-much, bleach with the bones of the dead sage, or molder under the high-lined junipers.

In 1928 Leopold quit the Forest Service, and in 1933 he was appointed professor of game management in the Agricultural Economics Department at the University of Wisconsin–Madison. Leopold taught at the University of Wisconsin until his death in 1948. While in Wisconsin, he purchased a worn-out farm and began to experiment with ways of reinvigorating the soils and of managing the site as a cohesive ecosystem. His efforts may have been the first such ecological preservation effort in the United States.

While working in Wisconsin, Leopold wrote his best-known work, *A Sand County Almanac.* A volume of nature sketches and philosophical essays, the *Almanac* is now recognized as one of the enduring expressions of an ecological attitude toward people and the land. Within its pages, Leopold penned a concept known as the land ethic, which was rooted in his perception of the human's need to see itself as one component in a larger environment. Ultimately the land ethic simply enlarges the boundaries of the human community to include soils, waters, plants, and animals or, collectively, the land.

Conclusion: Inspiring Modern Environmentalism

Contemporary environmentalists use this ethic as a way of measuring the impact and implications of human activity on the environment. They ask: Can we say that we are giving equal standing to soils, bugs, and the air and water? Can we give standing to an element of nature—such as a species of snail or a fish—when it slows or limits human opportunity and development? The land ethic strives to give more equal standing to these other elements of nature. Contemporary environmentalists who are able to fully commit to this ethic often refer to themselves as deep ecologists. Throughout the late 20th century, though, some variation of Leopold's land ethic could be found in much of the environmental policies that became a normal part of American life.

While Leopold's writing would construct a new environmental ethic for generations to come, federal programs had begun applying some of the lessons of the new ecology by the 1930s. However, these early efforts rarely reached far enough to meet with approval from Leopold and other idealists. The basic lessons, however, formed the foundation for environmental thought moving forward to consider larger systematic issues related to ecology. By the end of the century, chief among these issues would be climate change.

Brian C. Black

Further Reading

Andrews, R. N. L. *Managing the Environment, Managing Ourselves.* New Haven, CT: Yale University Press, 1999.

Cronon, William, ed. *Uncommon Ground: Rethinking the Human Place in Nature.* New York: Norton, 1996.

Gottleib, R. *Forcing the Spring: The Transformation of the American Environmental Movement.* Washington, DC: Island Press, 1993.

Leopold, A. *A Sand County Almanac, and Sketches Here and There.* 1948; reprint, New York: Oxford University Press, 1987.

Karliner, Joshua. "A Brief History of Greenwash." CorpWatch, March 22, 2001, http://www.corpwatch.org/article.php?id=243.

Kirk, Andrew. *Counterculture Green.* Lawrence: University Press of Kansas, 2007.

Nash. *Wilderness and the American Mind.* New Haven, CT: Yale University Press, 1982.

Opie, J. *Nature's Nation.* New York: Harcourt Brace, 1998.

Price, J. *Flight Maps.* New York: Basic Books, 2000.

Rothman, H. K. *The Greening of a Nation.* New York: Harcourt, 1998.

Rothman, H. K. *Saving the Planet: The American Response to the Environment in the 20th Century.* Chicago: Ivan R. Dee, 2000.

Worster, D. *Nature's Economy.* New York: Cambridge University Press, 1994.

Emiliani, Cesare (1922–1995)

A world-renowned figure in paleoclimatology (the study of past climate) and paleoceanography (the study of oceans in the geological past), Cesare Emiliani was born on December 8, 1922, in Bologna, Italy, to Luigi and Maria Emiliani. The younger Emiliani studied geology at the University of Bologna and received his DSc degree in 1945. As a micropaleontologist during 1946–1948 at Societa Idrocarburi Nationali, Florence, he began to publish research papers. His doctoral research was conducted at the Department of Geology at the University of Chicago during 1948–1950, and for the subsequent six years he remained there as a research associate. While working as an associate in the Geochemistry Laboratory of the Enrico Fermi Institute for Nuclear Studies under the famous scientist Harold Urey, Emiliani worked on the isotopic composition of carbon and oxygen in shells of marine organisms. In 1957 he began his career at the University of Miami and continued there until his retirement in 1993.

Emiliani's research in the 1950s was concerned with the paleoclimatology of the Pleistocene epoch and threw considerable light on the glacial cycles of the late Cenozoic era. The foraminifera (microscopic shells) deposited in ancient sediments of the ocean floor held many mysteries. Emiliani investigated the ratio of certain oxygen isotopes (specifically ^{18}O, as compared to the more common ^{16}O) of foraminifera within samples taken from ocean sediments to deduce the past temperatures of the ocean. The oxygen isotope cycles of long sediment cores gave evidence of many glacial and interglacial periods. The earlier accepted theory that there were 4 major glacial cycles during the Pleistocene epoch proved to be incorrect. Emiliani proved that instead there were 36 glaciations during the last 3 million years. He also showed that the temperature of the deep ocean had decreased considerably during the Cenozoic era (65 million years ago to the present) and that the temperature of the ocean was not constant, as had been earlier believed. Emiliani's study of Pleistocene glaciations and his contribution to knowledge about climatic cycles were milestones in paleoceanography.

Emiliani found that the Institute of Marine Science at the University of Miami was an ideal place for conducting research. His work required ships and technical personnel for drilling long sediment cores from the bottom of the sea. At the institute, Emiliani revamped the marine geology and geophysics programs and expanded the laboratory for isotope geology. His proposal to the U.S. National Science Foundation was accepted, and a ship called *Submarex* was provided to drill long cores on the Nicaraguan Rise. After a successful drilling, it became an established fact that drilled cores from ocean sediments would be a great help for studying the history of the ocean. Consequently, the program of the Joint Oceanographic Institutions for Deep Earth Sampling (JOIDES) was established, which undertook successive drilling work, first near Jacksonville, Florida (1966), and later the Deep Sea Drilling Project (1967–1983) and the Ocean Drilling Program (1984–2003).

Ocean floor samples collected all over the world have shown that there have been many sudden advances and retreats in global temperature and that Earth's climate has rarely been stable. Emiliani's technique became a standard procedure for interpreting the cyclic ice ages. In addition to demonstrating that research on deep-sea organisms can establish the history of Earth's climate, his work was a crucial contribution to the general understanding of climate variability. He was the recipient of many awards, including the Vega Medal from Sweden (1983) and the Alexander Agassiz Medal of the National Academy of Sciences (1989). A species of coccolithophorid algae, *Emiliania huxleyi,* the most abundant algae inhabiting Earth's oceans, was named in honor of Emiliani and Thomas Huxley.

Emiliani was not an armchair scientist but was concerned about contemporary problems faced by humankind, particularly the population explosion and its impact on the environment. He desired a rapport between scientists and the general public and authored useful books meant for the public as well as scientific researchers. For example, his *Dictionary of the Physical Sciences* (1987) defined various terms in astronomy, chemistry, physics, the geological sciences, cosmology, and many other scientific fields and was useful for students, teachers, and persons with a general interest in science. His book titled *The Scientific Companion* (1988) also contained a general survey of science. Another book, *Planet Earth* (1992), covered many branches of science within a framework of evolution and had the agenda of finding the means to mitigate the dangers faced by human beings and the environment. This informative and readable book encompassed almost everything from atoms to human beings. It also covered the origin of geoclines, theories pertaining to midocean ridges, the role of viruses in evolution, the origin of the moon, and many other topics. After going through the book, the reader would have a fair idea about mathematics, physics, chemistry, biology, cosmology, mineralogy, petrology, seismology, plate tectonics, and oceanography.

Apart from being a renowned scientist, Emiliani was well versed in many languages as well as in classical literature and history. He was against all sorts of doctrine. Attempts were made by him to eliminate the chronology based on BC ("Before Christ") and AD ("In the year of our Lord"), wishing to free the calendar of religious connotations. Truly a remarkable scientist and person, Emiliani died suddenly due to heart failure on July 20, 1995, in his home at Palm Beach Gardens, Florida.

Patit Paban Mishra

Further Reading

"Cesare Emiliani." Department of Geological Sciences, University of Miami, http://www.as.miami.edu/geology/emiliani.

"Cesare Emiliani (1922–1995): The Founder of Paleoceanography." National Oceanography Centre, Southampton, http://www.noc.soton.ac.uk/soes/staff/tt/eh/ce.html.

Emiliani, Cesare. *Dictionary of the Physical Sciences: Terms, Formulas, Data.* New York: Oxford University Press, 1987.

Emiliani, Cesare. *Planet Earth: Cosmology, Geology, and the Evolution of Life and Environment.* New York: Cambridge University Press, 1992.

Emiliani, Cesare. *The Oceanic Lithosphere.* New York: Wiley, 1981.

Emiliani, Cesare. *The Scientific Companion: Exploring the Physical World with Facts, Figures, and Formulas.* New York: Wiley, 1995.

Emiliani, Cesare, et al. *Earth Science.* Orlando, FL: Harcourt Brace Jovanovich, 1989.

Emissions Trading Scheme

Launched in 2005, the European Union (EU) Emissions Trading Scheme (or Emissions Trading System) (ETS) was the first multilateral emissions trading system in existence and at the time of this writing remains the largest. Under the scheme, all major emitters of greenhouse gases (GHGs) are required to monitor and report their CO_2 emissions and to return equivalent emission credits/allowances to the government. Surplus allowances can be sold on European and international exchanges and, by the same token, must be purchased to cover any deficits. Under the ETS, the governments of the EU member states agree on national emission caps, which have to be approved by the European Commission. Governments of the member states then allocate allowances across the relevant sectors and track and validate the actual emissions accordingly. To date, ETS trading periods (multiyear allocation periods have been agreed on due to annual weather fluctuations) are considered in three distinct phases, each of which have slightly different conditions: Phase I (2005–2007), Phase II (2008–December 2012, covering the period of the Kyoto Protocol), and Phase III (January 2013 onward). For the first two EU ETS phases, the total quantity of GHG credits allocated by each member state is defined in each member state's National Allocation Plan (NAP). The European Commission has oversight of the NAP process as per Annex III of the Emission Trading Directive (EU Directive 2003/87/EC), which takes the respective Kyoto targets into account.

In Phase I (2005–2007), the allocations were made freely via NAPs and covered around 12,000 installations in the energy and industrial sectors, accounting for about 40 percent of the EU's total CO_2 emissions. (All units in relevant sectors with a net heat excess of 20 megawatts [MW] or more were included.) This phase was originally created to operate apart from international mechanisms such as the Kyoto Protocol (and in fact slightly predated it). The EU subsequently agreed to incorporate some level of "flexible mechanism certificates" under the Kyoto Protocol into the ETS via a Linking Directive. These notably include Article 6 (Joint Implementation [JI] projects), Article 12 (Clean Development Mechanism [CDM] projects), and Article 17 (International Emissions Trading certificates) of the Kyoto Protocol. In effect, therefore, the units and mechanisms of the ETS were made compatible with the designs agreed through the United Nations Framework Convention on Climate Change. One EU Allowance Unit of one tonne of CO_2, or EUA, was designed to be identical (fungible) with the Assigned Amount Unit (AAU) of CO_2 defined under Kyoto, making them in principle tradable within the same system (which in practice was not possible before 2010). In the treaty's first year, 362 million tonnes of CO_2 were traded on the market for a sum of €7.2 billion, including in the form of futures and options. The price of allowances increased to its peak level in April 2006 of about €30 per tonne CO_2, but the market collapsed in the face of the generous emissions caps being offered by some countries. The price declined to €0.10 in September 2007. Verified emissions have actually seen a net increase over the first phase of the scheme, leading to accusations of governments buckling under industry pressure. This phase was subsequently defined as a "learning phase" by supporters of the EU ETS.

In Phase II of the EU ETS (2008–December 2012), the operators within each member state must surrender their allowances for inspection by the EU before they are retired. The second phase also expands the scope of the scheme considerably by operationalizing CDM and JI credits through the EU's Linking Directive, including aviation emissions and also including three non-EU members (Norway, Iceland, and Liechtenstein).

Ultimately the goal of the ETS is to include all GHGs and all sectors, including aviation, maritime transport, and forestry. For the transport sector, implementation is planned either as a cap and trade system for fuel suppliers or a baseline and credit system for car manufacturers. The National Allocation Plans for Phase II, the first of which were announced on November 29, 2006, were meant to result in a cut of nearly 7 percent below the 2005 emission levels. The use of offsets from CDM and JI projects, however, meant that in practice no actual cuts were needed to meet the Phase II cap. According to verified EU data from 2008, the ETS saw an emissions reduction of 3 percent, or 50 million tons. At least 80 million tons of "carbon offsets" were bought as part of the scheme.

In the face of this, the European Commission proposed various changes in a January 2008 package, including the abolishment of NAPs beginning in 2013 and auctioning a far greater share (60 percent in 2013, increasing thereafter) of emission permits. From the start of Phase III (January 2013) there will be a centralized allocation of permits rather than individual National Allocation Plans, with a greater share of auctioning of permits. The main changes proposed are setting an overall EU cap (with allowances then allocated to EU members), setting tighter limits on the use of offsets, allowing the banking of allowances between Phases II and III and continuing the use of CDM and JI credits to ensure flexibility, shifting from a stress on allowances to one on auctioning, and increasing coverage (of about 100 million metric tons of CO_2 equivalent) by adding the petrochemical, ammonia, and aluminum industries and further GHGs (including nitrous oxide and perfluoro compound emissions).

The UK Climate Change Committee, in its first report in 2008, estimated that to achieve a 20 percent cut in EU emissions (relative to 1990 levels), the reduction in total emissions required would be around 36 million tonnes per annum. In its second report in 2009, it projected a carbon price in 2020 of around 22 euro per tonne of CO_2. But these projections are subject to great uncertainty (e.g., over future fossil fuel prices), and many commentators are pessimistic about the future, predicting business-as-usual rates of emission. And, of course, at the time of this writing, these are proposals rather than agreements and may not all be implemented, particularly in the face of the ongoing economic difficulties.

Ranjan Chauduri

Further Reading

Betz, Regina, and Misato Sato. "Emissions Trading: Lessons Learnt from the 1st Phase of the EU ETS and Prospects for the 2nd Phase." *Climate Policy* 6 (2006): 351–359.

Committee on Climate Change. *Building a Low-Carbon Economy: The First Report of the [UK] Committee on Climate Change.* London: TSO, 2008. http://www.theccc.org.uk/pdf/TSO-ClimateChange.pdf.

Committee on Climate Change. *Meeting Carbon Budgets: The Need for a Step Change.* Progress Report to the UK Parliamentary Commission on Climate Change, October 12, 2009. http://www.official-documents.gov.uk/document/other/9789999100076/9789999100076.pdf.

Ellerman, A. Denny, and Barbara K. Buchner. "The European Union Emissions Trading Scheme: Origins, Allocation, and Early Results." *Review of Environmental Economics and Policy* 1(1) (2007): 66–87.

European Commission. "The EU Climate and Energy Package." European Commission, 2011, http://ec.europa.eu/clima/policies/package/index_en.htm.

European Commission. "Proposal for a Directive of the European Parliament and of the Council Amending Directive 2003/87/EC So as to Improve and Extend the Greenhouse Gas Emission Allowance Trading System of the Community." Brussels, January 23, 2008.

Energy Conservation and Efficiency

Energy efficiency can be accomplished in many ways, the most simple of which is through conservation, or just using less energy. This means, for example, turning off lights when not in use, not idling the car, keeping the thermostat low in the winter, or air-drying clothes instead of using a dryer. Energy efficiency also has many commercial and industrial applications, whether it is having fewer lights in a store, for example, or altering

> The challenge we face in international development is this. Can countries continue to grow and prosper in a way which uses energy and resources in a different way? Can energy be used more sparingly? Can cleaner ways be found to generate it? Can agricultural techniques, building designs, social support systems, insurance packages—be developed to help the world's poorest people cope with more extreme weather and natural disasters? Can governments develop the incentives that could unleash the transformative power of the private sector?
>
> This should be the most inspiring, exciting and overwhelming series of challenges to today's generation of bright, young people. As well as to wise, experienced, older hands. We are at the threshold of nothing less than a new industrial, agricultural and technological revolution.
>
> We know from previous industrial revolutions, that investment flows to where the leadership is. Whichever country seizes the opportunity presented by low-carbon growth, will reap the economic reward. The same is no less true of companies and citizens.
>
> *Andrew Mitchell, member of Parliament, United Kingdom, 2010*

a manufacturing process so that it runs more efficiently. As technology advances, so too are ways of conserving energy, and the methods are becoming more elegant and intricate.

Lighting

While there are many ways to conserve energy and use it efficiently, oftentimes lighting is the go-to example that is used because it can clearly illustrate the progression of energy efficiency. Conventionally, many consumers use incandescent lightbulbs, which work by using electricity to heat a tungsten filament in the lightbulb until it glows. Most of the energy consumed by this type of lightbulb is given off as heat, which gives it a low lumens (a measure of the power of light as perceived by the human eye) per watt (a measure of the rate of energy conversion, defined as one joule per second) rating, about 10–17 lumens per watt. They also have a fairly short operating life, lasting around 750 to 2,500 hours. While they are the least expensive to buy, incandescent bulbs are generally more expensive to operate.

Fluorescent lamps, including compact fluorescent lightbulbs (CFLs), use a quarter to a third of the energy that an incandescent lightbulb uses to provide the same amount of light. Fluorescent bulbs have a 30–110 lumens per watt rating, and they also last about 10 times longer, about 7,000 to 24,000 hours. Fluorescent lightbulbs work by an electric current that is conducted through mercury and inert gases. To regulate the current while operating and to provide a high start-up voltage, these lamps have a ballast, which makes this type of lighting more expensive. When CFLs were first introduced, they were criticized for becoming bright too slowly and emitting a harsh white light, but technology advances have led to a quicker start-up time and a softer white light that is more like the light output seen with incandescent bulbs.

Light-emitting diodes (LEDs) are the most energy-efficient and quickly developing lighting technology. They use at least 75 percent less energy than incandescent lightbulbs and last 25 times longer. If LEDs became a widespread technology, by 2027 they could save about 348 terrawatt hours of electricity, which is equivalent to the annual output of 44 large electric power plants. LEDs produce light in the form of photons, and the color of the light is determined by the energy of the photon. LEDs can have a wide range of applications, from stadium lighting to a status indicator on a television. Those that are designed for room lighting are more expensive than CFLs or incandescent bulbs.

There are also other technologies in development, including sulfur lamps, quantum dots, and organic light-emitting diodes. The advancement of the technology in this area is clear: there is the main focus on energy efficiency because each light source is using less energy and lasting longer while striving to produce the same light output both in brightness and in color.

Life-Cycle Analysis

In terms of energy efficiency and conservation, a life-cycle analysis or assessment (LCA) is employed to measure the energy uses at each stage of the life cycle of a process, service, or product from the gathering of raw materials at the beginning to when those materials are returned to the earth. This is also known as a cradle-to-grave approach. Full life-cycle analyses measure many different environmental impacts, including energy, water, and materials, and then evaluate potential environmental impacts to show a complete picture. In this way they are useful because they recognize and account for the transfer of pollution, such as from water to air. A life-cycle analysis is helpful for manufacturers, for example, when they are trying to make a more efficient product, so that they know which stage is the most energy intensive and focus attention on that stage to make it more efficient. When evaluating different potential products or processes, it also helps the company to see which is most efficient. Conducting an LCA, however, can be resource and time intensive, and obtaining data can be a problem.

To conduct a life-cycle analysis, first the goal should be defined and the scope of the project should be determined. This will ensure that the results are what was wanted at the beginning. Next, a life-cycle inventory (LCI) is conducted. This is when all of the necessary data, including all inputs and outputs to the product or process, is collected and organized. The next step is to carry out a life-cycle impact assessment. This is a way of processing the data effectively to determine relative energy impacts at each life-cycle stage and a way to know which impacts are worse, both within a life-cycle stage and between stages. Finally, life-cycle interpretation serves as the last step and is the step in which conclusions are reached and recommendations can be provided. The results are also organized in a way that is understandable and presentable.

For example, when Proctor and Gamble did an LCA on its detergent, it found that the energy used to heat the water in washing machines was by far the biggest energy user, so the company created a detergent designed to work in cold water. As long as the consumer used cold water, this effectively cut out entirely the largest energy input in the life cycle of the detergent. As another example, Stonyfield Farms did an LCA on its yogurt containers and found that the largest environmental impact occurs in the manufacture and transportation stages, so the company put its yogurt in No. 5 plastic containers, which is more difficult to recycle but can make the containers thinner and lighter than No. 2 containers. By conducting an LCA, the company was able to reveal where exactly was the largest energy input that might have otherwise remained hidden because it can seem counterintuitive to design packaging that cannot be recycled in many communities.

ENERGY STAR

The ENERGY STAR program is a national project, supported by the U.S. Environmental Protection Agency (EPA) and the U.S. Department of Energy. Established in 1992, it is a voluntary labeling program that qualifies products that are energy efficient, with the overall goal of reducing greenhouse gas emissions. The program identifies and promotes products that are qualified as ENERGY STAR. The program first started with computers and monitors and now qualifies more than 60 product categories, including major appliances, office equipment, lighting, and home electronics as well as buildings, including new homes and commercial and industrial buildings.

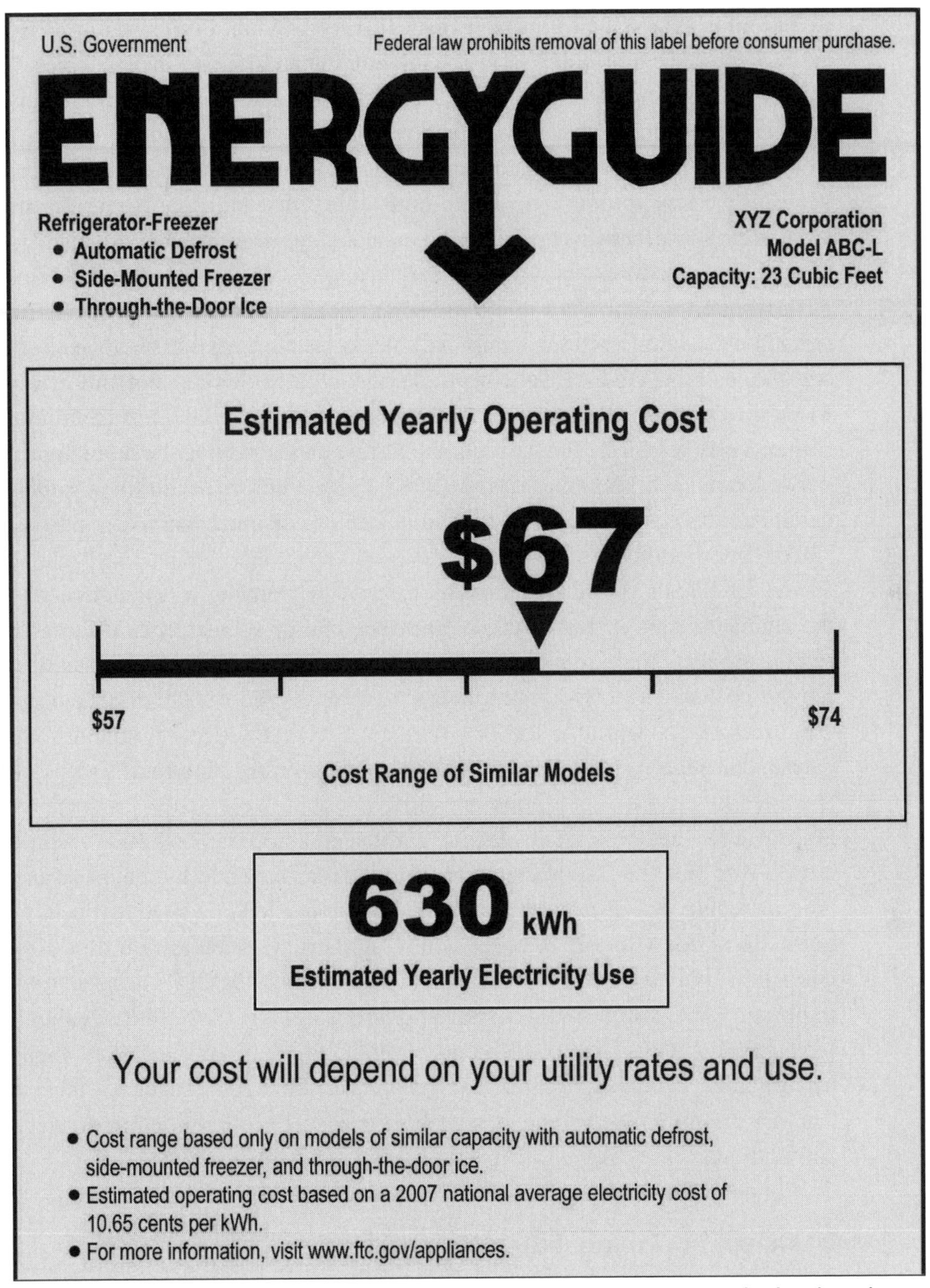

Consumer information is an important tool for energy conservation. The bright yellow EnergyGuide label appears on Energy Star–qualified appliances and shows the estimated yearly operating cost of the appliance in dollars per year. The information provides consumers with a context to compare the energy efficiency of different appliance models. (Federal Trade Commission)

The EPA establishes the specifications of a product qualification based on certain guiding principles. These include the following: the product categories must contribute significant energy savings nationwide; the qualified products must deliver the same features and performance found on non–ENERGY STAR–qualified products; if the product costs more than a non–ENERGY STAR product, then the consumer must be able to recoup the cost through utility bill savings in a reasonable amount of time; the energy-efficiency procedures can be achieved through widely available technologies offered by more than one manufacturer; the energy consumption and performance of the product must be able to be measured through testing; and the ENERGY STAR labeling will be visible to purchasers and effectively differentiate those products from non–ENERGY STAR products. The EPA can also revise the standards for a certain product category for different reasons, such as if there is a change in the federal minimum efficiency standards, technology changes, product availability, or performance or quality issues, among others.

The ENERGY STAR program partners with many different public and private entities looking to improve their own energy efficiency or making or selling products that are more energy efficient. From those selling products, rebates on ENERGY STAR products are often offered to further encourage consumer participation in the program. ENERGY STAR also supports its partners through its expertise, online tools, and training resources.

Different product categories use different ways to qualify products as ENERGY STAR. For example, for a refrigerator to be ENERGY STAR qualified, it must be 20 percent more efficient than required by the minimum federal standard. For a cordless phone to be ENERGY STAR certified, it must use under a certain wattage when on standby, depending on the features of the phone. The ENERGY STAR qualification for CFLs is based on a minimum lumens/watt rating, depending on the wattage of the CFL. Features for an ENERGY STAR–qualified home include effective insulation, high-performance windows, tight construction to minimize air leaks, efficient heating and cooling equipment, and ENERGY STAR–qualified appliances and lighting.

Leadership in Energy Efficiency and Design

Leadership in Energy Efficiency and Design (LEED) is an international certification system that provides third-party verification that a building or community was designed and built using methods so that it is energy efficient, conserves water, reduces greenhouse gas emissions, and improves indoor environmental quality. LEED was developed by the U.S. Green Building Council (USGBC) and introduced in 1998. Since then, it has registered more than 19,500 projects in all 50 states and 91 countries as of 2009. It was created as a way to define "green buildings," promote the whole-building approach in design, recognize environmental leadership in the industry, foster green competition, and raise awareness among the public of green building benefits.

The certification program standards are always being revised; the third version was released on April 27, 2009. This way, LEED standards can stay on top of emerging technology and not become outdated. The standards are developed through an open consensus-based process led by LEED committees. Each committee is made up of volunteers who are practitioners and experts representing the construction industry. There are rating systems for new construction, existing buildings for operations and maintenance, commercial interiors, core and shell construction, and schools, retail, health care, homes, and neighborhood development. The rating systems are point-based, so, for example, in the new construction category if the building is within a half-mile walking distance of a rail station or a quarter-mile walking distance of a bus station, then the project gets 6 points. If the building has on-site renewable energy, the project gets between 1 and 7 points, depending on the percentage of renewable energy that the building uses (ranging from a minimum of 1 percent to at least 13 percent renewable energy). There are also some aspects that are required, such as a minimum energy performance. To become LEED certified, a project needs 40–49 points, with 100 total points available. The rating system then has different levels for projects that want to go beyond 49 points, so to become LEED silver certified, the project must have 50–59 points, to get gold the project needs 60–79 points, and for platinum the project must have at least 80 points. The points are awarded by a LEED-accredited professional. There is also a LEED online tool that is used to manage the LEED registration and certification processes, and the USGBC offers support through this online tool.

The LEED ratings system is not without its problems, however. It is a design tool and not a performance measurement tool, so the LEED certification does not take into account what happens after the building construction is completed. It also is not yet climate-specific, and the rating system is not either, so buildings in Vermont and in Nevada, for example, would receive the same points for projects despite the fact that some projects would be more valuable in one case and not the other.

Rebound Effect

Improving energy efficiency can sometimes lead to greater use, which is a phenomenon termed the rebound effect. Horace Herring argues that to reduce energy use, energy efficiency is a means and not an end to this goal because of this rebound effect. Herring differentiates the terms "efficiency" and "conservation" such that the former is the ratio of energy services out to energy input, and the latter is reduced energy consumption through lower quality of energy services (for example, turning down the thermostat in winter). This can mean that increased energy efficiency at the individual level leads to a reduction of energy use at this level and also an increase of energy use at the national level. This is due to increased consumption of a product, because an improvement of efficiency in a product can cause the implicit price to fall. There is no consensus as to how much energy savings from the efficiency improvements goes to the rebound effect, but it is roughly estimated at 10 percent to 20 percent.

There are three types of rebound effect. The first is direct rebound effect, which is the increased use of energy services due to the price reduction of a product because of its increased energy efficiency. There is also the indirect rebound effect, which is caused by the reduction in the cost of energy services so that the consumer has a little more money to spend on goods and services. Finally, there are general equilibrium effects, which involve both the producer and the consumer and are the results of supply and demand across all sectors.

Herring presents lighting as a case study. In the early 1900s, incandescent bulbs with tungsten filaments replaced those with carbon filaments. These new bulbs used about one-quarter of the electricity of the old bulbs. The immediate result was a sharp drop in electricity use. However, soon a mass market was created from the increase in efficiency, and electricity took over gas for lighting. Herring goes on to predict that the same will happen with compact fluorescent lightbulbs because it will allow electric lighting for those unable to afford it otherwise and will also allow those with lighting to light more areas and to keep areas lit longer. Looking at data of public lighting in Great Britain in the 20th century, there has been more than a thirtyfold increase in energy consumption and twentyfold increase in efficiency in public lighting. When Herring overlaid this data with the mileage of construction of principal roads, he concluded that some of the increase in energy efficiency was taken by more miles of roads lit and in higher levels of illumination on roads.

With this information, Herring promotes conservation over efficiency as more effective and as the ethical choice. Improved efficiency can rationalize consumption and a wasteful lifestyle. This approach, however, has not caught on yet with Western culture. In the meantime, though, he recognizes the value of efficiency and notes that it should still be promoted, whatever the rebound effect might be.

Carbon Footprint

A carbon footprint is a subset of a life cycle analysis whereby only the energy inputs and outputs are measured. More recently, it has become more prevalent in pop culture as a way for individuals or households to measure how much energy they use personally each year and thus how much greenhouse gases are being emitted, which correlates to the size of the footprint. A simple search online will yield a number of websites with a carbon calculator where one puts in data about energy use from utility bills, car mileage and fuel efficiency, recycling habits, airplane flights, public transportation use, and diet and purchasing habits.

In the results, the calculator will display your footprint size in tons of carbon dioxide equivalent per year and also list the national average carbon footprint. Some calculators also list the footprint of developed countries and that of developing countries or what the target footprint should be to reach a carbon dioxide equivalent emissions goal to slow down or stop climate change. This way, it is very easy to compare one's personal footprint with the average and also with friends or family who also use the calculator. This allows for straightforward comparisons between energy users. Some calculators will advertise carbon offsets to buy. Many will offer suggestions at this point for ways to reduce this footprint, and this advice is customized based on the information put into the calculator. This information is most useful when the data put into the calculator are accurate. However, even then the results are a ballpark figure only.

Standby Power

Standby power, also called phantom power loss, vampire power, or leaking electricity, is a term used to describe when electronics are not shut off completely when the power button on them is switched off. Instead, these electronics—which include televisions, microwaves, and air conditioners, among many others—still draw a small but steady and constant supply of electricity, often unknown to the consumer. When the fact that the average American home has 40 devices constantly drawing power is considered, this small amount of electricity can add up. For example, the average CD player draws about 5 watts continuously, which translates to 45 kilowatt-hours per year. This equals about $5 per year on an electricity bill. In most developed countries, standby power is estimated to use 5–10 percent of residential energy use and is roughly responsible for 1 percent of global carbon dioxide emissions.

There are different ways that standby power is used. The black cubes that convert alternate current into direct current still draw electricity, as do the circuits and sensors in electronics with a remote control, soft keypads, and various displays, including those products with an LED status light. Sometimes it is not obvious that a product is using electricity when turned off, and a meter can be used to measure the electricity current.

There are ways to reduce standby power. If the product is not frequently used, it can be unplugged. This is the simplest way to ensure that power is not being used by this product. If the device is used more frequently, plugging it into a power strip with an on-off switch and placing the power strip in a convenient location makes it easy to reduce standby power. This is especially helpful for groups of products used together, such as a TV or computer setup, so that they can all be turned off at once. There are also power strips that can completely turn off peripheral electronics when the main product is turned off. These are also useful for TV or computer setups and avoid the necessary personal behavior to turn off the power strip every time. Another option is when shopping for products in the first place, look for those that have a low standby power draw, which includes all ENERGY STAR products. A watt meter can also be used in the home to check out the home's own standby power draw with the specific products in it. This can be very useful so that it is known what electronics have the highest standby power use.

Smart Grid

The electric grid currently is the largest interconnected machine on Earth and consists of more than 9,200 electric generating units with more than 1 million megawatts of generating capacity connected to more than 300,000 miles of transmission lines. The generation of electricity is responsible for about 40 percent of all the carbon dioxide emissions in the United States. Since 1982, growth in peak demand for electricity has outpaced transmission growth by almost 25 percent every year, but from 1973 to 2000 there was a decrease of funding in transmission. If the electric grid was 5 percent more efficient at transmitting electricity, the energy savings would be equivalent to permanently removing the fuel and greenhouse gas emissions of 53 million cars.

The Smart Grid, if adopted, will impact every part of the electricity delivery system, from generation and transmission to distribution and consumption.

It will increase the possibilities of distributed generation, which means that points of generation will be closer to consumers, such as solar panels on the roof of one's home. The shorter the distance that an electrical current has to travel, the more efficient it is. The Smart Grid will be automated and have a two-way flow of electricity and information and will be capable of monitoring power plants, customer preferences, and individual appliances. It will also give electricity generators and operators a better way to manage peak demand—those times of the day or year when electricity demand is highest. Currently, to supply extra power during these peak demand periods, generators turn on extra plants. These machines sit idle for most of the year and are sometimes older and less efficient. With the Smart Grid, operators can control loads to minimize or even eliminate this peak demand.

The Smart Grid can also allow for real-time pricing, which are energy prices that are set for a specific time period and can fluctuate from one set period to the next. Customers are notified ahead of time (such as a day or an hour) of the price change so they can change their electricity consumption accordingly. Residents can also set their preferences for comfort, price, and environment so that the Smart Grid can automatically adjust their electrical consumption to match these preferences. This extra knowledge can lead to tremendous consumer-side energy efficiency. The Smart Grid can also manage electricity for plug-in hybrid or solely electric vehicles, where they can act as batteries for the system by charging at night or other times when there is low electric demand and then selling the power back to the grid when it is needed.

There are already localized instances of the Smart Grid working at various stages in Hawaii, Illinois, West Virginia, California, and Colorado. In Hawaii at a substation in Maui, they are using an automated system with optimal dispatch of electricity that enables consumers to take control based on their preferences. In Illinois, they have developed a model of an electric system that cannot fail to deliver electricity to the consumer with the creation of microgrids. California is also studying microgrids at a substation in San Diego. These microgrids can isolate from the utility easily in times of power disturbance and then reconnect later just as easily.

A smart meter can play an important part in the Smart Grid. These devices provide a two-way communication between households and the utility. As of April 2010, worldwide there are approximately 76 million smart meters already installed, and 10 million of those are in the United States, most of them in California. Smart meters allow households to view their energy use by month, day, and hour online, while the meter itself will only show the cumulative electricity use. Households can also sign up online to get alerts when their electricity use is pushing them to a higher-cost tier so they can change their electricity consumption or be aware of this new pricing tier. With the introduction of this new technology, there have also been complaints about potential privacy issues. Smart appliances that could be introduced to work with these new meters can also tell energy generators what appliances are being used as part of the real-time data exchange that smart meters will facilitate. Some are also concerned about potential health risks because of the electromagnetic emissions when the smart meters send and receive information wirelessly.

Energy Assessment

Home energy assessments (also known as energy audits) are used to see how much energy a household consumes and to see what can be done to increase the efficiency of the house. A simple energy assessment can be done by individuals, or professionals can carry out a more thorough assessment. A do-it-yourself assessment can include checks for air leaks and drafts using incense sticks, since the smoke will show signs of air movement. Insulation can then be installed in those areas that have a draft. The water pipes leaving the water heater should also be insulated. If the home has a forced-air furnace, the air filter should be changed regularly, about every other month. The wattage of the lightbulbs in the household can also be reviewed to determine if lower-wattage or compact fluorescent lightbulbs can be substituted.

Professional energy assessments can go into great detail. The professional will go through the house room by room, review past utility bills, and ask questions about the thermostat setting, use of rooms, and when residents are home, the answers to which might reveal some easy ways to reduce energy consumption. The professional will use a blower door test to see how airtight the building is. A blower door is a large fan that fits into an exterior door frame. When it is on, it sucks air out of the house to create a lower air pressure that will pull outside air into the home through cracks and other openings. The professional may use a smoke pencil to find drafts. A calibrated blower door will also allow the professional to determine the overall tightness of the building. The professional will probably also use thermography to detect air leaks in the building. These infrared scanners measure surface temperatures by determining the light in the heat spectrum so that white areas on the scanner correspond to warmer areas in the building and black areas correspond to colder areas. These are usually done from inside the house so the source of the leak can be seen. Using this tool, the professional can also see how effective the current insulation in the home is.

Transportation

In 2008, the transportation sector in the United States emitted 1,930.1 million metric tons of carbon dioxide, which is 95.6 million metric tons lower than in 2007 but 343.2 million metric tons higher than in 1990. This translates to just about a third of greenhouse gas emissions (95 percent of which is carbon dioxide) in the United States, and this sector is second only to electricity generation in terms of volume and rate of growth of greenhouse gas emissions. Within transportation, highway travel accounts for 72 percent of transportation energy use and carbon emissions. Passenger cars alone account for 36 percent of transportation energy use, and air travel is next, at 10 percent, followed by modes of transportation with smaller percentages, such as marine, rail, and pipelines.

The fuel economy of passenger cars and light trucks increased from 13.1 miles per gallon (mpg) in 1975 to 21.9 mpg in 1988. However, from 1988 to 2004, the fuel economy for passenger cars and light trucks actually decreased to 19.3 mpg. This stagnation is due to the increasing popularity of light trucks, which have lower fuel economy standards. From 2004 to 2010, the fuel economy was on the rise again, with a final fuel economy in 2010 of 22.5 mpg. In commercial air travel, passenger mpg has increased 150 percent from 1975, which is mostly due to a near doubling of aircraft energy efficiency and because there are more passengers in the planes. Over this same time, the energy required to move a ton of freight by rail has decreased by half.

There are different ways to reduce these large emissions, including strengthening fuel economy; using alternative fuels such as liquefied natural gas, biomass, hydrogen, and electricity generated from renewable sources; developing a more efficient system on a large scale in terms of freight, mass transit, and roads; and reducing transportation activity. One way to strengthen fuel economy is to build engines that are not as powerful. In 2000, the engine of an average passenger car had more than eight times the power than it needed to cruise on a highway. Other ways include transmission efficiencies, reductions in aerodynamic drag, reducing the rolling resistance of tires, decreasing vehicle weight, and manufacturing and selling more hybrid vehicles. Some ways to develop a more efficient system are taking more direct routes from origins to destinations, increasing vehicle occupancy rates, shifting traffic to more efficient modes, and improving the efficiency of vehicles through better maintenance and driving behavior.

Hybrid vehicles are those cars in which the internal combustion engine is complemented by an electric motor powered by a nickel-metal hydride battery. The engine is downsized and mostly operates at its maximum efficiency.

The motor supplies peak power for acceleration and allows for the engine to be shut down during times of idling, deceleration, or traveling at very low speeds. The batteries are charged during regenerative braking. The first hybrid model is the Toyota Prius, introduced globally in 2001; more than 2 million of these cars had been sold as of September 2010.

Industrial Ecology

Industrial ecology looks at the material and energy flows through industrial systems to try to understand how industrial systems at various scales interact with the environment and has the goal of minimizing impact on the environment through waste reduction. This type of study is still relatively young, gaining popularity only in 1989. It has grown quickly, though, and there are now two journals devoted to this subject: the *Journal of Industrial Ecology* and *Progress in Industrial Ecology*.

The regional level is the most practical level for industrial ecology principles to work so that the exchange of materials can remain feasible. When considering forming an industrial ecology network, there are three main aspects that should be considered: collective problem definition and innovation activities, search at the intersectoral interface, and increased interorganizational interactions in environmental problem solving networks.

A case study to consider is in Landskrona, Sweden. An industrial ecology network was formed in this heavily industrial city in 2002. Synergistic connections that arose naturally were already in place, but a formal agreement was made to look more into the matter and identify further collaborative opportunities between different industrial sectors, including chemicals, waste management, metals processing and recycling, printing and printed packaging, motor vehicle components, agricultural seeds, transport, and logistics. Together, they numbered more than 20 companies, and there were also 3 public organizations involved. The group constructed a flow chart illustrating ways that waste at one company can become a useful input at another. For example, wastewater from car glass manufacturing was substituted for clean drinking water that had been in use at a wet scrubber removing volatile organic compounds from a printing company's flue gases.

Elisa Abelson

Further Reading

Department of Energy. "Energy Efficiency and Renewable Energy: Energy Savers; Home Energy Assessments." U.S. Department of Energy, 2010, http://www.energysavers.gov/your_home/energy_audits/index.cfm/mytopic=11160.

Department of Energy. "The Smart Grid: An Introduction." U.S. Department of Energy, 2008, http://www.oe.energy.gov/DocumentsandMedia/DOE_SG_Book_Single_Pages%281%29.pdf.

Environmental Protection Agency. "Life Cycle Assessment: Principles and Practice." U.S. Environmental Protection Agency, 2006, http://www.epa.gov/nrmrl/lcaccess/pdfs/600r06060.pdf.

Greene, David L., and Andreas Schafer. "Reducing Greenhouse Gas Emissions from U.S. Transportation." Pew Center on Global Climate Change, 2003, http://www.pewclimate.org/docUploads/ustransp.pdf.

Herring, Horace. "Energy Efficiency: A Critical Review." *Energy* 31 (2006): 10–20.

Energy Sources and Shortages

In the United States, federal discussion about energy use had begun to change during the 1970s. Stimulated by the 1973 Arab Oil Embargo, American leaders began to perceive our reliance on petroleum (acquired from Middle Eastern countries in increasing amounts by 1970) as a problem. Combining this political reality with the environmental initiatives of the 1970s, a new ethic of resource use and management took shape around the idea of conservation. Although not a culture-wide shift from the era of intense consumption, learning to live within natural limits during the late 1900s took a variety of forms.

Oil Shocks

Although the embargo had economic implications, it had begun as a political act by the Organization of Petroleum Exporting Countries (OPEC). Therefore, in 1974 the Richard Nixon administration determined that the embargo needed to be dealt with on a variety of fronts, including, of course, political negotiation. These negotiations, which actually had little to do with the petroleum trade, needed to occur between Israel and its Arab neighbors, between the United States and its allies, and between the oil-consuming nations and the Arab oil exporters. Convincing thc Arab exporters that negotiations would not begin while the embargo was still in effect, the Nixon administration leveraged the restoration of production in March 1974. Although the political contentions grew more complex in ensuing decades, the primary impact of the embargo came through the residual effects that it had on American ideas of energy.

Now, a panicked public expected action. Nixon, by this point increasingly embattled over the growing problem of the Watergate Scandal but having nonetheless been reelected in 1972, appeared before Americans on November 7, 1973, to declare an "energy emergency." He spoke of temporary supply problems:

> We are heading toward the most acute shortages of energy since World War II. . . . In the short run, this course means that we must use less energy—that means less heat, less electricity, less gasoline. In the long run, it means that we must develop new sources of energy which will give us the capacity to meet our needs without relying on any foreign nation.
>
> The immediate shortage will affect the lives of each and every one of us. In our factories, our cars, our homes, our offices, we will have to use less fuel than we are accustomed to using. . . .
>
> This does not mean that we are going to run out of gasoline or that air travel will stop or that we will freeze in our homes or offices any place in America. The fuel crisis need not mean genuine suffering for any Americans. But it will require some sacrifice by all Americans.

In his speech, Nixon went on to introduce Project Independence, which, he asserted, "in the spirit of Apollo, with the determination of the Manhattan Project, [would] . . . by the end of this decade" help the nation to develop "the potential to meet our own energy needs without depending on any foreign energy source."

In reality, Nixon's energy czar, William Simon, took only restrained action. Rationing was repeatedly debated, but Nixon resisted taking this drastic step on the federal level. Although he had rationing stamps printed, they were kept in reserve. In one memo, Nixon's aid Roy Ash speculated that "In a few months, I suspect, we will look back on the energy crisis somewhat like we now view beef prices—a continuing and routine governmental problem—but not a Presidential crisis." Nixon's notes on the document read "absolutely right," and overall, his actions bore out this approach. He refused to be the president who burst the American high of energy decadence.

Considering Reform in the United States

Of course, any argument for a conservation ethic to govern American consumers' use of energy was a radical departure from the postwar American urge to

Cars line up in two directions on December 23, 1973 at a gas station in New York City at the height of the energy crisis of 1973. Motorists waited in long lines and paid high prices for gas when the Organization of Petroleum Exporting Countries (OPEC) declared an oil embargo on the United States in October 1973. (AP/Wide World Photos)

resist limits and to flaunt the nation's decadent standard of living. Although this ethical shift did not take over the minds of all Americans in the 1970s, a large segment of the population began to consider an alternative paradigm of accounting for our energy use and needs. The national morals were shifting. The American public became interested in energy-saving technologies, such as insulation materials and low-wattage lightbulbs, as well as limits on driving speeds that might increase engine efficiency. As a product of the 1970s crisis, some Americans were even ready and willing to consider less convenient ideas of power generation, such as alternative fuels.

One conduit for such research would be the Department of Energy (DOE) that President Jimmy Carter created at the cabinet level. Similar crises of energy supplies might be avoided, it was thought, if in the future one agency administered strategic planning of energy use and development. The DOE's task was "to create and administer a comprehensive and balanced national energy plan by coordinating the energy functions of the federal government." The DOE undertook responsibility for long-term high-risk research and development of energy technology, federal power marketing, energy conservation, the nuclear weapons program, energy regulatory programs, and a central energy data collection and analysis program.

Just as with any government agency, however, the mandate and funding varies with each presidential election. During the next few decades, the DOE moved away from energy development and regulation toward nuclear weapons research, development, and production. Since the end of the Cold War, the DOE has focused on environmental cleanup of the nuclear weapons complex, nonproliferation and stewardship of the nuclear stockpile, and some initiatives intended to popularize energy efficiency and conservation. As the crisis faded, so too did the political will to strategically plan the nation's energy future.

Although the DOE did not necessarily present Americans with a rationale for less polluting fuels, new ethical perspectives moved through the public after the 1960s and created a steady appreciation for renewable energy. Interest in these sustainable methods as well as in conservation helped spur the public movement in the late 1960s that became known as modern environmentalism.

Looking for Energy Alternatives

After the 1970s, some policy initiatives specifically focused on the suspected causes of climate change, particularly on the use of hydrocarbons for fuel. Although some policies had by the 1970s begun to recognize pollution and other implications of the use of fossil fuels to create energy, regulations forced energy markets to reflect neither the full environmental nor economic costs of energy production, including potential implications for altering climate. Policy historian Richard Andrews writes that the 1973 embargo initiated three types of policy change related to energy. First, there was an emphasis on tapping domestic supplies or energy. Second, there was a new recognition that energy conservation was an essential element of any solution. And third, electric utility companies were forced to accept and pay fair wholesale rates for electricity created by any producer.

The Public Utilities Regulatory Policy Act of 1978 opened the electric grid to independent producers, including energy generated from renewable sources. Eventually the Energy Policy Act of 1992 expanded these possibilities nationally by allowing both the utilities and other producers to operate wholesale generating plants outside the utility's distribution region. Such initial steps often did not have the intended outcome. Andrews writes that "in effect it thus severed power generation from the 'natural monopoly' of electric transmission and distribution."

Carter and others had put the moral imperative for change in front of the American people at the end of the 1970s. Animating the health of Earth, much as a species of tree or animal that is impacted by human activity, many environmentalists spent the closing decades of the 20th century expanding the scale of our ability to consider the broader implications of our lifestyle. Most significant, after scientists of the 1980s presented findings that suggested a connection between the burning of fossil fuels and the rise in temperatures on Earth, Americans could begin to appreciate the implications of our trends in lifestyle. In short, more Americans than ever before came to see their everyday tendencies as having moral implications for the planet.

If significant change in energy use was to occur, however, personal transportation would be one of the most difficult alterations for Americans to accept. The dynamics of these new scientific understandings and the ethic of environmentalism altered personal transportation in small pockets as the 20th century closed. The rise of environmental concerns was focused in California in the late 20th century, and it is therefore not surprising that so did the development of electric vehicles. The California Air Resources Board (CARB) helped to stimulate CALSTART, a state-funded nonprofit consortium that functioned as the technical incubator for America's efforts to develop alternative-fuel automobiles during the 1990s. Focusing its efforts on the project that became known as the electric vehicle, this consortium faced auto manufacturers' onslaught almost singlehandedly. Maintaining the technology during the mid-1900s, however, had been carried out by a variety of independent developers.

Absent of governmental support and despite the contrary efforts of larger manufacturers after World War II, independent manufacturers continued to experiment with creating an electric vehicle that could be operated cheaply and travel farther on a charge. The problems were similar to those faced by Thomas Alva Edison and earlier tinkerers: reducing battery weight and increasing range of travel. Some of these companies were already in the auto business, including Kish Industries of Lansing, Michigan, a tooling supplier. In 1961 it advertised an electric vehicle with a clear bubble roof known as the Nu-Klea Starlite. Priced at $3,950 without a radio or a heater, the car's mailing advertisements promised "a well designed body and chassis using lead acid batteries to supply the motive energy, a serviceable range of 40 miles with speeds on the order of 40 miles an hour." By 1965, another letter from Nu-Klea told a different story: "We did a great deal of work on the electric car and spent a large amount of money to complete it, then ran out of funds, so it has been temporarily shelved." Nu-Klea was not heard from again.

No longer manufactured, the Hummer 2 (H2), was a sport-utility vehicle designed by General Motors. The H2 is modeled after the original Hummer (H1) which was developed for military use and carried a price tag of $100,000. (Richard Scherzinger/iStockphoto)

In 1976, the U.S. Congress passed legislation supporting the research of electric and hybrid vehicles. Focused around a demonstration program of 7,500 vehicles, the legislation was resisted by government and industry from the start. Battery technology was considered to be so lacking that even the demonstration fleet was unlikely. Developing this specific technology was the emphasis of the legislation in its final rendition. Historian David Kirsch writes that this contributed significantly to the initiative's failure. "Rather than considering the electric vehicle as part of the automotive transportation system and not necessarily a direct competitor of the gasoline car, the 1976 act sponsored a series of potentially valuable drop-in innovations." Such innovations would allow electric technology to catch up to gasoline, writes Kirsch. However, "given that the internal combustion engine had a sixty-year head start, the federal program was doomed to fail."

Various efforts to create electric vehicles with mass application followed, but nothing caught on with the American public. In fact, in one of the great moral rejections in energy history, most Americans spent the 1990s purchasing larger and heavier vehicles than ever before: SUVs and pickup trucks became the most popular vehicles on American roads. It appears, however, that this tendency was a residual holdover from previous eras in our high-energy existence; genuine cultural and social changes were indeed occurring in the very fabric of American lifestyles. Rising gas prices and the failure of the American automobile industry ultimately appear to have helped to alter American vehicle preferences.

Conclusion: Energy Transitioning

Will energy shortages spur transition to renewable sources? It appears that the shortages of the 1970s did spur a slow shift from fossil fuels. Rising prices in the 21st century moved the shift even closer to reality. In the United States, however, such transitions will take considerable time. Many critics complain that the vested interests of energy companies and the lobbying that they support have mightily slowed the American transition from fossil fuels. Consumers also seem to lack the willingness to prioritize information such as sustainability and emissions over the economics of their personal energy use. And clearly, political shifts in the United States have given limited support to the construction of legal regulations that would help stimulate such a transition.

In many other developed nations, however, the shortages of the 1970s began a process of transition that does continue today. Strategic thinking about energy security for many nations has evolved from ensuring access to petroleum and other fossil fuels to designing their own energy infrastructure to mitigate reliance on such sources.

Brian C. Black

Further Reading

Andrews, R. N. L. *Managing the Environment, Managing Ourselves.* New Haven, CT: Yale University Press, 1999.

Bradsher, K. *High and Mighty: SUVs; The World's Most Dangerous Vehicles and How They Got That Way.* New York: PublicAffairs, 2002.

Doyle, J. *Taken for a Ride: Detroit's Big Three and the Politics of Air Pollution.* New York: Four Walls Eight Windows, 2000.

Horowitz, D. *Jimmy Carter and the Energy Crisis of the 1970s.* New York: St. Martin's, 2005.

Roberts, Paul. *The End of Oil.* New York: Mariner Books, 2005.

Energy Technology Innovation

Energy technology innovation refers to the innovative processes associated with developing and deploying energy technologies. Energy technology innovation occurs within a broad, complex sociotechnical system that involves much more than the technological advancements made by scientists and engineers. Many interconnected social, economic, and political factors both influence and are influenced by technological advancement. Despite broad agreement that governments have a critical role to play in technology innovation, the politics associated with determining the appropriate and adequate role of government in supporting, encouraging, and facilitating technological innovation varies in different countries and different technology sectors.

In considering energy technology innovation, useful distinctions to consider are the three discrete but clearly intertwined phases of technological innovation, each of which is embedded in the larger sociotechnical system: (1) basic research and development (R&D), whereby technological details of new ideas are explored and advanced; (2) demonstration, whereby new technologies are piloted and tested; and (3) deployment, whereby new technologies are adopted at scale, implemented, and commercialized. Interactions and feedback among these three discrete phases of technological innovation are frequent, and social and technical learning occurs at each phase.

Beyond deployment, both R&D and demonstration involving learning by doing play important roles in the process of technological change. Major changes in existing technologies are unlikely without explicit financial support for R&D, but any new technological concept will not advance without learning by doing in demonstration projects and early deployment opportunities. It is often through accumulated experience and the learning by doing associated with demonstrating or deploying a technology that substantial cost reductions are realized.

Due to the environmental, economic, and geopolitical instability associated with current energy systems, energy technological innovation has been identified as particularly critical to transitioning to a more sustainable society. The need for facilitating accelerated energy technology innovation is becoming increasingly urgent, as a large portion of the human population has limited access to energy services, as the dangers associated with a societal dependence on oil are becoming more apparent, and as the risks associated with climate change continue to grow. Despite growing widespread acknowledgment of the need for energy technology innovation, the politics and policies associated with technology innovation within complex sociotechnical energy systems are complicated by a diversity of competing priorities that are economic, geopolitical, and environmental in nature.

Investing in new energy technologies is a well-recognized critical component of confronting the climate change problem. Energy technology innovation involves both private and public organizations and financial support. Prioritizing limited resources, however, for investments in the R&D, demonstration, and deployment continuum of emerging technologies is a difficult challenge. In some countries, government support for many components of energy technology innovation have been declining rather than increasing in recent decades, but in several countries more public attention and support for energy technology innovation are being recognized. With regard to the private sector, investment in energy technology innovation is difficult to assess, but it is clear that innovation in energy systems provides both opportunities and challenges for the private sector.

Jennie C. Stephens

Further Reading

Gallagher, K. S., J. P. Holdren, et al. "Energy-Technology Innovation." *Annual Review of Environment and Resources* 31 (2006): 193–237.

Holdren, J. P. "The Energy Innovation Imperative, Addressing Oil Dependence, Climate Change, and Other 21st Century Energy Challenges." *Innovations, Technology, Governance & Globalization* 1(2) (2006): 3–23.

Sagar, A. D., and B. van der Zwaan. "Technological Innovation in the Energy Sector: R&D, Deployment, and Learning-by-Doing." *Energy Policy* 34(17) (2006): 2601–2608.

Enhanced Geothermal Energy

Geothermal energy is an energy source of vast proportions. In the United States alone, there are enough recoverable geothermal resources to meet all our energy needs for more than 1,000 years. These geothermal resources are located at depths of 2,000 to 10,000 meters below the surface of Earth, where temperatures often exceed 200°C. This heat is produced by the radioactive decay of naturally occurring radioisotopes of potassium, uranium, and thorium that occur throughout Earth and from the slow conduction of heat away from Earth's core and mantle.

In any geothermal system, a medium is required to move the heat from the rock deep within Earth to the surface, where the energy can be converted into electricity. This can be done with water, but this also requires that there exist a reservoir of water that permeates the hot rock. And even more importantly, as this hot water is pumped to the surface, additional water must recharge the reservoir, and the new water can also permeate through cracks and fissures in the rock to be heated up in a reasonably quick time frame. Unfortunately, in most places in the United States and in the world these vast stores of heat do not have sufficient water resources and/or the underlying rock is not permeable to the recharging water. This is when enhanced geothermal systems (EGSs) are required. In an EGS, the rock can be fractured using hydraulic shearing, which is similar to a drilling technology borrowed from the shale gas industry in which rock is hydrofactured (or fracked) to allow trapped natural gas to be removed. Because commercial-scale power plants need a high flow rate of hot water (20 million gallons per day for a 50 megawatts electric [MWe] geothermal power plant), most EGSs also employ an artificial water recharging system in which an injection well is drilled so that water can be pumped into the fractured rock to replace the hot water that is removed. The operation and success of any EGS or more common geothermal power plant is nearly completely determined by the flow rate of water into and back out of the hot underlying rock.

Like common geothermal power plants, EGS plants can use a binary generator or a flash generator. In a flash generator, the hot water is pumped out of the ground at a temperature above 200°C. At this temperature, water is normally in a steam phase, but due to the immense pressure within Earth, this water is removed as a liquid. When the superheated water exits the production well, it flashes to steam from the reduction of pressure and is used to turn a steam generator to produce electricity. In a binary generator, the hot water is pumped out of the ground and is used to heat a secondary fluid to produce hot vapor that can be used to generate electricity. In both cases, the water can then be injected back into the ground to recharge the hot reservoir.

EGSs require many wells to be drilled into the underlying rock structure, and since the hot rock utilized by EGSs would not normally be porous to injected water, a series of spaced-out injection wells and production wells must be designed and all of these wells stimulated (hydrosheared) to create the necessary cracks and fissures to allow the water to move within this engineered geologic heat exchange system. The best geothermal resources in the United States are located west of the Mississippi River, most notably west of New Mexico and Colorado where water resources are fairly scarce. This is a real problem for EGSs, since it is estimated that during the 30-year lifetime of a 50 MWe EGS power plant, a total of nearly 10 billion gallons of water would be needed. Some of this water is used for stimulating (hydroshearing)

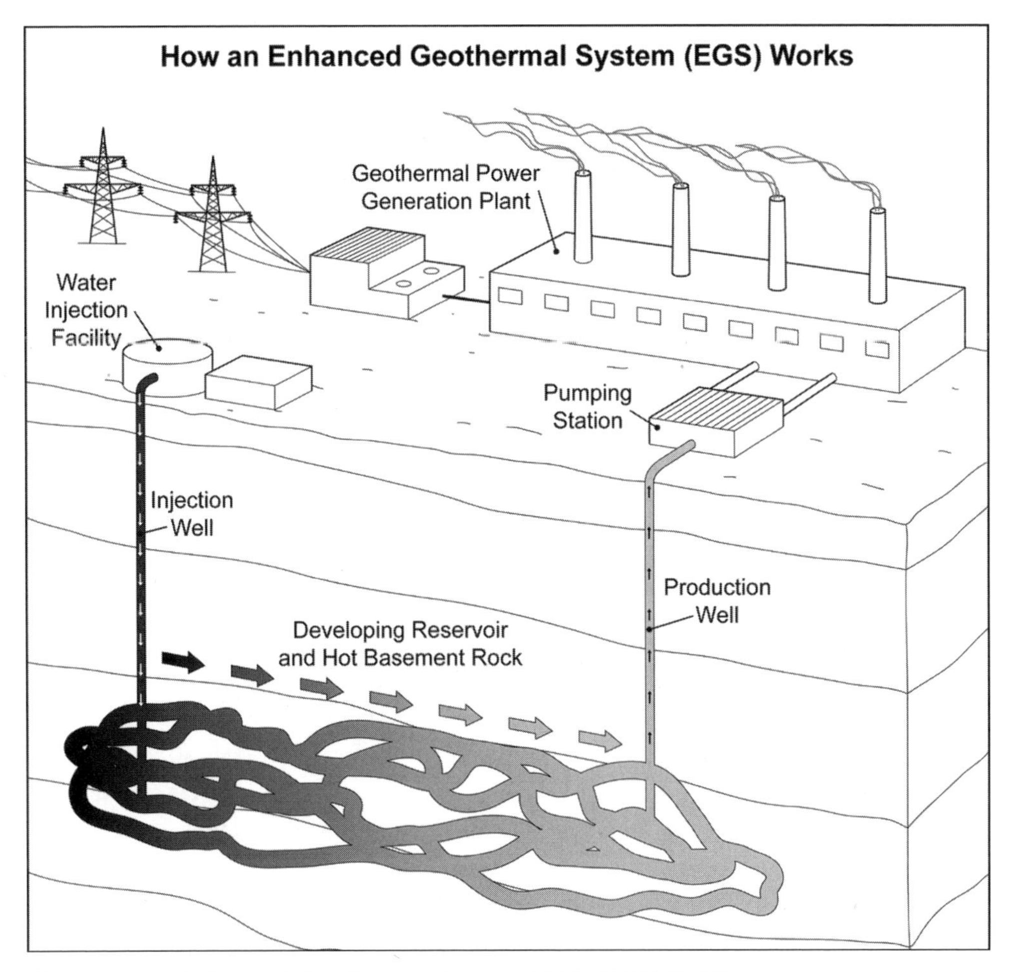

Illustration of Enhanced Geothermal System (EGS). (ABC-CLIO)

and drilling the many wells, but the vast majority of the water is to make up for small losses between the production and injected water volumes. Putting this into perspective, EGSs are expected to consume slightly more water per kilowatt-hour of electricity generated than a natural gas combined cycle power plant.

Geothermal electricity does not emit any CO_2, and since it is renewable it will never run out until Earth's center finally cools in a billion years or so. In some cases, geothermal releases pollutants such as sulfur oxides that are the primary acid rain–causing pollutant. Although evidence is limited to only a few sites that have extensive drilling, small earthquakes (up to magnitude 3) have been observed as a result of the hydraulic stimulation of the wells.

There are currently no commercial EGS facilities in the world that are operational. There are several commercial operations that are in various levels of development as well as a variety of research and development projects. It is expected that another decade of continued research will be needed before large-scale commercial EGSs will be practical. Until recently, government funding for EGSs was very small (just $1 million in 2000); this increased to $20 million in 2009. But even at this level, funding is inadequate for the full development of EGSs. A report led by MIT estimates that about $1 billion in government funding or incentives is needed to fully develop EGSs and that even with this level of funding, it would be 15 years before EGSs are commercially viable. The federal government provides financial incentives and funding for the commercial generation and research and development of many forms of energy. For example, in 2007 there was more than $3 billion in assistance given to coal, slightly more given to ethanol, $700 million for wind, and only $14 million given to geothermal (most of which was not EGSs). And yet of all the renewable energy sources available, EGS technology shows the best promise of providing inexpensive energy on demand regardless of weather conditions. The MIT report concludes that the cost of generating electricity from geothermal resources could drop from the current range of $0.10–$0.30 per kilowatt-hour (kWh) to $0.03–$0.09 per kWh with an investment of about $1 billion, mainly to be used to demonstrate and prove that new EGS technology works. With that investment (or subsidy), the reduced cost of geothermal would make it economically competitive with other fossil fuel commercial electricity without even considering any impacts on the environment or human health from other fossil fuels. If the environmental impacts of fossil fuels are somehow considered with tough pollution controls or a carbon tax, then electricity from EGS technology can easily become less expensive than from coal, natural gas, or nuclear. Many experts believe that all EGS technology needs in order to become a mainstay of the U.S. energy mix is an initial commitment of research and development resources.

Richard Flarend

Further Reading

Massachusetts Institute of Technology. *The Future of Geothermal Energy: Impact of Enhanced Geothermal Systems (EGS) on the United States in the 21st Century.* Cambridge: MIT Press, 2006. http://geothermal.inel.gov.

Energy Information Agency. *Federal Financial Interventions and Subsidies in Energy Markets, 2007.* Washington, DC: Energy Information Agency, 2008.

Enron and Energy Trading

Expansive use of energy is fed by a number of aspects of American society. During one particularly troubling episode, the economic structure of American business seemed to be compounding our addiction to fossil fuels. In the process, the Enron debacle made it seem all but impossible for our energy industry to address implications related to climate change.

Seen as one of the nation's most innovative companies, Enron became the leading example of a late 1990s precedent for corporate corruption and greed. Even when it was doing business as usual, though, Enron represented a new development in the commodification of energy. Taking the concepts that were at the root of John D. Rockefeller's Standard Oil Trust, Enron and others created the field of energy trading, which was now possible with electronic stock trading.

A Houston Powerhouse

After its founding in 1985, Enron, which was based in Houston, Texas, became the nation's seventh-biggest company in revenue by emphasizing the transmission and distribution of electricity and gas throughout the United States and the development, construction, and operation of power plants and pipelines worldwide. Following the trend (particularly in states such as California) toward energy deregulation, Enron became the essential middleman between energy producers and consumers. As a result, Enron was named "America's Most Innovative Company" by *Fortune* magazine for five consecutive years, from 1996 to 2000. Viewed as a great technological and business innovator, Enron, and its chairman Kenneth Lay, were credited with creating the energy markets that it grew to dominate.

For most of the 20th century, utility companies had generated, transmitted, and sold electricity as state-regulated monopolies. They also built and maintained the electrical grid—the network of transmission wires that carries electricity to homes and businesses. With deregulation, companies needed to make sure that their supply of power on the grid would remain consistent. This odd role of energy trading—without necessarily owning power plants or supplies of raw material—was Enron's route to success and ultimately its road to failure.

With its transactions taking place entirely on paper, Enron's business was tantalizingly tempting to illegal manipulation. For instance, the company has been accused of manipulating energy prices to create shortages in California in the summer of 1999. These practices, however, made Enron great sums of money, which, of course, made the company even more enticing to investors.

Traders at Enron were among those who took advantage of California's poorly constructed deregulation law and helped to bring about the state's energy crisis of 2000–2001. They concocted schemes to manipulate electricity markets and to maximize Enron's profit, using names such as Fat Boy and Death Star to describe the strategies. Some bantered casually in 2000 about how they were "stealing" from California and sticking it to "Grandma Millie" by overcharging for power, according to audiotapes of their conversations that have been made public.

The Bottom Falls Out

In a six-week downward spiral during 2000, Enron disclosed a stunning $638 million third-quarter loss, the Securities and Exchange Commission opened an investigation into the partnerships, and the company's main rival backed out of an $8.4 billion merger deal. After a series of scandals involving irregular accounting procedures bordering on fraud involving Enron and its accounting firm Arthur Andersen, Enron filed for protection from creditors on December 2, 2001, in the biggest corporate bankruptcy in U.S. history.

Enron's stock, worth more than $80 in 2000, tumbled to less than $1 per share. Enron's collapse left investors burned and thousands of employees out of work with lost retirement savings. In addition, Enron, which had 20,000 employees, barred them from selling Enron shares from their retirement accounts as the stock price plunged, saying that the accounts were being switched to a new plan administrator. Former Enron chief financial officer Andy Fastow was indicted on November 1, 2002, by a federal grand jury in Houston on 78 counts including fraud, money laundering, and conspiracy. He was sentenced to serve a 10-year prison sentence and forfeit $23.8 million.

The swift fall of this corporate giant caught investors by surprise. Maybe the best symbol of how quickly corporate entities could come and go was the new baseball stadium opened in Houston in 2000, which was named Enron Field. After the company collapsed, the Houston Astros baseball team needed to pay the company $5 million to have its name removed from the stadium. The team cited the need to do so in order to avoid the negative publicity associated with the former model of corporate success.

Ultimately Enron's failure was not tied directly to the actions of traders, who made hundreds of millions of dollars for the company. But company traders were speculating on energy prices, and aggressive accounting of risky long-term energy contracts made Enron even more susceptible to a blowup.

When Enron filed for bankruptcy in 2001, it was the largest filing in American history. Many employees and their families, including Osme Garcia shown protesting here in Houston, Texas, found themselves without jobs and with worthless stocks in their retirement accounts. (AP/Wide World Photos)

Therefore, even though Enron was a symbol of society's angst about corporate ethics, its business of trading on energy prices was a product of a new era in resource management and has outlived the company that perfected it.

Conclusion: Energy Trading on Scarce Resources

Enron, once the country's seventh-largest company, introduced its modern trading floor on national television. Today, companies such as Centaurus have learned from Enron's example. Since its 2002 founding, Centaurus has amassed $1.5 billion in assets under management. But it is doing so with a low profile. Energy trading—given a bad name by Enron—is springing to life again.

Volatile energy markets and record-high commodity prices are prompting renewed interest from investors. This has pushed banks and a growing number of hedge funds to hire more energy traders to maximize their ability to profit from the volatile sector. Whether Americans agree with such ethics, many appreciate the growth that it enables in their investment portfolios as energy continues to be recognized as an essential part of human society.

Brian C. Black

Further Reading

McLean, Bethany. *The Smartest Guys in the Room.* New York: Portfolio, 2003.

Munson, Richard. *From Edison to Enron: The Business of Power and What It Means for the Future of Electricity.* London: Praeger, 2005.

Swartz, Mimi. *Power Failure.* New York: Crown, 2004.

Environmental Health and Preventive Medicine

During the past 50 years, the subject of climate change has come to be considered one of the most significant problems facing humanity and has been debated extensively. Scientists, the Intergovernmental Panel on Climate Change, special interest groups, and government lobbyists have collected extensive information to determine both the degree of the problem and what should be done to mitigate its effects. A few historians of medicine and environmental health have contributed to the current debates. These include Vladimir Janković, in *Confronting the Climate: British Airs and the Making of Environmental Medicine* (2010), and Naomi Oreskes and Erik Conway, in *Merchants of Doubt: How a Handful of Scientists Obscured the Truth on Issues from Tobacco Smoke to Global Warming* (2010).

A consideration of historical examples before the 20th century can help us understand how climate change threats have long impacted human society as well as how society has sought to respond to them. There were important milestones in environmental health and preventive medicine as reactions to the impact of climate change phenomena between 1200 and 1900. Understanding the responses of human beings working in earlier environmental health traditions and foundations of preventive medicine provides useful background for current debates surrounding global warming.

It may be regarded as a long-standing belief in medicine and epidemiology that the different forms of disease affecting a given population are highly influenced by that population's social living conditions and biological environment. Historically, this belief can be traced as far back as the early observations made by ancient Greek physicians. For example, Hippocrates of Cos (ca. 460–370/377 BCE) wrote works on medical subjects such as physical diagnoses, dietetics, and hygiene while expressively stating in his work *Of Air, Water and Situation* (ca. 400 BCE) the following:

> [W]hen a physician comes a perfect stranger to a City, he should consider well the situation of it, how it stands with respect to the Winds, and the risings of the Sun. . . . He should also consider what sort of Water they use, whether standing and soft, or hard and from high grounds and rocks, or salt and such as will not boil well. The Land should be consider'd too, whether it be naked and dry, or woody and watery; whether in a bottom and suffocating place, or upon an eminence and cold. Nor should the way of Living they are most fond of be forgot; as whether, for instance, they love drinking, feasting, and idleness, or are given to exercise and labour, or are lovers of eating but not of drinking.

As documented in the ancient writings of the physicians and philosophers from the fifth century BCE, it became known that various human diseases required specific temperatures or moisture conditions in order to affect the human body. Other forms of diseases could only be transmitted to people if particular carrier animals were present. A good historical example is endemic malaria that survived in hot and humid climate zones and necessitated its transmission through mosquitoes. This was still the case in the southern parts of Europe throughout the late medieval and early modern period before malaria eventually became eradicated in these regions during the 19th and 20th centuries following major environmental health projects that involved the drying of swamps, the moving of population groups, and then later on the application of detergents and insecticides. Furthermore, diseases could be endemic to specific populations and become adjusted to these climates and environmental conditions.

Consider, for example, the devastating infections of the North American native peoples when Old and New World populations came into direct contact with them after the major discoveries and explorations of the 15th and 16th centuries. The enormous susceptibility of native populations to human viruses such as smallpox has been interpreted as a lack of genetic immunity, based on the former eradication of the smallpox virus (due to the small number of inhabitants and low exposure rates) and the harsh climate conditions on the North American continent. These had likewise posed crucial environmental constraints on viruses and bacteria from the time that the native peoples had immigrated during the Pleistocene Epoch over the Beringia land bridge from East Asia approximately 20,000 years ago. This early case of the human migration tendency, leading to altered disease and susceptibility patterns, shows the tight interrelationship among climate conditions, environmental impacts, and human diseases.

Another important example of changes in population health conditions is found in the era of European climate history often termed the Medieval Warm Period. This period lasted from approximately 1000 CE to the beginning of the 1300s, when the Northern Hemisphere, and particularly southwestern France and Spain, were influenced by relatively moist, warm airstreams from the Atlantic Ocean. The weather changes led to a more positive environment for grain agriculture and coincided with the invention of three-field crop rotation as a more intense way to cultivate European soils. In the following years, between the 11th century and the end of the 12th century, the overall population in Europe grew by one-third to one-half, totaling about 50 million inhabitants.

During the early 14th century, however, the average temperatures dropped—as testified by crop changes, lowered river water levels, the reduction of cultivated vineyards, and so on. As a result, Europe experienced a prolonged period of colder weather for about half a millennium, which historians later called the Little Ice Age. There is good evidence in the archaeological and historical epidemiology literature, such as in Sheldon Watts's *Epidemics and History: Disease, Power, and Imperialism* (1999), that poverty increased substantially in Europe between the 13th and 16th centuries. Brian M. Fagan, in *The Little Ice Age* (2000), has argued that the climate changes of the 14th and 15th centuries greatly impacted the economic and social trends of the period. Certain historians of medicine such as Alfred Jay Bollet, in *Plagues and Poxes* (2004), have argued that the devastating conditions of the long series of plague epidemics since 1347 (leading to an estimated net loss of more than 75 million people worldwide) could only have arisen as a result of the drastically changed climatic conditions.

During the Little Ice Age, mean temperatures changed frequently, with the coldest periods recorded directly after 1650, 1770, and 1850. It is suggestive that these periods coincide with major European conflicts, including the Thirty Years' War, the Seven Years' War, and the mid-19th-century Bourgeois revolutions.

In Western Europe during the late 18th century, crop damages caused by drought, hailstorms, and exceptionally cold winters resulted in the steep rise of grain prices in major cities and starvation in wide areas of the countryside. For a major part of the 18th century, France had experienced a significant increase in its agricultural productivity, with the prices for agricultural goods remaining high due to the ongoing population growth. With the devastating harvests after 1770, however, the lower social classes experienced considerable undernourishment, leading to an increased political frustration with the ruling aristocracy. Neither the powers of the absolutist military nor the commission work of the Royal Academy of Sciences in France—such as Pierre-Joseph Desault (1738–1795) and later Joseph Claude Récamier (1774–1852) and François Magendie (1783–1855)—in conjunction with Parisian municipalities, in their efforts to enhance nutritional values, could remedy the food shortages and living conditions. Political support for King Louis XVI (1754–1793) dwindled dramatically, and the French Revolution of 1789 ensued, with one of the major demands being the decrease of bread shortages and the lowering of food prices, something that the Jacobin government made as one of its major priorities, followed by state security and egalitarian liberal rights.

On the other side of the English Channel following the English Civil War, a national social security system had been developed. Between the late 1600s and the early 1800s, what came to be known as the Old Poor Law protected the poorest members of British society from food scarcity and led to an increase in the prospect for work and the establishment of poor houses in major cities and municipalities. The law had its origins in central government legislation and was significantly funded through mandatory local property taxes and administered on the clerical parish level. The 18th century then saw a particular growth in hospitals for patients with chronic diseases and mental illness, as exemplified by the Bedlam asylum in London.

When looking further into the historical changes of climate and its impact on man-made environments and public health during the 18th century, note that the agricultural economy of the Mediterranean regions also were severely affected. The wine crop failed in the late 1770s for many annual cycles, and with the decline in grain productivity, the gross income of farmers became substantially reduced. This in turn led to forms of considerable policing of peasants and day laborers and the strengthening of obsolete noble prerogatives that gave rise to public discontent in the autonomous north Italian regions and the French-Italian territories and islands.

After the end of the 17th century, ecological contexts were significantly altered through the worldwide phenomena of population growth—particularly in Western Europe and China. Population growth was made possible through the application of innovative and successful farming methodologies and irrigation systems. General living conditions also changed. In Western Europe, for example, city growth and industrialization led to significant increases in population densities, instigating a relocation of European soldiers, traders, and colonists to the Americas and other continents. From this significant transformation in population and environmental dynamics also emanated new routes for the proliferation of epidemics and diseases. A most remarkable example was the onset of a number of cholera epidemics that reached Europe from East Asia in the 1810s. After the latter half of the 18th century, circumscribed epidemics began to devastate many regions in the world, leading, for example, to a typhus epidemic in London in 1750 that decimated the densely populated poor districts of the English metropolis. On the other hand, the increase of agricultural productivity and political stability in Europe and North America constrained the spread of cholera and other epidemics and fortunately decreased incidence rates and disease lethality in these regions of the world.

The incidence rate of epidemics declined following the altered environmental contexts that reduced the natural habitats of viruses and bacteria. With the application of the three-field crop rotation, the agricultural economies of Western Europe morphed into intensive forms of cattle farming that in turn led to a decrease in the prevalence of malaria in the Mediterranean countries. Due to these new agricultural processes, the mosquitoes now primarily fed on cattle, transmitting the plasmodia parasites to them so that the life cycle of malaria transmission in humans eventually became disrupted. By the middle of the 18th century, environmental health conditions became more and more intentionally adjusted, originating in the requirement for public health programs and city planning projects (sewage, sanitation, and water supplies). The new public health tradition led to the implementation of technological and planning reforms, giving rise to the opening of additional city hospitals, decentralized health care departments, and the provision of clean water. During the early 19th century, public health programs in major European and North American cities and metropolitan regions, such as Boston and Philadelphia, became more and more efficient in containing the spread of disease and in improving general health and nutrition. The formerly serious epidemics of plague, cholera, and typhoid fever faded out.

At the same time, however, when the basic living conditions improved in major European and subsequent American metropolises, the industrial pollution (coal dust, smoke, oil sewage, and so on) ensued. In the area of environmental health, New York City physicians Burrill Bernard Crohn

(1884–1983), Leon Ginzburg (1898–1988), and Gordon D. Oppenheimer (1900–1974) had, for example, described an inflammatory condition of the small intestine that came to be known as Crohn's disease. After 1932, the reporting of cases of Crohn's disease allowed for the calculation of the prevalence and incidence rates in major cities around the world. Modern epidemiological studies after the 1950s have revealed that the incidence rates in developing countries increased with the emerging economic development. Before the landmark investigations of Crohn and his collaborators, cases of inflammatory diseases of the small and large intestines had been recognized in England, Scotland, Ireland, Germany, and Poland and were recorded in medical journals, textbooks, and hospital records. A review of modern medical literature reveals that there were many attempts to review historical case descriptions. As in other instances of retrospective diagnoses, the undertaking is difficult because of the similarities between Crohn's disease and ulcerative colitis, cancer, and dysentery and could have been influenced by the medical knowledge, diagnostic tools, and procedures available at the time of the occurrence. The emergence of Crohn's disease in the British Isles as economic development progressed parallels the breathtaking development of industrialization as an important factor in historical epidemiology and particularly so in that of urban mortality changes.

It was not until the late 19th century that scientists suggested that human emissions of greenhouse gases might be able to change the global climate. The Irish physicist John Tyndall (1820–1893) had explored the consumption of heat in different gases, and Svante Arrhenius (1859–1927), a Swedish scientist, calculated in 1896 that doubling the atmospheric CO_2 would lead to increased surface temperatures of 5–6 degrees Celsius. Arrhenius thought that the impact of global warming would eventually lead to the reduction of snowfalls, ice coverage, and glacial activity and might have beneficial health effects.

At the end of the 20th century, a consensus position emerged acknowledging that human activity had led to gradual warming phenomena, as reflected in the reports of the Intergovernmental Panel on Climate Change. The effect of global warming on public health and the response required by public health officials has been discussed in the research literature. For example, a roundtable is sponsored by the United Nations Foundation, the National Institute of Environmental Health Sciences, the National Center for Environmental Health, and the Centers for Disease Control and Prevention concerning the intersection of climate change, environmental health, and public health programs worldwide.

Frank W. Stahnisch

Further Reading

Bollet, Alfred Jay. *Plagues and Poxes: The Impact of Human History on Epidemic Disease.* New York: Demos Medical Publishing, 2004.

Crosby, Alfred. *Ecological Imperialism: The Biological Expansion of Europe, 900–1900.* Cambridge: Cambridge University Press, 1986.

Edelstein, Ludwig. "The Hippocratic Physician." In *Ancient Medicine: Selected Papers of Ludwig Edelstein,* edited by O. Temkin and C. L. Temkin, 87–110. Baltimore: Johns Hopkins University Press, 1967.

Epstein, Paul R. "Climate and Health." *Science* 285 (1999): 347–348.

Epstein, Paul R. "Climate Change and Human Health." *New England Journal of Medicine* 353 (2005): 1433–1436.

Fagan, Brian M. *The Little Ice Age: How Climate Made History, 1300–1850.* New York: Basic Books, 2000.

Fleming, James Roger. *Historical Perspectives on Climate Change.* New York: Oxford University Press, 1998.

Grove, Jean M. *The Little Ice Age.* London: Methuen, 1988.

Imbrie, John, and Katherine P. Imbrie. *Ice Ages.* Hillside, NJ: Enslow Publishers, 1979.

Janković, Vladimir. *Confronting the Climate: British Airs and the Making of Environmental Medicine.* New York: Palgrave, 2010.

McIvor, Arthur, and Roland Johnston. *Miners' Lung: A History of Dust Disease in British Coal Mining.* London: Ashgate, 2007.

Mitman, Gregg. "Where Ecology, Nature, and Politics Meet: Reclaiming the Death of Nature." *Isis* 97 (2006): 496–504.

Mitman, Gregg, Michelle Murphy, and Christopher Sellers, eds. *Landscapes of Exposure: Knowledge and Illness in the Making of Modern Environments. Osiris,* Vol. 19. Miami: History of Science Society, 2004.

Mooney, Graham, and Simon R. S. Szreter. "Urbanisation, Mortality and the Standard of Living Debate: New Estimates of the Expectation of Life at Birth in Nineteenth-Century British Cities." *Economic History Review* 40 (1998): 84–112.

Newman, Simon. "Dead Bodies: Poverty and Death in Early National Philadelphia." In *Down and Out in Early America,* edited by B. G. Smith, 41–62. University Park: Pennsylvania State University Press, 2004.

Oldstone, Michael B. A. *Viruses, Plagues, and History.* Oxford: Oxford University Press, 1998.

Oreskes, Naomi, and Erik Conway. *Merchants of Doubt: How a Handful of Scientists Obscured the Truth on Issues from Tobacco Smoke to Global Warming*. London: Bloomsbury, 2010.

Stahnisch, Frank. *Ideas in Action: Der Funktionsbegriff und seine methodologische Rolle im Forschungsprogramm des Experimentalphysiologen François Magendie (1783–1855)*. Muenster: LIT-Press, 2003.

Voegele, Joerg. *Urban Mortality Change in England and Germany, 1870–1910*. Liverpool: Liverpool University Press, 1998.

Watts, Sheldon. *Epidemics and History: Disease, Power, and Imperialism*. New Haven, CT: Yale University Press, 1999.

Environmentalism of the 1970s

The intellectual roots of American environmentalism are most often traced back to 19th-century ideas of romanticism and transcendentalism and thinkers such as Henry David Thoreau. These ideas of conservation or preservation are important components of what scholars today refer to as modern environmentalism; however, today's environmentalism composes a social movement with clear political outcomes. Most scholars include modern environmentalism with other reform efforts growing out of the 1960s.

Grounding the Environmental Paradigm of the 1970s

Overall, the 1960s counterculture contributed to the development of many institutions that would change basic relationships in American life. The American relationship with nature was one of the most prominent shifts. Much of what became known as the modern environmental movement was organized around groups and organizations that prospered with the influence of 1960s radicalism; however, the real impact of these organizations came during the later 1960s and 1970s when their membership skyrocketed with large numbers of the concerned and not-so-radical middle class. These growing organizations demanded a political response from lawmakers. Whether the issue was pollution, the need for alternative energy, or the overcutting of rain forests, concerned citizens exerted an environmental perspective on a scale not seen previously.

Contrasted with the conservation movement of the late 19th century, the social landscape of 20th-century environmentalism had changed a great deal. For instance, many of these environmental special interest groups would evolve into major political players through lobbying. Nongovernmental organizations (NGOs) broadened the grassroots influence of environmental thought; however, they also created a niche for more radical environmentalists. The broad appeal as well as the number of special interest portions of environmental thought stood in stark contrast to 19th-century environmentalism. Whereas early conservationists were almost entirely members of the upper economic classes of American society, the new environmentalists came mostly from the middle class that grew rapidly after World War II.

During the 1970s and 1980s, these NGOs helped to bring environmental concerns into mainstream American culture. Some critics argue that American living patterns changed little; however, the awareness and concern over human society's impact on nature had reached an all-time high in American history. These organizations often initiated the call for specific policies and then lobbied members of Congress to create legislation. By the 1980s, NGOs had created a new political battlefield as each side of environmental arguments lobbied lawmakers. Environmentalists were often the ones offering their perspective as the moral high road that would help rein in human development and expansion with an ethic of restraint.

The American public often financially supported organizations that argued for their particular perspectives. Even traditional conservationist organizations such as the Sierra Club (founded in 1892), the National Audubon Society (founded in 1905), the National Parks and Conservation Society (founded in 1919), the Wilderness Society (founded in 1935), the National Wildlife Federation (founded in 1936), and the Nature Conservancy (founded in 1951) took much more active roles in policy making. The interest of such organizations in appealing to mainstream middle-class Americans helped to broaden the base of environmental activists. It also contributed to the formation of more radical-thinking environmental NGOs that disliked the mainstream interests of the larger organizations. In fact, many devout environmentalists argued that some of these NGOs were part of the establishment that they wished to fight.

One of the first writers to take advantage of this increased interest among middle-class Americans was the scientist and nature writer Rachel Carson. She began writing about nature for general readers in the late 1950s. In 1962, Rachel Carson's *Silent Spring* erupted onto the public scene to become a best seller after first being serialized in the *New Yorker.* Carson's scientific findings brought into question basic assumptions that Americans had about their own safety and many of the chemicals that they used to create their comfortable standard of living.

By the end of the 20th century, environmentalism became a diverse movement. In this photo, members of the conservation group Friends of the Earth stage a protest against the oil industry at the United Nations Framework Convention on Climate Change in Buenos Aires, Argentina in November 1998. (AP/Wide World Photos)

Overall, the cultural attitude toward progress predicated on cheap energy and manufactured chemicals was beginning to lose its dominant hold. In the case of Carson's work, her exposé of the health impacts of chemicals helped to disrupt the paradigm that supported Americans' trust, more generally, in technological progress. In a single summer, chemical science and blind confidence in technological progress had fallen from its unchallenged pedestal. Here is a portion of what Carson wrote:

> The "control of nature" is a phrase conceived in arrogance, born of the Neanderthal age of biology and philosophy, when it was supposed that nature exists for the convenience of man. The concepts and practices of applied entomology for the most part date from that Stone Age of science. It is our alarming misfortune that so primitive a science has armed

> itself with the most modern and terrible weapons, and that in turning them against the insects it has also turned them against the Earth.

Carson's story and her words would inspire an entire portion of the American population to reconsider our society's living patterns.

Following Carson, in 1968 Garrett Hardin wrote an article that developed the ecological idea of the commons. This concept and his argument of its tragic (undeniable) outcome in depletion gave humans a new rationale with which to view common resources such as the air and the ocean. He wrote:

> The tragedy of the commons develops in this way. Picture a pasture open to all. It is to be expected that each herdsman will try to keep as many cattle as possible on the commons. Such an arrangement may work reasonably satisfactorily for centuries because tribal wars, poaching, and disease keep the numbers of both man and beast well below the carrying capacity of the land. Finally, however, comes the day of reckoning, that is, the day when the long-desired goal of social stability becomes a reality. At this point, the inherent logic of the commons remorselessly generates tragedy.
>
> As a rational being, each herdsman seeks to maximize his gain. Explicitly or implicitly, more or less consciously, he asks, "What is the utility to me of adding one more animal to my herd?" This utility has one negative and one positive component. . . .
>
> Adding together the components . . . the rational herdsman concludes that the only sensible course for him to pursue is to add another animal to his herd. And another. . . . But this is the conclusion reached by each and every rational herdsman sharing a commons. Therein is the tragedy. Each man is locked into a system that compels him to increase his herd without limit—in a world that is limited. Ruin is the destination toward which all men rush, each pursuing his own best interest in a society that believes in the freedom of the commons. Freedom in a commons brings ruin to all.

This essay marked a crucial moment in Americans' ability to apply the scientific ideas of ecology, conservation, and biology to human life—to assume that we possibly were *not* the exceptional species on Earth. Or, even if we were exceptional, maybe this status came with a responsibility for stewardship and management instead of for expansion. Such new perspectives were critical to preparing the cultural soil that would be able to conceive of the concept of global warming when it was presented by scientists. Most important, for the first time Americans were learning to view the world around them not as indestructible but instead as quite volatile and composed of limited resources.

Connecting Emissions to Larger Ecological Problems

In a variety of examples during the 1970s, new scientific findings—or at least findings that were newly made public—brought complex implications of human life into the public mind. As Carson and Hardin demonstrated, humans' high-energy existence exerted impacts on Earth's natural systems. Environmentalism helped to create a culture that might consider such issues and their implications in an effort to correct these problems. In hindsight, we find that these issues of the 1970s were baby steps toward grasping the issue of climate change.

The new appreciation of the environmental impact of the internal combustion engine was just the beginning of the problems that would face the brokers of America's high-energy existence. A number of developments during the late 1970s and early 1980s started to move global warming into the political limelight. As a general backdrop, a number of environmental issues became important to the general public and to American politics. One of the first was the problem of urban smog, caused by the release of smoke and sulfur dioxide (SO_2) from various human activities. (Volcanoes are a secondary source of SO_2 but, of course, do not explain urban smog problems.) Sulfur dioxide reacts with water vapor to form sulfuric acid and other sulfates, which hang in the atmosphere as aerosols. Automobile and industrial emissions also release nitrogen oxides, which combine with various organic compounds to create yet more particulate matter in the atmosphere. Although smog had already been a political issue in the early part of the 20th century, urban pollution continued to be a visible problem in the early 1970s, and the American press renewed concern for the issue. Public awareness gave the issue political traction, and legislation, which started with the Clean Air Act of 1963, continued to be passed regularly by Congress, up to the Clean Air Act of 1990.

It was also during the 1970s that concern grew over the environmental issue known as acid rain. The problem was that the various sulfur and nitrogen compounds formed as aerosols in the atmosphere came down with precipitation. This created acidic soil conditions (low pH values), which were not good for plant growth, agriculture, and aquatic life. The Acid Rain Program was an outgrowth of the Clean Air Act of 1990. The program allowed the Environmental Protection Agency (EPA) to head a market-based system in which industries (such as coal-burning power plants) were encouraged to reduce the release of sulfur dioxide and nitrogen oxides by establishing emission quotas and then allowing the companies to buy and sell emission allowances, depending on their economic circumstances and needs.

The growing complications related to air pollution grew when scientists' view of smog was no longer limited to local areas, such as the city of Los Angeles. New computer modeling combined with better understanding of the functioning of various layers of Earth's atmosphere to make clear that something was rapidly depleting the planet's protective ozone layers. In addition, heat was becoming trapped in Earth's atmosphere at an alarming rate, suggesting that the natural greenhouse effect might be increasing. Finally, as discussed previously, by the 1990s scientists concluded that Earth was warming at a pace without historical precedent.

Considering the American Standard of Living

It did not take long for this new way of viewing the human condition to focus itself on the ethics behind Americans' high-energy lifestyle. The thinker most often given credit for making this transition in thought is E. F. Schumacher, a British economist who, beginning in 1973, wrote a series of essays titled *Small Is Beautiful.* Another title, *Small Is Beautiful: Economics as if People Mattered,* became a best seller. Like all the books in the series, this one emphasized the need to consider a different view of progress than the expansive energy-intensive American approach. Building from the idea of limits that the embargo had reinforced, Schumacher emphasized a philosophy of enoughness in which Americans designed their desires around basic human needs and a limited appropriate use of technology. Later, this approach was termed "Buddhist Economics."

Schumacher particularly faults the conventional economic thinking that failed to consider sustainability and instead emphasized growth at all costs and a basic trust in the idea that bigger is better. The key, he argues, was in the conception of new technologies—when inventors and engineers were choosing why they pursued an innovation. He writes:

> Strange to say, technology, although of course the product of man, tends to develop by its own laws and principles, and these are very different from those of human nature or of living nature in general. Nature always, so to speak, knows where and when to stop. Greater even than the mystery of natural growth is the mystery of the natural cessation of growth. There is measure in all natural things—in their size, speed, or violence. As a result, the system of nature, of which man is a part,

> tends to be self-balancing, self-adjusting, self-cleansing. Not so with technology, or perhaps I should say: not so with man dominated by technology and specialisation. Technology recognises no self-limiting principle—in terms, for instance, of size, speed, or violence. It therefore does not possess the virtues of being self-balancing, self-adjusting, and self-cleansing. In the subtle system of nature, technology, and in particular the super-technology of the modern world, acts like a foreign body, and there are now numerous signs of rejection.
>
> Suddenly, if not altogether surprisingly, the modern world, shaped by modern technology, finds itself involved in three crises simultaneously. First, human nature revolts against inhuman technological, organisational, and political patterns, which it experiences as suffocating and debilitating; second, the living environment which supports human life aches and groans and gives signs of partial breakdown; and, third, it is clear to anyone fully knowledgeable in the subject matter that the inroads being made into the world's nonrenewable resources, particularly those of fossil fuels, are such that serious bottlenecks and virtual exhaustion loom ahead in the quite foreseeable future.
>
> Any one of these three crises or illnesses can turn out to be deadly. I do not know which of the three is the most likely to be the direct cause of collapse. What is quite clear is that a way of life that bases itself on materialism, i.e. on permanent, limitless expansionism in a finite environment, cannot last long, and that its life expectation is the shorter the more successfully it pursues its expansionist objectives.

Although Schumacher's points may have been extreme, they presented a new paradigm of restraint in energy management that appealed to some intellectuals. It was organized by a new ethic of restraint that was tailor-made for the findings of climate science.

One of the most noticeable spokespeople of this alternative energy paradigm was the economist Amory Lovins, who published an article titled "Soft Energy Paths" in *Foreign Affairs* in 1976. In his subsequent book, Lovins contrasted the "hard energy path," as forecast at that time by most electrical utilities, and the "soft energy path," as advocated by Lovins and other utility critics. He writes:

> The energy problem, according to conventional wisdom, is how to increase energy supplies . . . to meet projected demands. . . . But how much energy we use to accomplish our social goals could instead be considered a measure less of our success than of our failure. . . . [A] soft

> [energy] path simultaneously offers jobs for the unemployed, capital for businesspeople, environmental protection for conservationists, enhanced national security for the military, opportunities for small business to innovate and for big business to recycle itself, exciting technologies for the secular, a rebirth of spiritual values for the religious, traditional virtues for the old, radical reforms for the young, world order and equity for globalists, energy independence for isolationists. . . . Thus, though present policy is consistent with the perceived short-term interests of a few powerful institutions, a soft path is consistent with far more strands of convergent social change at the grass roots.

Lovins's ideas moved among intellectuals but found immediate acceptance with neither political leaders nor the general public. The shift, though, seemed to arrive in the form of President Jimmy Carter.

With additional instability in the Middle East by the late 1970s, Carter elected to take the ethic of energy conservation directly to the American people. Despite some rapid success in the area of energy conservation, the American environmental movement of the 1970s came out of the decade under siege. The 1990s brought a governmental backlash to many of the initiatives.

Nature Strikes Back and Presses a Tipping Point

In the 1970s, many Americans favored the establishment of environmental regulation because they felt that the American ethic for development had pushed nature too far. If any additional evidence was needed to prove the need for such legislation, nature seemed to be in open rebellion against humans in 1969. These events, which became large-scale media events, functioned to radically change the role of nature in everyday American life.

Together, these environmental catastrophes changed the American conscience. Each of the following events cycled directly into the new concern for the environment and became prominent headlines in American newspapers and news media.

Lake Erie Is Dead

In the 1960s, with little visible life in its waters, Lake Erie was declared dead. Scientists quickly learned that the opposite was true; the lake was full of life but not the correct balance of life forms. Primarily, an excessive amount of algae, created by pollution and excessive nutrients, had created eutrophication in Lake Erie. As the excessive algae expanded, it soaked up the lake's supply of oxygen, making it impossible for other species to survive.

Lake Erie's situation was particularly acute because it is the shallowest and warmest of the five Great Lakes. It had been abused extensively for decades, enduring runoff from agriculture, urban areas, industries, and sewage treatment plants along its shores. The worst of the pollutants coming into the lake was phosphorus, which entered from agricultural fields. Unfortunately, the phosphorus did just what the farmers intended when they spread the fertilizer on their fields—only it also stimulated vegetative growth once it ran off into Lake Erie. This phosphorus was one of the primary reasons for the growth of the algae.

In response to Lake Erie's "death," the Great Lakes Water Quality Agreement (GLWQA) was signed by the United States and Canada in 1972. The agreement emphasized the reduction of phosphorus entering Lakes Erie and Ontario. Coupled with the U.S. and Canadian Clean Water acts, the GLWQA made significant progress toward reducing the phosphorus levels in Lake Erie. Through GLWQA, two nations for the first time committed themselves to creating common water quality goals.

The Cuyahoga River Is on Fire

Throughout American industrialization, many rivers were exploited and left heavily laden with pollutants. Rivers near industrial centers such as Pittsburgh and Cleveland became symbols of this degradation. This legacy attracted national attention in 1969 when the Cuyahoga River did something that it had done many times before: caught on fire. This time, however, the American public viewed the occasion as a profound statement about the impact of pollution on our natural environment.

The fire was actually very brief in duration. It began at 11:56 a.m. and lasted for approximately 20 minutes. The area of the river that caught fire was fairly out of the way: just southeast of downtown Cleveland. The flames damaged two railway bridges and were estimated to reach heights of roughly five stories. The cause was nothing out of the ordinary for the Cuyahoga: a slick of highly volatile petroleum derivatives had leaked from one of the refineries located along the river. That turned out to be the real issue—national disdain that such an event was fairly routine for such industrial locales.

The river fire served as a reminder of the importance of continued support for cleanup of Lake Erie and the Cuyahoga River. On the day after the fire, Cleveland mayor Carl Stokes stood on the damaged Norfolk and Western Bridge and called for the public to rally support for an effort to clean up the

Cuyahoga and promised to sue the State of Ohio and the individual polluters. Stokes referred to the polluted state of the river as "a long-standing condition that must be brought to an end."

The Cuyahoga fire was most important as a symbol for a new era. The river fire received major national media attention, with a *Time* magazine article on August 1, 1969 (approximately one month after the fire). In October 1969, federal officials passed a bill that would grant states more assurances that projects aimed at improving water quality would receive federal support.

Oil Spills off Santa Barbara, California

On the afternoon of January 29, 1969, a Union Oil Company platform stationed six miles off the coast of Summerland suffered what is referred to as a blowout. Oil workers had drilled a well down 3,500 feet below the ocean floor. Riggers began to retrieve the pipe in order to replace a drill bit when the mud used to maintain pressure became dangerously low. Natural gas blew out of the pipe, sending oil into the surrounding water. When the drillers successfully capped the well, pressure built up, and five breaks appeared in an east-west fault on the ocean floor, releasing oil and gas from deep beneath Earth.

For 11 days, oil workers struggled to cap the rupture. During that time, 200,000 gallons of crude oil bubbled to the surface to form an 800-square-mile slick. Incoming tides brought the thick tar to beaches from Rincon Point to Goleta, covering 35 miles of coastline and coastal life. The slick also moved south, tarring Anacapa Island's Frenchy's Cove and beaches on Santa Cruz, Santa Rosa, and San Miguel islands.

Many lessons were learned on this first oil spill of the environmental era. Rapid cleanup was crucial. In this spill, it took oil workers 11.5 days to control the leaking oil well. The cleanup had to be multifaceted: skimmers scooped oil from the surface of the ocean, while airplanes dumped detergents on the spill to try to break up the slick. Meanwhile, on the beaches and harbors, volunteers spread straw to soak up the tar and oil, and rocks were steam-cleaned.

Just days after the spill occurred, activists founded Get Oil Out (GOO) in Santa Barbara. Founder Bud Bottoms urged the public to cut down on driving, burn oil company credit cards, and boycott gas stations associated with offshore drilling companies. Volunteers also helped the organization gather 100,000 signatures on a petition calling for the ban of offshore oil drilling. Although drilling was only halted temporarily, new laws eventually tightened regulations a bit more on offshore drilling.

Three Mile Island and the Antinuclear Movement

Following World War II, nuclear technology served as a symbol of the capabilities of American know-how to master basic needs such as energy consumption. By the 1970s, a movement against nuclear technology, which began with weapons and expanded to energy production, became an important component of the emerging environmental perspective.

Following the use of nuclear weapons to end World War II in Hiroshima and Nagasaki, J. Robert Oppenheimer and other scientists voiced serious concerns about humans' future with atomic technology. In the United States, President Dwight D. Eisenhower launched the domestic atomic era during the 1950s when he led the push to use atomic technology not just for weapons but also for energy production. During this same era, however, scientists disclosed the dangerous implications of any atomic technology—primarily, the release of radiation. A growing number of international critics began to publicly demand the control and discontinuation of atomic development.

During the late 1940s, the United States remained in the experimental phase of its nuclear technology, staging tests (primarily at the Bikini Atoll in the Pacific Ocean) that also demonstrated its power to the world. David Bradley worked as a scientist measuring radiation levels after the explosion. He became so disgusted with the results—and the government's refusal to release them publicly—that he wrote a book, *No Place to Hide,* to announce his findings to the world. Although the book was initially banned, his findings initiated the call for more information that grew into a full-blown protest movement.

The antinuclear movement was strongest in Europe, where groups including the H-Bomb National Campaign, the National Council for the Abolition of Nuclear Weapons Tests, and the Direct Action Committee (DAC) protested any further development or testing of atomic weapons. Possibly the most influential organization, however, was the Campaign for Nuclear Disarmament (CND). Founded in February 1958, the CND—along with the DAC—made international headlines several weeks later with a protest march from London to a British nuclear weapons facility. The CND's annual Easter marches soon became major events, with thousands of participants.

The CND's protest effort created an international symbol that reached well beyond the issue of nuclear disarmament: the peace sign. Designed in 1958 by Gerald Holtom, the peace symbol was unveiled for the 1958 Easter weekend protest. Five hundred cardboard lollipops on sticks were produced that were half black and half green. Eventually the design was put on badges and worn for other protests. In addition, the peace sign immediately crossed to the United States to be used to protest the Vietnam War.

The CND became an important political actor in Britain by demanding that Britain unilaterally get rid of its nuclear weapons. In Britain's 1964 general elections, the Labour Party, after CND pressure, campaigned on a platform to cancel government plans to buy U.S. Polaris nuclear missile submarines. Similar antinuclear groups, such as the Dutch Interchurch Peace Council, also became influential. Other protest groups found members in East and West Germany, which functioned as ground zero in any possible Cold War confrontation.

During the 1960s, American environmental organizations used information such as *No Place to Hide* and began to protest nuclear testing. Most protests took place outside the United States; the veil of Cold War support and secrecy limited many Americans' willingness to speak out on this issue. When the U.S. government made plans for a test at Amchitka Island, Canadian and American members of the Sierra Club created a protest of watershed importance for the internationalization of the protest movement. After the protests, Paul Watson joined a number of other protestors and established the Don't Make a Wave Committee. It was this small group that sponsored the voyage of the *Greenpeace I* in 1971.

The *Greenpeace I* was a retired Canadian fishing boat that set out from Vancouver, British Columbia, bound for Amchitka Island. The activists' intention marked a new idea in antinuclear protests: they would disregard warnings and guard ships and sail directly into the test site. On board *Greenpeace I* were 13 volunteers, including Robert Hunter, Rod Marining, and Lyle Thurston. Three decades later, these three would still be sailing with Watson on campaigns for a new organization: Sea Shepherd. Thanks to the intervention of *Greenpeace I,* the test at Amchitka was postponed. The nuclear test had been delayed to foil the voyage of the *Greenpeace I;* however, the U.S. Atomic Energy Committee advanced the next blast date to avoid the *Greenpeace Too,* which was the second ship in the protest fleet.

The controversy that the *Greenpeace* voyages generated led to the decision to cancel further tests. The detonation of November 1971 was the last nuclear test to take place at Amchitka. In 1972, the Don't Make a Wave Committee took the name of the two ships from the first campaign and renamed themselves the Greenpeace Foundation. This organization marked the beginning of a truly international movement to support the environment and other causes. Greenpeace continues to aid in raising awareness about global warming and many other issues with international consequences. The first issue, though, to reach across national borders and effect all humans was atomic weapons.

Weapons were one thing; it was pretty simple for people to understand their inherent danger. However, nuclear power had been marketed to the American public as a safe power source. Nuclear power lost its futuristic promise for Americans with one incident: the Three Mile Island accident. A power plant near Harrisburg, Pennsylvania, Three Mile Island ran two very effective nuclear reactors for approximately five years. At 4:00 a.m. on March 28, 1979, while the Unit 2 reactor was operating at 97 percent power, it experienced a relatively minor malfunction in the secondary cooling circuit. This malfunction caused the temperature of the primary coolant to rise, which triggered an automatic shutdown of the reactor. The problem arose when a relief valve failed to close and instrumentation did not reveal this fact to operators. Due to this sequence of events, so much of the primary coolant drained away that the residual decay heat in the reactor core was not removed. The core suffered severe damage as a result. In short, the primary problem revealed at Three Mile Island was twofold: first, the potential volatility of what the public had been told was a benign technology, and second, the industry experienced difficulty in properly monitoring, maintaining, and reacting to such potential difficulties. Modern environmentalism had found yet another reason to suspect technology and industry's ability to regulate it.

Environmental Protection in the 1970s

Although many Americans remained skeptical about the federal government's ability to regulate the natural environment and about the prescience of any controls over industrial development, these events helped to convince politicians and voters that action was needed. Many Americans had simply had enough. Others very clearly felt shameful at the present state of America's natural environment. During the next decade the social and cultural change initiated by the 1960s and fed by a barrage of demonstrations of environmental degradation and the need for conservation created a deluge of environmental legislation.

The public outcry would be so severe that even a conservative such as Richard Nixon might be deemed "the environmental president" as he signed the National Environmental Protection Act in 1969, creating the EPA. The act reads in part as follows:

> (a) The Congress, recognizing the profound impact of man's activity on the interrelations of all components of the natural environment, particularly the profound influences of population growth, high-density

urbanization, industrial expansion, resource exploitation, and new and expanding technological advances and recognizing further the critical importance of restoring and maintaining environmental quality to the overall welfare and development of man, declares that it is the continuing policy of the Federal Government, in cooperation with State and local governments, and other concerned public and private organizations, to use all practicable means and measures, including financial and technical assistance, in a manner calculated to foster and promote the general welfare, to create and maintain conditions under which man and nature can exist in productive harmony, and fulfill the social, economic, and other requirements of present and future generations of Americans.

(b) In order to carry out the policy set forth in this Act, it is the continuing responsibility of the Federal Government to use all practicable means, consistent with other essential considerations of national policy, to improve and coordinate Federal plans, functions, programs, and resources to the end that the Nation may—

1. fulfill the responsibilities of each generation as trustee of the environment for succeeding generations;
2. assure for all Americans safe, healthful, productive, and aesthetically and culturally pleasing surroundings;
3. attain the widest range of beneficial uses of the environment without degradation, risk to health or safety, or other undesirable and unintended consequences;
4. preserve important historic, cultural, and natural aspects of our national heritage, and maintain, wherever possible, an environment which supports diversity, and variety of individual choice;
5. achieve a balance between population and resource use which will permit high standards of living and a wide sharing of life's amenities; and
6. enhance the quality of renewable resources and approach the maximum attainable recycling of depletable resources.

(c) The Congress recognizes that each person should enjoy a healthful environment and that each person has a responsibility to contribute to the preservation and enhancement of the environment. . . .

The Congress authorizes and directs that, to the fullest extent possible: (1) the policies, regulations, and public laws of the United States shall be interpreted and administered in accordance with the policies set forth in this Act, and (2) all agencies of the Federal Government shall—

(A) utilize a systematic, interdisciplinary approach which will insure the integrated use of the natural and social sciences and the environmental design arts in planning and in decisionmaking which may have an impact on man's environment;

(B) identify and develop methods and procedures, in consultation with the Council on Environmental Quality established by title II of this Act, which will insure that presently unquantified environmental amenities and values may be given appropriate consideration in decisionmaking along with economic and technical considerations;

(C) include in every recommendation or report on proposals for legislation and other major Federal actions significantly affecting the quality of the human environment, a detailed statement by the responsible official on—

(i) the environmental impact of the proposed action,
(ii) any adverse environmental effects which cannot be avoided should the proposal be implemented,
(iii) alternatives to the proposed action,
(iv) the relationship between local short-term uses of man's environment and the maintenance and enhancement of long-term productivity, and
(v) any irreversible and irretrievable commitments of resources which would be involved in the proposed action should it be implemented.

The public entrusted the EPA as its environmental regulator to enforce ensuing legislation monitoring air and water purity, limiting noise and other kinds of pollution, and monitoring species in order to discern which required federal protection. The public soon realized just how great the stakes were.

During the 1970s, nearly every industrial process was seen as having environmental costs associated with it. From chemicals to atomic power, long-believed technological "fixes" came to have long-term impacts in the form of wastes and residue. Synthetic chemicals, for instance, were long thought to be advantageous because they resist biological deterioration. In the 1970s, this inability to deteriorate made chemical and toxic waste the bane of many communities near industrial or dump sites.

Among the assorted catastrophes, Love Canal stood out as a new model for federal action. The connection between health and environmental hazards became obvious throughout the nation. Scientists were able to connect radiation, pollution, and toxic waste to a variety of human ailments. The smoking gun, of course, contributed to a new era of litigation in environmentalism. Legal battles armed with scientific data provided individuals armed with only NIMBY convictions with the ability to take on the largest corporations in the nation.

Rapidly this decade instructed Americans, already possessing a growing environmental sensibility, that humans—just as Carson had instructed—needed to live within limits. A watershed shift in human consciousness could be witnessed in the popular culture as green philosophies infiltrated companies wishing to create products that appealed to the public's environmental priority. Recycling, requiring the use of daylight savings time, carpooling, and environmental impact statements became part of everyday life after the 1970s.

Green Party History

The political impact of environmentalism became very great by the end of the 20th century. Most often, members of each major American political party might have an environmental commitment. By the 1980s, though, a growing number of Americans continued an international trend and initiated a party organized around environmental and other humanitarian principles: the Green Party.

The U.S. Green Party traces itself to the European Greens, who first organized as an antinuclear propeace movement at the height of the Cold War. It was the German Greens, organized by Petra Kelly, who were most specifically influenced by the U.S. environmental movement of the early 1970s. The Green Party's primary impact, though, came elsewhere in the world.

As a flexible group that could be formed in any nation, the Green Party primarily pursued principles of sustainable governance, including social justice, nonviolence, grassroots democracy, conservation of diversity, and ecological wisdom. Greens believe that the exercise of these principles leads to world health. Expanding through New Zealand and Tasmania in the 1970s, the Green Party may have found its most significant inroads in Germany over the issue of nuclear power. The German Green Party, as well as the Finnish Green Party, experienced political successes in the 1980s and 1990s. Many European nations as well as others with democratic systems saw the concerns of the 1970s environmental movement evolve into political parties by 2000.

Political Outcomes in the United States

Instead of creating an entire new American political party, the political impact of 1970s environmentalism came primarily through policies. As the timeline below portrays, the 1970s brought a clear connection between identification of problems or issues and expectations that the federal government would react by creating laws, regulations, or agencies to confront the challenge. The priority was to ensure a safe and sustainable American standard of living.

Timeline: The Stirring of Environmental Action in the 1960s–1970s

- 1962: Rachel Carson writes *Silent Spring*
- 1964: Barry Commoner's group, formed in the 1950s to oppose the development of atomic energy, begins publishing the journal that will become known as *Environment.*
- 1965: Commoner Ralph Nader leads a critique of the American economic system and the ability of science to question development.
- 1965: The Sierra Club files a lawsuit to protect Storm King Mountain in New York from a power project. The U.S. Supreme Court rules in favor of the club and of noneconomic interests in a conservation case.
- 1966: The Sierra Club opposes building two dams that would have flooded the Grand Canyon and publishes newspaper ads that say "This time it's the Grand Canyon they want to flood. The Grand Canyon." The dam plan is defeated.
- 1969: The Santa Barbara oil spill attracts public attention to polluted beaches.
- 1969: Congress passes the National Environmental Policy Act (NEPA) to mandate consideration of environmental issues prior to major public decisions.
- 1970: The first Earth Day occurs on April 22.
- 1972: Congress passes the Clean Water Act, the Coastal Zone Management Act, the Federal Environmental Pesticide Control Act, the Marine Mammal Protection Act, the Ocean Dumping Act, and the Federal Advisory Committee Act to require representation of public interest advocates on committees.
- 1972: Oregon passes the first bottle-recycling law.
- 1972: The Club of Rome publishes *Limits to Growth* in 1972.
- 1973: Congress passes the Endangered Species Act.
- 1974: Congress passes the Safe Drinking Water Act.

- 1977: Love Canal, New York, is identified as a chemical waste site, requiring evacuation of all inhabitants.
- 1979: The Three Mile Island nuclear reactor near-meltdown occurs in Pennsylvania.
- 1980: Congress passes the Superfund Bill to help identify and pay for cleaning up abandoned toxic waste sites.

Conclusion: Earth Day 1990

While new expectations formed from new scientific understandings to shape a movement to create environmental law and regulation during the 1970s, the major development that supported all of this shifting was a change in the public. The environmental awakening of the 1970s changed American culture and created a social movement that would be known from this point forward as environmentalism. While political winds shifted over the next few decades, this cultural shift endured. The greatest symbol of it may be the most significant cultural act of 1970s environmentalism: Earth Day.

Earth Day 1970, which took place on April 22, 1970, involved 20 million people worldwide. In the United States, the public activities ranged from cleaning up rivers to symbolically burying cars. School groups took on their first environmental community service. And politicians publicly considered the state of their local environments. The focus of attention on environmental causes was a revolutionary change. Earth Day became an annual day to do something on behalf of Earth and also provided us with a way to measure change in environmental awareness over time.

How far could the influence of environmental thought come in 20 years? April 22, 1990, may have seemed like an ordinary day; however, instead of the regular television schedule listings, the *New York Times* listed "TV and the Environment." Most impressive, there were a host of programs to delineate. However, many Americans were not taking time to watch TV; instead, they performed cleanup enterprises on behalf of Mother Earth. Internationally, more than 40 million humans marked some kind of celebration on Earth Day 1990.

For American culture, though, Earth Day 1990 marked a day of broader recognition. American society celebrated a new relationship with the natural world surrounding it. Today, polls reveal that nearly 70 percent of Americans refer to themselves as environmentalists. Such developments are simply the latest in a watershed shift in Americans' awareness of the human impact on the natural environment. The 1990s marked a maturing period for the environmental movement, which had been evolving in the United States since the 19th century.

As part of that maturing process, the policy and science that converged in the 1970s also made it possible for Americans to consider an entirely new scale of problems facing Earth. Primary among these was the issue of climate change.

Brian C. Black

Further Reading

Andrews, R. N. L. *Managing the Environment, Managing Ourselves.* New Haven, CT: Yale University Press, 1999.

Ashworth, W. *The Late, Great Lakes: An Environmental History.* New York: Knopf, 1986.

Boyer, Paul. *By the Bomb's Early Light.* Chapel Hill: University of North Carolina Press, 1994.

Carson, Rachel. *Silent Spring.* New York: Mariner Books, 2002.

Hardin, Garrett. "The Tragedy of the Commons." *Science* 162 (1968): 1243–1248.

Liddington, Jill. *Long Road to Greenham.* New York: Virago, 1990.

Lovins, Amory. *Soft Energy Paths.* New York: HarperCollins, 1979.

Nash, R. *Wilderness and the American Mind.* New Haven, CT: Yale University Press, 1982.

Opie, J. *Nature's Nation.* New York: Harcourt Brace, 1998.

Rothman, H. K. *The Greening of a Nation.* New York: Harcourt, 1998.

Rothman, H. K. *Saving the Planet: The American Response to the Environment in the 20th Century.* Chicago: Ivan R. Dee, 2000.

Steinberg, T. *Down to Earth.* New York: Oxford University Press, 2002.

Williams, Terry Tempest. *Refuge.* New York: Vintage, 1992.

Environmental Justice and Climate Change

The environmental justice movement calls for a redefinition of the term "environment" to include the places where people live, work, and play, that is, cultural environment as well as physical or natural environment. The Environmental Protection Agency (EPA) defines environmental justice as the fair treatment and meaningful involvement of all people, regardless of race, color, national origin, or income, with respect to the development, implementation, and enforcement of environmental laws, regulations, and policies.

Some scholars define environmental justice as equal distribution of environmental "goods" and environmental "bads" regardless of race, class, or gender. However, there is some disagreement on this point. The Energy Justice Network, for example, distinguishes between environmental equity ("poison people equally") and environmental justice ("stop poisoning, period"). Environmental justice advocates span academic disciplines, from ecology to communication to political science. It is important to note, however, that the movement primarily consists of community organizers, grassroots activists, and nongovernmental organizations (NGOs).

Origins of the Environmental Justice Movement

Accounts of where the environmental justice movement began conflict somewhat, which is unsurprising given the nebulous nature and grassroots organization that defines new social movements. Many scholars argue that environmental justice emerged from the civil rights movement, because activists began recognizing that environmental hazards were yet another front in the struggle for social justice and racial equity. Local environmental justice movements emerged in response to the presence of toxins in homes, schools, or places of work; problematic models of risk assessment; and discriminatory zoning and land-use practices.

According to most accounts, the environmental justice cause emerged as its own movement in the national arena during the summer of 1982 in Warren County, North Carolina. This small low-income and predominantly African American community was selected by the state as the site for a toxic waste landfill. Local citizens, who felt that their community was chosen for political reasons rather than ecological ones, protested and staged numerous demonstrations. As a result of these activists' efforts and his own participation in the Warren County protests, Congressman Walter Fauntroy requested in 1983 that the U.S. General Accounting Office conduct research in eight southern states to determine the correlation between the location of hazardous waste landfills and race or socioeconomic status. The study found that three of every four landfills were oriented near a minority community.

Additional studies confirmed these findings. For example, in 1987 the United Church of Christ's Commission for Racial Justice published *Toxic Wastes and Race in the United States* under the direction of Reverend Benjamin Chavis, who was also active in the Warren County protests. The report indicated that race was the single most important variable in determining the

Hot Spot

Madagascar

A survey conducted in late 2006 by the Wildlife Conservation Society revealed that rising sea temperatures—part of the worldwide trend of global warming—were having a disastrous effect on one of Madagascar's ecosystems, the unique coral reefs located off Madagascar's southwestern coast. While conducting the study, scientists discovered 4 species of coral previously unknown to science, 19 species of coral unknown in Madagascar's waters, and 164 coral species in all. In addition, they identified 3,865 species of fish living in the reefs. The survival of all of these species is in serious jeopardy due to the rising ocean temperatures, which are killing the algae that serve as the main food for coral. In some areas, 99 percent of coral cover has already disappeared.

---

siting of toxic facilities. Furthermore, it confirmed that this was not accidental but instead was the intentional consequence of land-use policy.

Elsewhere around the nation, grassroots environmental groups in minority communities began charging corporations with targeting their neighborhoods for dumping and pointing out government complicity in environmental inequity. While case studies are innumerable, the following is a short list of nationally discussed struggles for environmental justice at the end of the 1980s:

- 1986—After discovering dioxin contamination, the U.S. government evacuated and relocated residents of Times Beach, Missouri.
- 1988—The Mothers of East L.A., in conjunction with other Latino grassroots organizations, successfully stopped the construction of a toxic waste incinerator in their community. That same year, a Navajo community also successfully blocked the siting of a toxic waste incinerator in their community in Dilkon, Arizona.

Environmental Racism

Environmental racism refers to the phenomena whereby minority communities are disproportionately burdened by negative impacts while at the same time enjoying far fewer environmental goods. Chavis first coined the term during the Warren County protests and defined environmental racism succinctly in the conference proceedings of the First National People of Color Environmental Leadership Summit as "Racial discrimination in environmental policy making and the enforcement of regulations and laws, the deliberate targeting of people of color communities for toxic and hazardous waste facilities, the official sanctioning of the life-threatening presence of poisons and pollutants in [those] communities, and the history of excluding people of color from the leadership of the environmental movement."

Robert D. Bullard is a sociologist who is credited as both the "Father of Environmental Justice" and the lead campaigner working against environmental racism. However, as the movement grew, the term "environmental justice" was adopted instead to include economically disadvantaged white groups or international communities facing an unfair distribution of environmental "bads."

Principles of Environmental Justice

On October 24, 1991, delegates to the First National People of Color Environmental Leadership Summit in Washington, D.C., compiled and adopted a list of 17 principles of environmental justice. As the environmental justice movement grows and expands, these principles still serve as the defining canon. The document begins with the following preamble:

> We, the people of color, gathered together at this multinational People of Color Environmental Leadership Summit, to begin to build a national and international movement of all peoples of color to fight the destruction and taking of our lands and communities, do hereby reestablish our spiritual interdependence to the sacredness of our Mother Earth; to respect and celebrate each of our cultures, languages and beliefs about the natural world and our roles in healing ourselves; to ensure environmental justice; to promote economic alternatives which would contribute to the development of environmentally safe livelihoods; and, to secure our political, economic and cultural liberation that has been denied for over 500 years of colonization and oppression, resulting in the poisoning of our communities and land and the genocide of our peoples, do affirm and adopt these Principles of Environmental Justice:
>
> 1. Environmental Justice affirms the sacredness of Mother Earth, ecological unity and the interdependence of all species, and the right to be free from ecological destruction.

2. Environmental Justice demands that public policy be based on mutual respect and justice for all peoples, free from any form of discrimination or bias.
3. Environmental Justice mandates the right to ethical, balanced and responsible uses of land and renewable resources in the interest of a sustainable planet for humans and other living things.
4. Environmental Justice calls for universal protection from nuclear testing, extraction, production and disposal of toxic/hazardous wastes and poisons and nuclear testing that threaten the fundamen tal right to clean air, land, water, and food.
5. Environmental Justice affirms the fundamental right to political, economic, cultural and environmental self-determination of all peoples.
6. Environmental Justice demands the cessation of the production of all toxins, hazardous wastes, and radioactive materials, and that all past and current producers be held strictly accountable to the people for detoxification and the containment at the point of production.
7. Environmental Justice demands the right to participate as equal partners at every level of decision making, including needs assessment, planning, implementation, enforcement and evaluation.
8. Environmental Justice affirms the right of all workers to a safe and healthy work environment without being forced to choose between an unsafe livelihood and unemployment. It also affirms the right of those who work at home to be free from environmental hazards.
9. Environmental Justice protects the right of victims of environmental injustice to receive full compensation and reparations for damages as well as quality health care.
10. Environmental Justice considers governmental acts of environmental injustice a violation of international law, the Universal Declaration On Human Rights, and the United Nations Convention on Genocide.
11. Environmental Justice must recognize a special legal and natural relationship of Native Peoples to the U.S. government through treaties, agreements, compacts, and covenants affirming sovereignty and self-determination.
12. Environmental Justice affirms the need for urban and rural ecological policies to clean up and rebuild our cities and rural areas in balance with nature, honoring the cultural integrity of all our communities, and providing fair access for all to the full range of resources.

13. Environmental Justice calls for the strict enforcement of principles of informed consent, and a halt to the testing of experimental reproductive and medical procedures and vaccinations on people of color.
14. Environmental Justice opposes the destructive operations of multinational corporations.
15. Environmental Justice opposes military occupation, repression and exploitation of lands, peoples and cultures, and other life forms.
16. Environmental Justice calls for the education of present and future generations which emphasizes social and environmental issues, based on our experience and an appreciation of our diverse cultural perspectives.
17. Environmental Justice requires that we, as individuals, make personal and consumer choices to consume as little of Mother Earth's resources and to produce as little waste as possible; and make the conscious decision to challenge and reprioritize our lifestyles to ensure the health of the natural world for present and future generations.

Additional Environmental Justice Canonical Documents

At the Second People of Color Environmental Leadership Summit in 2002, further principles were developed that emphasized working together with other communities and offered advice about working productively with mainstream environmental organizations. *Principles of Working Together* emphasized the need to work from the ground up, beginning with grassroots organizing and activism. The document also argued for the need to recognize the value of traditional or indigenous knowledge and the importance of diversity. In addition, the document valued cooperation across generational, racial, and class boundaries.

Relationship with Mainstream Environmentalism

While the environmental and environmental justice movements might seem like natural allies, the environmental justice movement is often described as a critique of the more traditional environmental movement and conservancy organizations that were founded in response to climate change. Specifically, early environmental justice activists argued that environmental organizations, such as the Sierra Club and the Nature Conservancy, excluded poor people and people of color from positions of leadership. Furthermore, environmental justice advocates were dismayed that the mainstream environmental movement envisioned a strict boundary between the natural world and the human world, in essence favoring a pristine environment that needed to be protected from people. Many urban issues that environmental justice coalitions were worried about (neighborhood security, housing, etc.) were not deemed adequately environmental by powerful NGOs and were described instead as "public health" concerns. Environmental justice advocates remind us that the environment, or nature, is not separate from people and urges environmentalists to consider the interrelated social, economic, political, and ecological sustainability of present actions for future generations.

Gradually, however, mainstream environmental organizations have recognized the importance of social justice for environmental responsibility and climate change mitigation. On January 16, 1990, a coalition of environmental justice organizations working on the coast of the Gulf of Mexico sent a letter to the so-called Group of Ten mainstream environmental organizations. The letters accused the environmental movement of complicity in the unequal distribution of environmental hazards in poor communities and communities of color. Environmental justice activists cited "debt-for-nature" programs (whereby developing nations are invited to trade land rights for national debt reduction) as one example of the environmental movement's marginalization of communities of color from decision making. The letter-writing campaign was a success in raising awareness about the disconnect between the two environmental movements, and the Sierra Club now devotes resources to community partnership programs to explore linkages between environmental quality and social justice.

At the second People of Color Environmental Leadership Summit in 2002, the following "Principles of Alliance with Green Groups" were adopted to ensure mutually beneficial relationships moving forward:

1. Prior to forming an alliance, green groups shall show respect for the history, culture, traditions, and capacity of the community.
2. Green groups, their staff, members and volunteers must actively seek to gain a substantive understanding of environmental racism and environmental injustice, as defined by the Environmental Justice Movement.
3. There must be a good faith belief amongst all alliance members that the alliance will maximize the opportunity to create new solutions, agreements and decisions for greater value.
4. All members of an alliance, regardless of resources and status, should be treated as full and equal partners.

5. Green groups professional and decision-making staff, board, membership, and vendors will reflect the full richness and racial diversity of America.
6. Prior written consent should be obtained before the use of a partner organization's name in connection with endorsements of initiatives, policies or funding.
7. There must be a prior agreement on how resources should be sought, administered, allocated, and how costs should be borne.
8. There should be notification of meetings, press opportunities, and inclusion in key meetings and negotiations.
9. Given that there is a continued disparity in funding, alliances should build the long-term capacity of EJ [Environmental Justice] organizations which have not had the benefit of the 20 to 30 years of continuous funding that Green groups have experienced.
10. Thc alliance should work together to educate government agencies, foundations, and other funding sources on the value of directly funding Environmental Justice groups.
11. Fundraising for collaborative projects must be a joint effort.
12. Alliances should create a space where conflict can be respectfully resolved.
13. Members of an alliance should not engage in unilateral decision-making or actions. There should be a process in place that ensures input from all alliance members.
14. Green groups should only work on Environmental Justice issues at the invitation of or with the consent of the community group.
15. Green groups will respect the right of the community groups to set the agenda, including identifying the problem, determining the goals, and defining success.
16. Credit for outcomes should be shared.

The Environmental Protection Agency

In response to increased public concern, the EPA created the Office of Environmental Justice in 1992 and made an effort to integrate environmental justice concerns into EPA programs and policies. In 1993 the National Environmental Justice Advisory Council (NEJAC) was established as a forum for addressing national environmental justice issues. Membership is composed of scholars, community leaders, industry, NGOs, and tribal governments or indigenous groups. The NEJAC was renewed in 2008 with an EPA charter with the mandate that it provide independent advice and recommendations to the governments about all issues related to environmental justice. Lisa Garcia, U.S. EPA associate assistant administrator, has compared the organization's vision of nationwide environmental equity to achieving world peace.

However, there is mixed opinion as to whether or not the EPA is the best respondent to environmental justice concerns. For example, in the current debate over hydraulic fracturing for natural gas in the Marcellus Shale gas play, many environmental justice activists assert that the EPA is more concerned with the needs and desires of the oil and gas industry than with community activists' points of view.

Globalizing Environmental Justice

Although the environmental justice movement originated in the United States during the 1980s, scholars and activists alike have argued that the effects of a globalized economy and colonialism result in environmental justice concerns all across the planet. Scientific data indicates that climate change is causing and will continue to cause a myriad of environmental shifts, such as flooding or desertification, that will in turn cause millions of people to leave their homes without the possibility of ever returning. This phenomena, called environmental displacement, disproportionately affects poor and nonwhite communities.

J. Timmons Roberts argues that improvement in electronic communications now allows for environmental justice groups around the globe to connect with one another in transnational solidarity. This is necessary because the world's poorest (and often nonwhite) regions face a triple threat of (1) depletion of environmental resources by richer nations, (2) the processing of those resources in polluting ways, and (3) outsourced manufacturing to lower-wage nations, resulting in severe exposures of workers and downwind communities. The following case in Bhopal, India, perhaps the most visible of countless examples of international environmental justice concerns.

In 1969, Union Carbide began manufacturing pesticides in a plant in Bhopal. This plant was marketed as a means to achieve India's green revolution because pesticides would aid in agricultural self-sufficiency. Ten years later, Union Carbide chose the Bhopal facility for the production and storage of a new, dangerous, and little-understood chemical, methyl isocyanate (MIC). Because health and safety regulations are often less strict in the developing world, the siting of hazardous facilities in the global South is common practice. Then, on December 23, 1984, a series of mitigating factors combined to cause the worst industrial disaster in history. A gas leak exposed thousands of workers and hundreds of thousands of nearby residents to MIC. While reports vary wildly, between 3,800 and 8,000 people died in the first weeks after the disaster, and another 8,000 are estimated to have since died as a result of their exposure. Some estimates by environmental justice organizations argue that more than 25,000 people lost their lives as a result of the accident. The Indian government released an affidavit in 2006 stating that the leak caused 558,125 injuries, many of which were permanently disabling.

Investigation after the disaster listed overfilled storage containers, safety systems switched off to save money, and poor plant maintenance as some of the factors contributing to the Bhopal disaster. Effects of the gas cloud were multiplied by packed living conditions in nearby slums. Reported health defects range from difficulty breathing to birth defects to neurological disorders.

Bhopal is often credited as the longest standing campaign against corporate crime and environmental injustice. The Indian government passed the Bhopal Gas Leak Disaster Act, allowing the government to represent victims in legal action. The case between Union Carbide and the Indian government was settled in 1989 for $470 million, less than a seventh of what the Indian government estimated was necessary to rehabilitate the community. Final compensation for personal injury was $830 and for death was $2,058. Though the decision was challenged, the U.S. Supreme Court ruled that the original settlement would stand. Indian officials charged Union Carbide chief executive officer Warren Anderson with homicide along with eight other executives, but Anderson cannot be extradited from the United States. Chemicals abandoned at the plant are still entering groundwater sources, although researchers are in dispute as to whether the chemicals are harmful. Union Carbide was later purchased by Dow Chemical Company, which claims that it is not its responsibility to clean up the Bhopal waste site or offer additional compensation to victims.

Just Sustainability

In 2003, Julian Agyeman, Robert Bullard, and Bob Evans reiterated that the challenge of environmental justice, with particular consideration for equity across racial, economic, gender, and sexual orientation barriers, must be integrated with the domain of environmental concerns if we are to respond productively to climate change. They define just sustainability as "The need to ensure a better quality of life for all, now and into the future, in a just and equitable manner, whilst living within the limits of supporting ecosystems."

The just sustainability paradigm insists that shortfalls of the environmental justice movement and the environmental movement can be overcome by acknowledging the complex and embedded nature of sociopolitical and economic relationships between the human and nonhuman worlds.

Katherine Cruger

Further Reading

Agyeman, Julian, Robert D. Bullard, and Bob Evans. *Just Sustainabilities: Development in an Unequal World.* Cambridge, MA: Earthscan/MIT Press, 2003.

Bullard, Robert D. *Dumping in Dixie: Race, Class and Environmental Quality.* Boulder, CO: Westview, 1990.

Bullard, Robert D. *The Quest for Environmental Justice: Human Rights and the Politics of Pollution.* San Francisco: Sierra Club Books, 2005.

Bullard, Robert D., ed. *Confronting Environmental Racism: Voices from the Grassroots.* Boston: South End, 1993.

Bullard, Robert D., ed. *Unequal Protection: Environmental Justice and Communities of Color.* San Francisco: Sierra Club Books, 1994.

Bullard, Robert D., and Beverly H. Wright. *Race, Place, and Environmental Justice after Hurricane Katrina: Struggles to Reclaim, Rebuild, and Revitalize New Orleans and the Gulf Coast.* Boulder, CO: Westview, 2009.

Energy Justice Network. "Environmental Justice Resources." http://www.ejnet.org/ej/.

Environmental Protection Agency. "Environmental Justice." http://www.epa.gov/environmentaljustice/.

Roberts, J. Timmons. "Globalizing Environmental Justice." In *Environmental Justice and Environmentalism,* edited by Ronald Sandler and Phaedra C. Pezzullo, 285–320. Cambridge, MA: MIT Press, 2007.

General Accounting Office. "Siting of Hazardous Waste Landfills and Their Correlation with Racial and Economic Status of Surrounding Communities." GAO/RCED-83-168, June 1, 1983, http://archive.gao.gov/d48t13/121648.pdf.

Newton, David E. *Environmental Justice: A Reference Handbook.* Santa Barbara, CA: ABC-CLIO, 1996.

Pezzullo, Phaedra. *Toxic Tourism: Rhetorics of Pollution, Travel, and Environmental Justice.* Tuscaloosa: University of Alabama Press, 2007.

Sandler, Ronald, and Phaedra Pezzullo, eds. *Environmental Justice and Environmentalism: The Social Justice Challenge to the Environmental Movement.* Cambridge, MA: MIT Press, 2007.

United Church of Christ Commission for Racial Justice. *Proceedings of the First National People of Color Environmental Leadership Summit.* Washington, DC: National Council of Churches USA, Eco.-Justice Program Office, 1991.

Wentz, Peter S. *Environmental Justice.* Albany: State University of New York Press, 1988.

Environmental Justice and Climate Change: Implications for Oil Producing Countries

The necessity of petroleum has created a new world order in the last few decades. Developed nations have come to view access to crude as a matter of national security, even worthy of going to war in some cases. On the other side of the ledger, nation's holding oil reserves have worked to discern how best to leverage it in order to ensure their nation's future. In some producing nations, particularly in Africa, developing oil resources has also become a threat to citizens' health and to political stability. These destabilizing effects stem directly from the centrality of petroleum in the modern world.

Blessing or Curse?

Petroleum as an issue of national security would have never occurred to the American public in 1950, even though World War II had made political leaders acutely aware of its strategic importance; however, today the concept is so common that most Americans find it impossible to recall a day when they controlled the bulk of the world's petroleum supply. Indeed, a new world order has evolved that is organized by the power emanating from crude—those who must have it but do not possess sufficient reserves themselves and others who possess it but do not need it for themselves. Political scientist Michael Klare writes that in this 21st-century era of haves and have-nots, the United States and other developed nations now exist in an era of "resource wars," which he describes in this fashion:

> For the American military establishment, this concern has particular resonance: while the military can do little to promote trade or enhance financial stability, it *can* play a key role in protecting resource supplies. Resources are tangible assets that can be exposed to risk by political turmoil and

> conflict abroad—and so, it is argued, they require physical protection. While diplomacy and economic sanctions can be effective in promoting other economic goals, only military power can ensure the continued flow of oil and other critical materials from (or through) distant areas in times of war and crisis. As their unique contribution to the nation's economic security, therefore, the armed forces have systematically bolstered their capacity to protect the international flow of essential materials.

Economists have also parceled the concept of resource wars into categories of new and old warfare. By doing so, they follow the logic of these conflicts to argue that oil and war have been linked since the start of the 20th century, as oil "was considered a key strategic commodity and security." Economists Mary Kaldor, Terry Lynn Karl, and Yahia Said explain that in new oil wars, the government connection has been eroded. "New wars," they write, "are associated with weak and sometimes ungovernable states where non-oil tax revenue is falling, political legitimacy is declining and the monopoly of organized violence is being eroded. In such wars, the massive rents from petroleum are used in myriad ways to finance violence and to foster a predatory political economy."

As a rentier war, conflict over oil is based only on crude's remarkable value. Interested parties express little or no interest in long-term development of the region or resource. In addition, they often care little about the global nature of the commodity—except that it will bring them revenue. Often they work with global oil corporations in an unfettered and unregulated arrangement that is seen as a major threat to the stability of crude as a commodity.

The necessity of a stable supply of petroleum to U.S. national security came of age in the second half of the 20th century and would only intensify. As American petroleum reserves grew depleted and increasing consumption left the nation no choice but to increase imports, the United States was drawn into the Middle East petroleum vortex that had involved the great imperial powers for a century of colonialism. Although Saudi Arabia remained the strongest American friend in the region, Iraq emerged as the most significant demonstration of the geopolitical concept of resource wars. With the erosion or entire retreat of colonial authority, many nations suddenly faced power vacuums that were seized by a variety of leaders. The late 20th century saw many examples of leaders of the developed world attempting to find diplomatic or military methods for managing relations with such individuals. These efforts grew more intense if the nation was of strategic importance because of its location or the resources that it possessed. In this fashion the postcolonial era saw nations categorized under a petroleum measuring stick:

haves or have-nots. Of course, the suddenly independent nations of the Middle East fell into the "have" category.

Possessing oil, however, did not result in an automatic economic shift within a nation. As journalist Peter Maass writes in *Crude World,* "One of the ironies of oil-rich countries is that most are not rich, that their oil brings trouble rather than prosperity." In Nigeria, government ministers clash with military generals, and civilians are entirely ignored in the effort to ease access to the nation's oil reserves. Ecuador's lack of concern over the behavior of oil developers contaminated a tributary of the Amazon River on which all life in the region depends. And in nations ranging from Russia to Venezuela and Guinea, government officials have used oil to consolidate political power and undergird their presence on the world stage.

The events in Iraq of the 21st century were set in motion by the embargo of the 1970s, which was directed toward the United States but carried with it a dramatic effect on European powers, particularly France and Britain. As a result, the petroleum-desperate United States expanded its relations with Saudi Arabia and Kuwait. In addition, the primary concessions that had begun in Persia in 1901 and tied the region to Europe had expired. A bold new day of American geopolitics in the region had begun with the leadership of the shah of Iran in the late 20th century. In addition, from the 1970s forward, a nation such as Iraq was not simply left to British trade and development. In fact, Iraq, under the command of Saddam Hussein, remained a largely autonomous wild card in the region. After supplying nations caused petroleum disruptions in the last decades of the century, it became apparent that ensuring control and supply could be used as a rationale for warfare. Competition for the remaining petroleum reserves focused the attention of many nations on the Middle East, which holds approximately two-thirds of the known supply. The ascendance of geopolitical power to the region emerged in the second decade of the 21st century in the form of Dubai in the United Arab Emirates, which grew from the desert much like a mirage.

Petro Dictators and Socialists Leverage Growing Scarcity

In an era of petroleum hoarding, the nations possessing crude—the haves—obviously experienced a significant increase in their global stature. The term "petro dictator" has been attached to a variety of leaders in locations ranging from Azerbaijan to Venezuela. Each leader or group uses the power of the petroleum commodity to his own advantage and to raise the international stature of his nation. Although part of the landscape that holds new oil wars, these leaders are not only using their oil for its rentier value.

Defining the form, Hussein, the Iraqi leader during 1979–2003, pressed the advantage of petroleum wealth more than any other (Iraq had nationalized its oil industry in 1972). In the end, most observers would claim that he overplayed his petroleum advantage; petro dictators who followed have learned from his example how to preside over the commodity in this new era of resource wars.

In his boldest move, of course, Hussein sought to function as the enforcer of the Organization of Petroleum Exporting Countries (OPEC) interest to limit production and thereby manage the global price. Making its own determination to appease the United States, Kuwait's leaders consistently overshot its production caps during the 1980s. Although this willingness enhanced the nation's relations with Western powers, OPEC leaders grew increasingly frustrated with its rogue production. In Iraq, Hussein rose to dictatorial power in 1979 and began building the region's largest military, which was first used in 1979 to invade Iran. In 1990 Hussein, with an additional eye toward Kuwait's access to the Persian Gulf, decided to be OPEC's enforcer, and the Iraqi Army invaded Kuwait. Hussein increased his oil reserves by 20 percent overnight as the world looked on and imagined the consequences if Hussein's campaign continued into the lightly armed Saudi Arabia and United Arab Emirates—which would then provide him with control of approximately half of the world's proven petroleum reserves. With its hand forced by a continued need for petroleum, the United States and allies drew a line in the sand.

The ensuing months were marked by efforts to use the United Nations (UN) to arrive at a nonmilitary diffusion to the situation. This was ultimately abandoned on January 17, 1991, when a massive multinational force authorized by the UN and led by the United States descended on the Persian Gulf with the goal of returning Hussein and his army to Iraq. Most of the fighting lasted just hours as Hussein's army suffered grave defeat at the hands of the world's most advanced military technology. In fact, the war itself symbolized the gap between developed and less developed nations as Hussein's army—dominant within the Middle East and African region—appeared primitive and hopelessly overmatched. This, however, did not mean that his forces could not exert great damage on the real source of the conflict: before retreating, Hussein's army set afire approximately 800 Kuwaiti oil wells, creating a modern environmental disaster.

Allowed to retreat to Baghdad, Hussein remained in power until another American president—George W. Bush—seized the moment in 2003 to commit a largely American force to dislodging the dictator. The logic of the war in 2003 was tied to the attacks on American soil on September 11, 2001. Action against leaders such as Hussein, President Bush argued, fell into a new strategy of preemptive warfare that was designed to head off future attacks or conflicts. Critics immediately claimed that it was a resource war designed to open Iraq's petroleum reserves to unfettered use and development by the United States. For the purpose of our consideration, hindsight demonstrates that at the very least access to crude was one of the fringe benefits to unseating Hussein. The instability that ensued after Hussein's fall, capture, and death thwarted hopes for immediate development of Iraqi oil; however, by 2010, Iraq's new petroleum order was clear.

Learning from Hussein's model, the next most obvious petro dictator is Venezuela's Hugo Chavez. A great admirer of Cuba's Fidel Castro, Chavez swept to political leadership in 1998 and ever since has sought to use his nation's enormous oil reserves to leverage international standing for himself and Venezuela. Internally, Chavez promised "revolutionary" social policies and constantly abused the "predatory oligarchs" of the establishment as corrupt servants of international capital. Internationally, he employed what he refers to as "oil diplomacy." Venezuela has "a strong oil card to play on the geopolitical stage," he explained. "It is a card that we are going to play with toughness against the toughest country in the world, the United States." In OPEC, Chavez has fought to keep prices high and has even publicly questioned whether or not barrel prices should still be measured on the basis of the American dollar. Whether speaking at the UN to demonize the United States or threatening to only sell Venezuela's oil directly to underprivileged populations in the United States, Chavez's international standing—whatever it might actually be—is based on his nation's vast supply of crude.

Russia has followed a different political model in recent years; however, petroleum has emerged as a major structuring agent for its base of national power following the fall of Communism. Oil production is no longer financed by the state budget; it is now financed by selling the output to other nations. In at least one region—western Siberia—just as the Communist government fell and Russia emerged as an independent region, the former Soviet Ministry of Oil petitioned Moscow to form a joint stock company known as Lukoil. Other petroleum resources were divided among workers and private companies in very complex and unclear arrangements during the early days of Russia's independence. Historian John D. Grace writes that "By the beginning of 1995, of the roughly three dozen original Soviet-era producers in Russia, over 20 were still wholly in state hands and 13 were listed as private companies. . . . The most important of these were Lukoil, Yukos, Surgutneftegaz, Slavneft, Sidanco, Kominift, Eastern Oil and Onako." In the Volga-Ural Basin, Grace added two companies that remained under the control of local governments: Tatarstan and Bashkortostan. As a few Russians took control of the nation's banking system, these oligarchs soon became major players in the new oil companies—-particularly in Lukoil.

By the early 21st century, Lukoil used Western oil and gas corporations as its model. It took over smaller companies and diversified into international operations beyond exploration and production, including refining, marketing, and the petrochemical industry. In 2002, Lukoil became the first Russian oil company to list its shares on a Western exchange (in London). Lukoil became an active player in Colombia and Iraq and also took over many of the major pipeline projects near the Caspian Sea. New trading arrangements were formed with Asian nations, particularly Japan and China, and poised Lukoil to take advantage of some of the world's fastest-growing oil markets in the 21st century. Whether the companies are truly independent or not, thanks to their rapid success the new Russia stands as a leader in production and distribution of oil today.

Petro dictators have often managed to maintain control of their nations for fairly lengthy regimes. There is growing evidence, however, that a lopsided emphasis on petroleum development by dictatorial powers does not end well. Regardless, petroleum supplies have emerged as the single most significant equalizer for nations on the less developed side of the gap.

Other nations have used a state-owned or socialist model to emphasize oil development to support infrastructural development. In developing the North Sea supply of oil, for instance, a group of European nations has carried out a joint initiative. The United Kingdom, Denmark, Norway, Germany, and the Netherlands formed joint tax and licensing regimes to develop the difficult North Sea offshore supply after 1968. In Norway, for instance, the state-owned Statoli corporation has helped the nation become the world's third-largest oil exporter and eighth-largest producer. Choosing not to join OPEC, Norway instead established the Petroleum Fund of Norway in 1990 to collect profits from sales and licensing fees. One of the largest public funds in the world, this fund is largely held to ensure the nation's economic stability when oil supplies diminish. Particularly because Norway's population stands at less than 5 million, critics in recent years have questioned whether or not it is necessary to create such a large savings fund. Recently many critics have called for the fund to be used more for internal improvements and national needs.

Environmental Justice and Modernization in Developing Africa

Decolonization brought changes to nations, whether or not they possessed valuable resources such as petroleum. In the case of many African nations, the decades after World War II witnessed a great expansion of agricultural capabilities. In many of the nations of Asia and Central and South America, the Green Revolution has been a terrific success in food production. In other areas such as Africa, it has been less successful.

Petroleum played a role in this agricultural shift, and some of these nations applied details of modern life seen in more developed societies. In most cases, however, African nations have found this productivity difficult to sustain. For this and other reasons, the initial modernization of many African nations after the colonial era has given way to destabilization and conflict. The revolution in agricultural productivity has resulted, at least at times, in conflict and even the destruction of nations.

Darfur, which is located in the African nation of Sudan, provides the preeminent example of how such destabilization might exacerbate existing ethnic or economic divisions. Arab nomads with livestock are pitted against African farmers who practice more sedentary plant cultivation. As the expanding population has resulted in ecological crisis and desertification, each group blames the other. In desperation, each group in Darfur fights the other for survival, quite literally: the nomads expand southward searching for new grazing land, and the farmers move northward to find arable land. The bloody atrocities of the conflict grow from ethnic differences; however, the stress of the pressures of development and sustainability create the pressure that ignites the conflict.

In Equitorial Guinea, the difficulties are quite different. Similar to the petro dictators discussed above, President Teodoro Obiang has used new petroleum discoveries to his political advantage; however, his country receives little benefit. Particularly in less developed nations, oil wealth can be stolen by a corrupt leader who accepts bribes or negotiates lucrative side deals to direct exploration and production contracts. Oil discoveries in the 1990s jettisoned Equitorial Guinea into prominence, and by the 21st century the nation was the third-leading producer of petroleum in sub-Saharan Africa. In short, Obiang's relationship with the developed world—particularly the United States—has done little to close the gap for his nation, even though he personally has reaped millions.

One of the last African nations to become independent in the 1960s, Equitorial Guinea, which is dispersed over a series of islands, had little to offer industrialists. Defined by profit margin, oil companies create harvest and production infrastructure in places such as Equitorial Guinea entirely off the grid and using only imported materials and labor. Applying the ethic behind earlier boomtowns, Marathon and other oil production companies moved easily into less developed environments and created the setting that they required for oil production. In Equitorial Guinea, the company built a concrete plant, for instance, that could be dismantled and moved elsewhere when the oil had played out. Asian workers at oil production plants lived in trailers that had been imported from abroad. And each of the company's facilities was joined by its own satellite communication network reaching directly back to headquarters in Houston, Texas. The work of oil producers was made simpler in nations such as Equitorial Guinea in which leaders made sure that there would be no dispute or discussion over plans for the access and development of oil reserves.

More problematic from a legal and ethical standpoint, Equitorial Guinea's citizens have also been insulated from the proceeds of the harvest of its oil reserves. Journalist Peter Maass writes that oil "not only offers itself as a treasure to be stolen; it can become a political amulet that protects thieves from abandonment or punishment." Obiang lived opulently on his proceeds from oil development, and additionally, he set up accounts at the Riggs Bank in Washington, D.C., that would provide easier deposits from the American companies with which he was working. The arrangement worked well enough to process nearly $1 billion in payments from oil companies, which were content not to concern themselves with what happened to the funds afterward. After being alerted to the arrangement, the U.S. Senate investigated and released the report "Money Laundering and Foreign Corruption: Enforcement and Effectiveness of the Patriot Act; Case Study Involving Riggs Bank." Although there had been great wealth to be had in oil for Equitorial Guinea, internal politics limited its value for the nation.

For African nations, crude has often only meant an extractive enterprise; proceeds have done little to close the development gap, while the crude is drained and used to help other nations prosper. In the case of Equitorial Guinea, Obiang has shifted today to doing most of his oil business with China.

Warring for Petroleum Access

From the perspective of developed nations on the other side of the gap, the use of petroleum supplies as a political weapon demanded an increasingly active culture of engagement. At his inauguration as U.S. president in 1989, George H. W. Bush seemed to speak directly to the Middle East and to petro dictators when he said that "They got a President of the United States that came out of the oil and gas industry, that knows it and knows it well." Bush's worldview teamed with his business experience to make him one of the first Western leaders who clearly—and openly—believed in the strategic importance of a U.S. influence in the OPEC-dominated Middle East, which was now responsible for producing two-thirds of the world's oil.

At this historic juncture, OPEC was wrestling with the idea of fixing petroleum prices for the good of all its members, but many individual nations were unwilling to limit production due to their own economic limitations. When Hussein invaded Kuwait in 1990, American president and oilman Bush orchestrated the joint action by UN forces to stop Hussein's progress and ultimately force Iraqi troops out of their neighboring nation. As Iraqi forces fled Kuwait, they lit on fire many of the nation's oil wells. This act of terrorism created an environmental hazard and debilitated Kuwait's immediate ability to produce oil. Most damaging, though, was Hussein's miscalculation that presented the United States with a military presence in the world's oil region. Bush accomplished his goal of creating a mutually dependent relationship between Persian Gulf nations and the United States. However, this did not necessarily mean that price stability would last. The late 1990s brought more problems related to underproduction. The production imbalance fed the tripling of gasoline prices in 1999–2000. As Kaldor et al. trace the roots of the 21st century war in Iraq, the United States, they argue, sought to fight an "old war" about oil by using the commodity to facilitate the decision to go to war. The conflict, it was argued, would be paid for and largely absorbed by Iraq's "oil revenues."

In a twist of historical fate, the presidential election of 2000 brought George W. Bush, son of the previous president, into office. Although Iraq's leader Hussein and Middle Eastern oil supplies were priorities of the younger Bush, oil prices remained somewhat low. Energy security, though, emerged in the public sphere with the 9/11 terrorist attacks. Although unrelated to Hussein, these attacks became a leveraging point with which President Bush could make unseating the Iraqi leader a mission of the American military in two fashions: first, Hussein's unreliability and possible dangerousness was compounded by his control of such significant oil reserves; second, if an invasion was carried out, it was argued, that revenue from petroleum sales would help quickly stabilize the new Iraq and decrease the financial resources necessary from the United States or any other occupying nation. Each of these components for war derived from the importance of petroleum.

Had the 2003 American-led invasion been about oil supplies? Wound tightly into our ecology of oil, developed nations required a steady supply of crude. And clearly, the need to preserve energy security abroad had steadily increased during the 20th century. Until the end of World War II, domestic supplies allowed Americans to watch European powers colonize and develop Middle Eastern supplies; the end of the war, though, brought the same realization to American foreign policy makers. The world that emerged in the 21st century clearly factored geopolitics into nearly every diplomatic interaction.

Protected from this reality by more than a half century of disinformation, American consumers clearly became the last remaining disconnect in comprehending the implications of our dependency. History may show that the invasion, occupation, and support of Iraq had crude at its core. Clearly, however, this war was a leading symbol of new world petroleum order organized by the haves and have-nots and various efforts to compensate for each nation's particular standing in the petroleum organization.

To make the new petroleum order even more clear, in 2004 Nancy Birdsall and Arvind Subramanian published "Saving Iraq from Its Oil" in the influential journal *Foreign Affairs.* The article sought to advise the United States and occupying nations how they might best make Iraq's petroleum a beneficial resource for developing a new nation. This argument was based on a simple yet remarkable main idea. Birdsall and Subramanian write of petroleum's "resource curse" that "Oil riches are far from the blessing they are often assumed to be. In fact, countries often end up poor precisely because they are oil rich. Oil and mineral wealth can be bad for growth and bad for democracy, since they tend to impede the development of institutions and values critical to open, market-based economies and political freedom."

The bitter irony of the Iraq War, of course, is that despite its origins, it became a new oil war in which nonstate actors took an active role in fomenting dissent and complicating American occupation of the country. In the process, these activities by terrorists neutralized the ability of profits from crude to assist the settlement of a new Iraq.

Conclusion: Dubai and Human Migrations for Crude

Symbols can be very important to entire societies. For instance, the events of September 11, 2001, demonstrated that for some in less developed nations, made to feel powerless by a lack of opportunity and access to developing their nation's resources, two identical office towers, known as the World Trade Center in New York City, marked the point at which—in some small or potentially significant way—the gap could be breached. Although the loss of thousands of American lives that day has resulted in expanded military activity by the United States as well as an intensified culture of domestic security, the world's skyline tells us that the gap has very likely not closed; however, the centers of power may have begun to shift.

Erupting into the sky with much more symbolism than utility, the Burj Khalifa opened for business in Dubai in early 2010. A symbol of an emerging world order, Burj Khalifa is a rocket-shaped edifice that soars 828 meters, or 2,717 feet. It is the world's tallest structure, with views that can reach 100 kilometers (approximately 60 miles). At a cost estimated at $1.5 billion, the Burj took five years to build, is more than 160 floors high, and has comfortably surpassed the previous record holder in Taipei. If the 1970s Oil Crisis marks the point where developed nations were forced to acknowledge their need for oil from less developed nations, Burj Khalifa marks the permanent institutionalization of this crude reality.

Dubai, the city that the Burj towers over, emerged in the first decade of the 21st century as a global phenomenon. It is one of the seven emirates of the United Arab Emirates, located south of the Persian Gulf on the Arabian Peninsula. The City of Dubai is sometimes called Dubai state to distinguish it from the emirate. Although the city grew with the petroleum industry at the close of the 20th century, at the start of the 21st century Dubai positioned itself as the economic center for an emerging global economy. Focused on organizing the financial development of projects in the Middle East and Southcast Asia, thc city also functions as an oasis for the diverse workforce that began to operate within the Persian Gulf.

The increasing importance of crude was not only noticeable in political leadership at the dawn of the 21st century. Harvesting crude, wherever it occurred, also created patterns of worker migration from neighboring regions. In the Persian Gulf region, population shifts connected to petroleum have become a defining characteristic for the entire region. One of the most significant labor influxes to the Persian Gulf since 1990 has been from Kerala, India. In 1998, for instance, nearly 1.4 million Keralans emigrated from India, approximately 95 percent of whom were destined for Arab countries of the Middle East. Nearly 40 percent of this total immigrated to Saudi Arabia, and 30 percent immigrated to the United Arab Emirates.

In the Middle East, generations of workers have moved with the oil industry. Facing low-paying jobs or unemployment at home, for decades many have worked abroad as laborers, taxi drivers, or food preparers and wire money to their families or return with it on occasional visits. Recent years have seen an increase in the level of professional workers, particularly coming from nations such as Egypt. Whereas in the past engineers and other professionals would take their training to find employment to Europe or the United States, they are finding increasing opportunities closer to home. Egypt has an estimated 5 million workers abroad, including 1.5 million based in the Persian Gulf region. Remittances from those in the United States, Europe, and the Persian Gulf region are a key source of foreign currency for Egypt. Egyptians sent home $8.56 billion in remittances in the 2007–2008 fiscal year, up from $6.32 billion a year earlier.

Much of this work is focused on a new city that has taken shape in the region, both in the work of international economic trading that takes place in it and in constructing the physical monument in which much of it will take place. Blending World Trade Center with Las Vegas decadence, Dubai now focuses around the Burj, with its mix of nightclubs, mosques, luxury suites, and boardrooms. In the Burj one finds the extravagant splendor of the world's first Armani hotel, the world's highest swimming pool (on the 76th floor), the highest mosque (on the 158th floor), and 54 elevators that can hit speeds of 65 kilometers (40 miles per hour). For the more than 12,000 people who occupy its 6 million square feet, the Burj is an oasis from the desert that surrounds it as well as from the overwhelming poverty of the majority of the public in the United Arab Emirate.

To make the irony more acute, a global economic slowdown in 2008 made Dubai and the Burj appear more as a symbol than a genuine accomplishment. All of the city's real estate market collapsed, and the Burj project was rocked just as it neared completion. At this point, the building was named Burj Dubai. To save the project and ensure its 2010 completion, the neighboring kingdom of Abu Dhabi propped up its financing. "Dubai not only has the world's tallest building, but has also made what looks like the most expensive naming rights deal in history," said Jim Krane, author of *City of Gold: Dubai and the Dream of Capitalism.* "Renaming the Burj Dubai after Sheikh Khalifa of Abu Dhabi—if not an explicit quid pro quo—is a down-payment on Dubai's gratitude for its neighbor's $10 billion bailout [in 2010]."

Economic observers have little doubt that the economy will recover, and when it does it will still be organized by certain basic global realities. Chief among these, of course, is the increasing need for petroleum emanating through the Burj, which makes this skyscraper a symbol of a new world order.

Brian C. Black

Further Reading

Birdsall, Nancy, and Arvind Subramanian. "Saving Iraq from Its Oil." *Foreign Affairs* 83(4) (July–August 2004): 77.

Black, Brian. *Crude Reality.* New York: Rowman and Littlefield, 2012.

Blum, Justin. "Chavez Pushes Petro-Diplomacy." *Washington Post*, November 22, 2005.

Grace, John D. *Russian Oil Supply.* London: Oxford University Press, 2005.

Kaldor, Mary, et al. *Oil Wars.* London: Pluto, 2007.

Klare, Michael. *Blood and Oil.* New York: Metropolitan Books, 2004.

Krane, Jim. *City of Gold: Dubai and the Dream of Capitalism.* New York: St. Martin's, 2009.

Maass, Peter. *Crude World.* New York: Vintage Books, 2009.

Environmental Justice and Climate Change: Influence on Native Americans

Environmental justice is defined by David Schlosberg as "the fair treatment and meaningful involvement of all people regardless of race, color, sex, national origin, or income with respect to the development, implementation and enforcement of environmental laws, regulations, and policies." Since Europeans arrived on the shores of North America, Native Americans have been denied this right on the policy level. As North America struggles to uncover new resources for industry in the face of a rapidly changing climate, Native Americans are left to bear a heavy burden from these choices. At times they have cooperated for economic reasons, but in many cases the tribes have worked to halt the exploitation of their reservation lands and the further marginalization of their people.

The Euro-American method of taking claim over native lands was through relocation of Native Americans to reservations. The justification for this mass removal was the highly popular ideology of Manifest Destiny. According to Manifest Destiny, it was divine providence that entitled whites to expand to the interior of North America. Native Americans were thought of as an obstacle to fulfilling this expansion. A reservation, by definition, is a tract of land set apart from its surrounding lands for a specific population. The Indian Removal Act of 1830 mandated the mass movement of Native Americans who occupied lands east of the Mississippi River to west of the river so as not to interfere further with settlement. The Indian Appropriations Act of 1851 further solidified white Americans' determination to relocate Native Americans by approving reservation lands in present-day Oklahoma.

Choosing Reservation Lands

Reservations tend to be land that is not desirable for industry or intensive agriculture. They also tend to be in areas with few local water sources. These factors combined make it very likely that Native American lands are

Hot Spot

Inuvik, Canada

In the Canadian Inuit town of Inuvik, 90 miles south of the Arctic Ocean near the mouth of the Mackenzie River, the temperature rose to 91°F on June 18, 1999, a type of weather unknown to living memory of anyone in the area. "We were down to our T-shirts and hoping for a breeze," said Richard Binder, age 50, a local whaler and hunter. Along the MacKenzie River, according to Binder, "Hillsides have moved even though you've got trees on them. The thaw is going deeper because of the higher temperatures and longer periods of exposure." In some places near Binder's village, the thawing earth has exposed ancestral graves, and remains have been reburied.

Inuit hunters in the northernmost reaches of Canada say that ivory gulls are disappearing, probably as a result of decreasing ice cover that affects the gulls' habitat. Sea ice in the Arctic in the year 2000 covered 15 percent less area than it did in 1978, and it has thinned to an average thickness of 1.8 meters, compared with 3.1 meters during the 1950s.

Bruce E. Johansen

among those that will be most affected by climate change in North America. Increasingly, these lands are under threat from many environmental hazards, including nuclear waste disposal, gas pipelines, and water pollution. Despite treaties and agreements in place to prevent environmental degradation, mistreatment still takes place.

Resistance and the Trail of Tears

Some tribes saw removal and relocation as imminent and signed treaties with the War Department. Other tribes fervently resisted. Perhaps the most famous act of resistance came from the Cherokees who established their own nation in western Georgia and took their case to the U.S. Supreme Court. While they did receive a favorable ruling, the Treaty of New Echota was signed in 1835, signaling the forced removal of the Cherokees. They were led on a march, known as the Trail of Tears, to Oklahoma, and thousands died from sickness and exhaustion along the way.

Agriculture

The relocation of Native Americans to lands farther west represented a severing of long-standing place-based agricultural knowledge. Lands on which they were relocated in many cases lacked sufficient water sources or had undesirable soil to grow traditional crops. Native American agricultural techniques include companion planning (e.g., the Three Sisters, consisting of corn, beans, and squash), minimum tilling, seed saving, and promoting biodiversity with their chosen crops. Currently more attention is being paid to these principles because the industrial farming system has caused mass loss of topsoil, biodiversity, and compromised soil health and has allowed the proliferation of pests and superdiseases. Nitrogen fertilizers seep into groundwater, compromising water tables. They also run off into nearby streams, rivers, and eventually oceans, causing dead zones and large numbers of fish deaths. A large percentage of nitrogen fertilizer goes into the air and is a large contributor to greenhouse gas emissions (GHGs). Intensive logging for timber threatens native populations of fish in nearby waters and nearby farmland. The main threat through logging is soil erosion.

Mining

In recent years, corporations have realized that Native American reservation land has many important resources for industry. This is especially true of the mining industry. Because Native Americans are a marginalized group, they have less political power to keep corporations from extracting coal and building plants near their land. Native Americans carry the brunt of the burden in their health and amenities, yet they benefit the least from the results of mining and extraction. Coal mining compromises the landscape and the health of those living in proximity to the mining. Air quality also suffers greatly as a result of being in the vicinity of a coal plant. Some tribes have allowed coal mining on their lands for economic reasons. For example, the Crow tribe approved an extension of a coal mine to be developed on its tribal lands. The taxes from this account for millions of dollars going to the Crow tribe each year. However, most Native Americans living on reservation lands are without power, despite the close proximity to plants and mining operations.

Energy

Nuclear energy is derived from uranium, a heavy metal that must be mined in order to be extracted. Nuclear energy is mostly used for electricity but is also used to produce nuclear weapons. A by-product of nuclear energy is radiation.

Sheila Watt-Cloutier, an Inuit activist in Canada, is a leader in raising awareness of global warming impacts. (AP/Wide World Photos)

Exposure to radiation can cause immediate death and/or cancers that will result in death down the road. Native American lands are particularly vulnerable to the ill effects of nuclear mining and waste management because there happens to be a significant amount of uranium beneath the surface of lands they reside upon, especially in the U.S. Southwest. Tribe members in the areas of conflict have fought against the installation of waste sites on or near their lands, using evidence of health hazards and environmental racism on the part of industry to back their arguments. Perhaps the most famous proposed nuclear waste site is Yucca Mountain, located 100 miles northwest of Las Vegas, Nevada. Nuclear waste must be buried underground for many years to cool. In the case of Yucca Mountain, the Department of Energy wanted to store upwards of 77,000 tons of waste in the area. After much debate, Yucca Mountain is no longer being considered for nuclear waste disposal and burial.

Conclusion

Native Americans unfairly bear the burden of industry's extraction and processing decisions. Reservations have very little in the way of electricity and

Hot Spot

Sachs Harbour, Canada

Born in 1954, Rosemarie Kuptana grew up in a traditional Inuit hunting society and spoke only Inuvialuktun (the Western Arctic dialect of the Inuit language) until the age of 8. Her home community of Sachs Harbour is a Banks Island village of about 120 people on the Beaufort Sea in the Arctic Ocean about 800 miles northeast of Fairbanks, Alaska. Born in an igloo, Kuptana has been an Inuit weather watcher for much of her life (she was 50 years of age in 2004). Her job was to scan the morning clouds and test the wind's direction to help the hunters decide whether to go out and determine what everyone should wear.

"We can't read the weather like we used to," said Kuptana. She said that autumn freezes now occur a month later than they did in her youth; spring thaws come earlier as well. Residents of Sachs Harbour still suffer through winters that most people from lower latitude would find chilling, with temperatures as low as –40°F. While such temperatures once were commonplace during the winter, however, they now are rare. "The permafrost is melting at an alarming rate," said Kuptana. Foundations of homes in Sachs Harbour are cracking and shifting because of the melting permafrost.

Kuptana said that at least three experienced hunters had recently fallen to their deaths through unusually thin ice. Never-before-seen species (including robins, barn swallows, beetles, and sand flies) have appeared on Banks Island. No word exists for "robin" in Inuktitut, the Inuit language. Growing numbers of Inuits are suffering allergies from white pine pollen that recently reached Banks Island for the first time. At Sachs Harbour, mosquitoes and beetles are now common sights where they were unknown a generation ago. Sea ice is thinner and now drifts far away during the summer, taking with it the seals and polar bears upon which the village's Inuit residents rely for food. Young seals are starving to death because melting and fracturing sea ice separates them from their mothers.

Bruce E. Johansen

clean running water, yet their land is most often used for providing these services to faraway areas. Reservations are intended as a space set aside for an independent group, yet industry and the federal government only follow these guidelines when convenient. The irony is that reservation lands were originally chosen because they were undesirable to white settlers. In the present day, driven by the need for natural resources, it comes to light that reservation lands are rich with exactly what today's society needs to continue functioning as it does, including a place to store the waste. Going forward, it is important to assess the energy habits of modern society and its insistence on dominance over the environment with the overarching Native American partnership with the environment.

Amanda West

Further Reading

"About Yucca Mountain Standards." Environmental Protection Agency, 2011, http://www.epa.gov/radiation/yucca/about.html.

"Energy Justice." Indigenous Environmental Network, 2011, http://www.ienearth.org/energy.html.

Fliessbach, Andreas, Adrian Müller, and Urs Niggli. "Organic Farming and Climate Change." Research Institute for Organic Agriculture, 2009, http://orgprints.org/view/projects/ch-fibl-sustainability-climate.html.

Foreman, G. *Indian Removal: The Emigration of the Five Civilized Tribes of Indians.* Norman: University of Oklahoma Press, 1953.

Grossman, Z. "Native and Environmental Movements." Native Council for Science and the Environment, November 1995, http://www.cnie.org/NAE/docs/grossman.html.

"Indian Removal." Public Broadcasting Service, n.d., http://www.pbs.org/wgbh/aia/part4/4p2959.html.

Klapp, J., J. Cervantes-Cota, and A. J. Chávez. *Towards a Cleaner Planet.* Berlin: Springer Berlin, 2007.

Moriarty, J. T. *Manifest Destiny: A Primary Source History of America's Territorial Expansion in the 19th Century.* New York: Rosen, 2005.

"Overview of Nuclear Energy." World Nuclear Association, March 2011, http://www.world-nuclear.org/education/intro.html.

Reijntjes, C., B. Haverkort, and A. Waters-Bayer. "Farming for the Future." Center for International Earth Science Information Network, 1992, Retrieved October 12, 2011, http://www.ciesin.org/docs/004-176a/004-176a.html.

Schlosberg, D. *Defining Environmental Justice: Theories, Movements, and Nature.* Oxford: Oxford University Press, 2007.

Topping, J. C., Jr. "Climate-Related Core Issues." Climate Institute, 2010, http://www.climate.org/topics/national-action/native-americans-climate.html.

Vander, Hook Sue. *Trail of Tears.* Edina, MN: ABDO, 2010.

Environmental Justice and Climate Change in China

Environmental harms result from extensive exploitation of natural resources and the disposal of foreign wastes in this globalizing world. China has established an extensive network of environmental protection legislative bodies and administrative authorities at all levels of government and the economy, reflecting the growing significance of environmental protection as a key national policy. Prior to the launching of striking economic reform, the Environmental Protection Leading Group was established under the State Council to supervise the environmental performance in 1974. In 1982, the Environmental Protection Agency (EPA) was set up under the Ministry of Urban and Rural Construction to strengthen environmental performance of economic activities. Six years later, the EPA was promoted to an agency directly under the State Council, which increased its authority to issue regulations and guidelines. In addition, the Environmental Protection Law became permanent in 1989. In 1998, the EPA's status was upgraded to a semiministry and eventually became the Ministry of Environmental Protection in 2008. In addition, laws and regulations are also formulated by the National Environmental Protection and Resources Conservation Commission under the National People's Congress. Furthermore, the National Climate Change Program was initiated in 2007 in order to mitigate and adapt to climate change.

However, as the world's largest developing country with nearly three decades of almost double-digit economic growth, China's systemic shift from state socialism to market economy has brought up the crucial topic of environmental justice and social equity. Such issues have been actively promoted by grassroots activists and academics both in China and overseas. Given the impact that environmental protection can have on a variety of social groups and regions, more procedures are needed than just simply controlling greenhouse gas emissions and charging environmentally harmful ventures. Moreover, the burden of environmental and climate hazards (such as desertification, soil erosion, and deforestation) has fallen disproportionately on poorer and less powerful individuals and regions. This increasing evidence illustrates that people and regions in China do not benefit from the same level and degree of protection from hazards.

The overall environmental quality has been worsening in China over the same time period, with environmental pollution equivalent in monetary terms to an estimated annual loss of 3–8 percent of gross domestic product (GDP). Searching for the appropriate equilibrium between economic

Hot Spot

China

Huge industrial plants of all types line the cities of China. The vast majority of these factories and utilities are powered by coal, which is plentiful in China. However, burning so much coal has led to severe environmental degradation. For example, in northern China, such cities as Beijing and Shenyang have poor air dispersal and low-level temperature inversions. As a result, they suffer some of the worst air pollution in the world. The rapid increase in the size of China's deserts—a process called desertification—has caused yellow dust storms to increase in intensity over the past decade. These dust storms, also called sandstorms, originate in China and disperse sand and toxic pollutants over Korea, Japan, and the United States, prompting international efforts to reduce China's industrial pollution.

In southern China, such areas as Sichuan, Guangxi, Hunan, Jiangxi, and Guangdong have growing acid rain problems. Again, China's heavy industry is being blamed for the pollution, because industrial emissions are increasing as the country continues to modernize. Toxic substances emitted into the air from coal burning have dramatically affected human health, with disproportionately high lung cancer and heart disease rates in China's cities. Acid rain from Chinese industry also crosses national borders, killing forests and lakes in neighboring nations.

Efforts to wean the country off of coal have created new environmental dilemmas. Construction of the Three Gorges Dam on the Yangtze River sparked an international outcry because the dam displaced more than 1 million people and submerged huge areas of natural beauty—and an untold bounty of cultural relics. The Yangtze is home to several unique species such as the Yangtze river dolphin, which has been threatened by the damming of the river. However, Chinese leadership considers the hydroelectric power generated by the project a fair trade-off for the environmental devastation it has caused.

growth and environmental/climate justice is an emerging and pressing issue in China's uneven socioeconomic and environmental transformation. Chinese society has essentially changed, with expanding income differentials and rising social structural gaps. Specifically, spatial inequality of limited resources (such as water, air, and land) has been formed across various social groups, accompanied by people's unequal exposure to the increasing environmental risks and climate hazards. In other words, China is experiencing dramatic spatial and social reconfiguration, differentiation, and polarization, which gives rise to different environmental benefits based on people's economic status with a simultaneous deconstruction of the socialist social order. Meanwhile, regional income inequality dramatically rose between urban and rural and between coastal and inland areas, despite the overall fast economic growth. China changed from a socialist regime of planned economy to a more market-based profit-seeking entrepreneurial state. Specifically, these politically correct and economically rewarding entrepreneurial activities are encouraged, admired, and pursued by government agencies and individuals. Hence, the well-being of the marginalized and disadvantaged people remains of little or no concern.

Heightened public awareness of environmental crises and demands for a more moral landscape have promoted the ethical discussion of environmental and climate issues in China, while the dilemma still exists between uneven development and environmental/climate change mitigation in this continued dynamic economic system. Regions compete intensively for investment capital, while officials collaborate with factory owners and developers in the hasty quest for economic growth as the career promotion criteria and sometimes due to a direct financial benefit or personal relationships. Since there is a lack of incentives to make environmental protection a priority, local officials prefer to emphasize economic growth instead of heeding environmental mandates from the central government. The implementation of environmental laws and regulations is especially problematic in poor areas. What worsens environmental injustice is widespread local protectionism. Because the local government benefits from those who pollute or violate environmental standards, ordinary people's health can be sacrificed and ignored.

It is evident that environmental/climate harms and mitigation are not equally distributed within China. Motivated by least-cost production and comparative advantage, the less fortunate, marginalized, and disadvantaged social groups in China are increasingly suffering from an unequal share of the environmental harms of trade relative to its financial gains, while key beneficiaries of these harmful production processes are overseas consumers and rich Chinese. In many hilly inland areas of China, for instance, local residents have been routinely exposed to toxic and sometimes radioactive materials left by extensive and exploitive mining. Similarly, in wealthy coastal regions such as Guangdong Province, several poor towns have engaged in retrieving precious metals from highly contaminated foreign electronic wastes, which greatly harm the health of the workers and fellow villagers as well as local ecosystems by releasing lethal substances into the air and into groundwater. All these activities benefit small groups of recipients at costs to a number of other citizens.

China's national assessment report on climate change warns that national average temperature will grow by 2–3°C over the next 50–80 years if the country continues on its current development style. Increases in extreme weather events (such as heat waves, wildfires, droughts, flooding, and hurricanes) in China are just the most apparent threats due to climate hazards and global warming. For instance, China suffers floods during the June–August rainy season every year, which often cause deadly landslides. However, it should not be assumed that the Chinese government has paid less attention to the environmental implications of climate change. As averting catastrophic climate change has risen on the domestic and international agenda, the Ministry of Environmental Protection in China has gained more significant impact on the decision-making process of national policy.

Environmental justice should be adopted by China as a principle to guide domestic environmental regulation in the context of an increasingly interdependent world. The philosophical case of social justice is to some extent less complicated, while the political mission is arduous. Specifically, the central government in China continues to stress and endorse the need for economic growth as the priority, which is considered to be the foundation of the sustained authority of the government and the stability of Chinese society. All these activities lead to continued growth in selected coastal regions, with fast expansion of energy supplies at the expense of the poor. However, linking the issue of climate change to the problem of development is more likely to achieve a win-win goal than will treating climate change separately from development.

Nevertheless, climate change is not just an environmental issue; it also a political one that is closely associated with the environmental equity and fairness across social hierarchies. Environmental injustice is in part generated by economic growth focusing single-mindedly on GDP growth. Environmental justice and protecting the environment are coming together to facilitate the policy making that puts people above economic growth. In other words, all social groups should have equitable rights to access a clean environment with equitable distribution of resources from the point of view of philosophy, law, and moral ethics. In addition to domestic environmental regulation, China has spent considerable effort in promoting the ideas of sustainable development and a harmonized society, reflecting its national strategy of balancing environmental protection and development, which gradually moved from command-control to market-based instruments.

Current climate hazards have more adversely influenced those in poorer parts of China or among the poorer sectors of society, who unequally carry the burdens of climate change, though Article 26 of the Constitution of the People's Republic of China declares that "the State protects the environment and natural resources and prevents and eliminates pollution and other hazards to the public." Minority groups and low-income populations among "the public" as well as marginalized regions in particular bear negative health and environmental impacts from climate hazards. For example, critical water shortages, expanding deserts, and more intense storms in poor western provinces have threatened the livelihoods of vulnerable citizens, who lack the administrative support and financial ability to deal with the consequences of these climate changes. Such situations contribute in a feedback loop to continued warming and accumulated climate injustice. While the climate injustice problem in China has not attracted the same level of attention as in other developed nations, environmental impacts are awakening Chinese citizens to seek their rights to environmental justice through laws and community movements. The costs of public health (the safety of the air they breathe, the water they drink, the food they eat) and the inability of the government to deal with the environmental situation are brewing social conflicts. Thus, innovations of China's environmental decision making and international cooperation, which mean both procedural and distributive justice nationally and internationally, are required to terminate this trend of exportation and displacement of environmental harms.

Considering global calls to combat climate change and promote environmental justice as well as China's uneven regional socioeconomic development, the transfer of polluting industries, wastes, and ecologically destructive practices should be prevented from developed coastal regions to undeveloped inland regions, from urban to rural, and from rich to poor. Some poor and ecologically sensitive areas still invite all possibilities for a cheaper path to industrialization, and little free public discussion is permitted regarding potential environmental harms or their implications. Therefore, a nationwide and systematic ecoservices scheme should be implemented based on the principle of common but differential responsibilities. Principles of responsibility and capacity should be applied. Furthermore, these principles encourage the efficient and equitable allocation of environmental resources across regions to balance poverty alleviation with environmental protection. That is, those who have created the pollution and are more able to bear the cost should pay for and be committed to reducing climate hazards. Hence, the burden of emissions reductions must lie with rich affluent polluters, who should also support developing regions to avoid the polluting path to development through financing and technology transfer. Then, developing regions can be enabled to adapt to the consequences of climate change. In the meantime, this nationwide pricing scheme should be used to reimburse the opportunity cost borne by those who adopt eco-friendly activities instead of polluting development. Moreover, pollution victims need a compensation fund to respond to health risks and lost income.

Nonetheless, empowering the public and various stakeholders to play an active role in combating climate change is vital in order to ensure sustainable development in the name of environmental justice. Compared to most developed states, civil society movements are relatively new and weak in China. No environmental nongovernmental organizations (NGOs) were officially registered until 1994, which is marked as the start year of Chinese environmental activism. Environmental education and biodiversity protection serve as the core task of NGO missions, while grassroots activism on climate change and popular pressure for government action on environmental issues are growing in both scale and extent. A variety of international NGOs have strong presence in China now, such as World Wildlife Fund (WWF), Friends of Nature, the Nature Conservancy, and Conservation International.

Environmental activists protest environmental pollution issues through various means such as letters, campaigns on the Internet, and newspaper editorials. However, NGOs need to be sponsored by government agencies before they can be officially registered and operate. In other words, government agencies will supervise NGOs' activities, membership, and funding sources. This governmental conduct significantly limits the range of action adopted by environmental NGOs. At the same time, they still largely rely on international funding, which is subject to criticism for being a source of foreign interference in China's domestic affairs. However, substantial progress can only be made gradually before China's young environmental NGOs gain greater influence by attracting funding from domestic sources as well as raising environmental justice awareness among a growing number of citizens. Given China's population scale and geographical complexities as well as its considerable contributions to global emissions, whether and how it meets the challenges of climate change will have massive implications not only for Chinese citizens but also for citizens worldwide.

Xinyue Ye

Further Reading

Brajer, V., R. Mead, and F. Xiao. "Adjusting Chinese Income Inequality for Environmental Equity." *Environment and Development Economics* 15 (2010): 341–362.

Economy, E. C. *The River Runs Black: The Environmental Challenge to China's Future.* Ithaca, NY: Cornell University Press, 2004.

Hong, Q. "Why Are There More and More Cancer Villages?" *Environmental Economy* (March 2005): 58–59.

Liu, L. "Sustainability Efforts in China: Reflections on the Environmental Kuznets Curve through a Locational Evaluation of 'Eco-Communities.'" *Annals of the Association of American Geographers* 98(3) (2008): 604–629.

Lu, Y. "Environmental Civil Society and Governance in China." *International Journal of Environmental Studies* 64(1) (2007): 59–69.

Ma, L. J. C. "From China's Urban Social Space to Social and Environmental Justice." *Eurasian Geography & Economics* 48(5) (2007): 555–566.

Mol, A. P. J. "Environmental Governance through Information: China and Vietnam." *Singapore Journal of Tropical Geography* 30(1) (2009): 114–129.

Rock, M., and D. Angel. "Grow First, Clean Up Later?" *Industrial Transformation in East Asia, Environment* 49(4) (2007): 8–19.

Smil, V. *China's Environment: Resilient Myths and Contradictory Realities.* Armonk, NY: M. E. Sharpe, 2000.

Tang, S., and X. Zhan. "Civic Environmental NGOs, Civil Society, and Democratisation in China." *Journal of Development Studies* 44(3) (2008): 425–448.

Wang, M., M. Webber, B. Finlayson, and J. Barnett. "Rural Industries and Water Pollution in China." *Journal of Environmental Management* 86(4) (2008): 648–659.

Wong, K. "Greening of the Chinese Mind: Environmentalism with Chinese Characteristics." *Asia-Pacific Review* 12(2) (2005): 39–57.

Environmental Justice and Climate Change in India

The focus on environmental justice as a social movement and as a scholarly field has its origins in the United States. The U.S.-based environmental justice movement emerged in the 1980s as a response to the unequal protection that communities of color received regarding environmental pollution by local, state, and national regulatory agencies. The Principles of Environmental

Justice were adopted at the 1991 First People of Color Environmental Leadership Summit. Subsequently, recent academic literature has expanded its reach to other national contexts, and many works now consider environmental inequality occurring across nation-states or within other nations. One important global arena where environmental justice scholars and activists have sought to analyze environmental policies and problems is in the field of climate change.

India and Climate Change Policy

India has long maintained that it has not been responsible for most of climate change and will only follow with efforts to control green house gas (GHG) emissions once the industrialized countries have made decisive first steps. For the majority of the last two decades, India's policy toward climate change has been guided by the idea of common but differentiated responsibility. This idea allows for a distinction to be made in emissions between states according to their level of industrialization. This idea gained momentum through the work of Anil Agarwal and Sunita Narain at the Center for Science and Environment in New Delhi, India. Agarwal and Narain made a distinction between what they believed constituted survival and luxury emissions. Agarwal and Narain argue that developing countries such as India should not pay mitigation costs or bear any responsibility for causing climate change, as they have historically produced only survival emissions.

This principle was strongly articulated again in the 2007 United Nations Human Development report *Fighting Climate Change: Human Solidarity in a Divided World.* The report examines how both the roots of climate change and its negative impacts are globally differentiated as well as internationally differentiated within the global North. For instance, according to the report, "the state of Texas (population 23 million) in the United States registers carbon dioxide emissions of around 700 Mt [metric tons] or 12 percent of United States' total emissions. That figure is greater than the total carbon dioxide footprint left by sub-Saharan Africa—a region of 720 million people." These types of statistics are often cited by poor nations such as India that want rich nations to act immediately and reduce their emissions. Furthermore, invoking principles of justice, they argue that a climate treaty should punish countries that have contributed the largest amount to GHG emissions.

In 1990 when negotiations on the United Nations Framework Convention on Climate Change (UNFCCC) began, India produced 0.6 billion Mt of carbon dioxide. Today, India's emissions have more than doubled to 1.3 Mt of carbon dioxide. India now ranks as the third-largest consumer of primary

Hot Spot

India

The environmental issues facing India are enormous. With more than 1 billion residents, the country is the second most populated nation on Earth after China but encompasses a land area only one-third the size of the United States. India's population continues to grow at a high rate—the nation's annual number of births exceeds its number of deaths by almost 3 to 1. With a growing number of people competing for a limited amount of natural resources, the country will likely be in a state of environmental crisis for years to come.

In such cities as Mumbai (formerly Bombay) and Kolkata (formerly Calcutta), there is inadequate housing for most of the population. Forty-six percent of all houses in Mumbai are less than 100 square feet—the same size as an average bedroom in a U.S. home. Several Indian families usually share one of these tiny homes, sometimes sleeping in shifts. Sewage and garbage disposal are also inadequate. Potable water is difficult to come by. Often an entire neighborhood will share a single tap, and people must carry water to their houses in buckets and jugs.

India's large industrial plants also pollute heavily. Waste is often poured directly into the rivers or the sea, and industrial accidents are common. One of the worst industrial accidents in the planet's history occurred at the Union Carbide pesticide plant in Bhopal, India, in 1984. Due to defective safety procedures, 40 tons of deadly methyl isocyanate leaked out of storage tanks, spewing a toxic gas cloud over large portions of the city. The official death toll from the accident was 2,352 people killed, although later counts put the number as high as 10,000.

Large areas of tropical and hardwood forest remain in India, but deforestation caused by slash-and-burn agriculture and timber harvesting continues. In the late 20th century, satellite photos revealed that India had lost about 386 square miles of its prime forests. Today, only about 10 percent of the country still has forest cover, and only 4 percent of India's land is protected within national parks and reserves. In the past few decades, the government has taken serious steps to improve environmental management and has established more than 350 parks, sanctuaries, and reserves.

energy supplies and is one of the top five emitters of GHG emissions. However, despite having a 4.23 percent share of global emissions, its per capita emissions are well below the world's average. India's per capita emissions are at 1.31 tons per capita, while China has a per capita emission rate of 4.91 and that of the United States is at 19.18.

Nevertheless, despite certain positive statistics, India's climate trajectory is hardly encouraging. Realizing sustained economic growth rates of 8 percent requires energy- and carbon-intensive materials. A 2008 report released by the Energy and Resources Institute in New Delhi argued that India would need to expand its primary energy supply threefold to fourfold and boost electricity supply by some five to seven times of current levels by 2031. Of India's more than 1 billion people, more than 800 million still subsist on less than $2 a day. More than 700 million people still cook on traditional stoves using crop waste and animal residue. More than 400 million still do not have access to electricity. Addressing the needs of the poor will no doubt drive up India's carbon emissions for the coming two decades. A U.S. Department of Energy report predicts that India's CO_2 emissions could increase between 72 percent and 225 percent by 2025, which would severely threaten the goal of curbing global temperature increases within 2 degrees Celsius.

Impact of Climate Change on India

While India now numbers among the top countries for global GHG emissions, it also figures among those countries most vulnerable to climate change. With an economy that is closely tied to its natural resource base and climate-sensitive sectors such as agriculture, water, and forestry, India will most likely face a major threat because of projected changes in climate. A 2010 report by the Energy and Resources Institute predicts that sugarcane production could go down by as much as 30 percent under climate change, severely impacting people's livelihoods. In addition, it appears that the increasing temperatures associated with climate change (combined with other pressures such as increasing population, a decline in soil fertility, and a decrease in genetic diversity of popular varieties) may lead to a reduction in crop production in India. Recent reports have shown that in eastern Uttar Pradesh the seasons have turned increasingly erratic, which has severely disrupted agriculture in the area.

Other possible climate change impacts in India are sea-level rise and increased severe storms and flooding. It is estimated that a 1-meter sea-level rise could displace up to 7 million people, with 5,764 square kilometers of land and 4,200 miles of road lost. Studies have shown that India is likely to suffer the highest gross domestic product loss from climate change if there is a 2.5 degree Celsius increase in global temperature. Early indication of economic losses is already becoming evident in some parts of the country. A recent news report raised concerns that rising temperatures may be affecting India's tea plantations in Assam. As Assam produces more than half of India's tea and India accounts for nearly one-third of the world's tea production, such rising temperatures could have dire consequences for India's export industry.

National Action Plan on Climate Change

In response to some of these concerns, India has acknowledged its responsibility in addressing threats arising from climate change. Prime Minister Manmohan Singh recently stated that "There should be no doubt in anybody's mind that we fully recognize not just how important this issue is to India but also our obligation to address it." India released its first National Action Plan on Climate Change (NAPCC) in June 2008. The plan outlines existing and future policies and programs addressing climate adaptation. Through the NAPCC, the Indian government has resolved to take steps to combat, mitigate, and adapt to various scenarios that may arise due to climate change. The plan does not impose quantitative emission targets on the country but instead focuses on efficiency targets. The report identifies measures that promote development objectives while yielding cobenefits for climate change. In doing so, the plan seeks to balance India's economic growth objectives with its development goals. Eight core national missions running through 2017 are identified in the following areas: solar, enhanced energy efficiency, sustainable habitat, water, Himalayan ecosystem, green India, sustainable agriculture, and strategic knowledge on climate change.

The NAPCC has been followed by the formulation of state-level Strategy and Action Plans on Climate Change. These state-level plans are meant to convert and adapt the national level policy imperatives to subnational level actions. All state governments were asked to prepare these action plans and submit them by March 31, 2011.

Environmental Justice Concerns

The environmental justice movement in India has had a vibrant history, from the Chipko movement in the 1970s to the contestations over the building of the Narmada Dam in the 1990s. The climate justice movement in India dates back to 2002, when the Eighth Conference of the Parties was held in New Delhi. Paralleling the UNFCCC meeting was the Climate Justice Summit attended by hundreds of activists from throughout the country. A resulting Delhi Climate Justice Declaration stated in part:

> We affirm that climate change is a rights issue—it affects our livelihoods, our health, our children, and our natural resources. We will build alliances across states and borders to oppose climate change inducing patterns and advocate for and practice sustainable development. We reject the market based principles that guide the current negotiations to solve the climate crisis: Our World is Not for Sale.

Despite this action, there appeared to be very little action on the climate justice front between 2002 and 2008. Since the NAPCC was released in 2008, there has been a significant civil society outcry about the lack of attention given to environmental justice concerns. September 2008 saw a large number of people's movements, progressive trade unions, people's science groups, forest groups, fish-workers federations, antiextractive struggles, and other progressive civil society organizations come together and form the Indian Climate Justice Forum. The forum consists of about 75 groups from all over India, many of them large umbrella organizations in their own region. The Indian Climate Justice Forum highlights the serious threat that climate change poses to the poor and takes the Indian government to task on various fronts.

First, the forum critiques the Indian government's stance on mitigation, which argues that India will "not allow its per capita GHG emission to exceed the average per capita emissions of the developed countries." The main problem with the government of India approach is that in setting the emissions bar so high (compared to that of the developed nations, including the United States), the Indian government is setting its achievement expectations pretty low. Environmental justice activists point out that elites in India have emission rates approaching American and European levels, whereas the majority of Indian citizens do not even have access to electricity. Activists argue that the country's touted low per capita emissions are on the backs of the poor. Indeed, a recent report by Green Peace demonstrated that

> While only 14% of the Indian population earns more than 8,000 rupees a month, they contribute 24% of the CO_2 emissions of the country. . . . [W]hen it comes to CO_2 emissions, a relatively small wealthy class of 1% of the population in the country is hiding behind a huge proportion of 823 million poor people. It is the country's poor with an income of less than 5000 rupees a month, who keep the average CO_2 emissions really low.

Second, environmental justice groups argue that adaptation to climate change is woefully underfunded, especially in key areas such as mangrove conservation, afforestation, and biodiversity conservation programs.

Third, activists argue that technology transfer to India should be carried out free of conditionalities and intellectual property rights restrictions. They also advocate for the adoption of local and sustainable technologies that are appropriate to people's needs, as opposed to large capital projects.

Finally, activists argue that industrialized countries should pay for adaptation funds for developing countries, that the funds should be monitored, and that their ultimate use should be decided in a transparent participatory manner—through consultation with state and local governments.

Conclusion

The environmental justice movement in India faces some significant challenges in the future, particularly with regard to furthering their propoor climate justice agenda. India's middle class constitutes a significant portion of the political population. Responsible for a significant rise in GHG emissions and waste production, they are also responsible for the circulation of new more authoritarian forms of environmental discourses in urban areas. Scholars such as Amita Baviskar and Ramachandra Guha term this development the rise of "bourgeois environmentalism." They point to a distinct social group—professionals holding jobs as university lecturers, bankers, and journalists—as perpetuating this bourgeois form of environmentalism and differentiate this type of environmentalism from the environmentalism of the poor. Environmental justice activists need to be cognizant of these trends and not only enlist the Indian state to further its cause but also make sure that their strategies target urban populations and make them their allies.

Sonalini Sapra

Further Reading

Agarwal, A., and S. Narain. *Global Warming in an Unequal World: A Case of Environmental Colonialism.* New Delhi, Center for Science and Environment, 1991.

Ananthapadmanabhan, G., K. Srinivas, and V. Gopal. *Hiding behind the Poor: A Report by Greenpeace on Climate Injustice.* Bangalore: Greenpeace India Society, 2007.

Baumert, Kevin, Timothy Herzog, and Jonathan Pershing. *Navigating the Numbers: Greenhouse Gas Data and International Climate Policy.* Washington, DC: World Resources Institute, 2005.

Baviskar, Amita. "The Dream Machine: The Model Development Project and the Remaking of the State." In *Waterscape: The Cultural Politics of a Natural Resource,* edited by Amita Baviskar, 281–307. New Delhi: Permanent Black, 2007.

Baviskar, Amita. "The Politics of the City." *Seminar* 516 (2002): 40–42.

Brenkert, A. L., and E. L. Malone. "Modeling Vulnerability and Resilience to Climate Change: A Case Study of India and Indian States." *Climatic Change* 72(1–2) (2005): 57–102.

The Energy and Resources Institute. *Predicting Local Impacts in Indian States.* 2010. http://www.teriin.org/themes/climate-change/maharashtra.php.

Government of India. *National Action Plan on Climate Change.* New Delhi: Prime Minister's Council on Climate Change, Government of India, 2008.

Government of India. *The Road to Copenhagen: India's Position on Climate Change Issues.* Washington, DC: Public Diplomacy Division, Ministry of External Affairs, Government of India, 2009.

Guha, Ramachandra. "The Authoritarian Biologist and the Arrogance of Anti-Humanism: Wildlife Conservation in the Third World." *Ecologist* 27 (1997): 14–20.

Human Development Report. *Fighting Climate Change: Human Solidarity in a Divided World; Human Development Report 2007/2008.* New York: Palgrave Macmillan, 2007. http://hdr.undp.org/en/reports/global/hdr2007-2008/.

India Resource Center. "Delhi Climate Justice Declaration," 2002, http://www.indiaresource.org/issues/energycc/2003/delhicjdeclare.html.

Integrated Research and Action for Development. *Gender Analysis of Renewable Energy in India: Present Status, Issues, Approaches, and New Initiatives.* Washington, DC: ENERGIA, 2009.

Kelkar, U., K. K. Narula, V. P. Sharma, and U. Chandna. "Vulnerability and Adaptation to Climate Variability and Water Stress in Uttarakhand State, India." *Global Environmental Change: Human and Policy Dimensions* 18(4) (2008): 564–574.

Manoj, C. G. "Climate Change: PM Asks States to Ready Their Action Plans." *Indian Express,* August 19, 2009, http://www.indianexpress.com/news/Climate-change—PM-asks-states-to-ready-their-action-plans/503761.

Michel, David. "Introduction." In *Indian Climate Policy: Choices and Challenges,* edited by David Michel and Amit Pandya, 1–18. Washington, DC: Henry L. Stimson Center, 2009.

Nordhaus, W. D., and J. Boyer. *Warming the World: Economic Models of Global Warming.* Cambridge, MA: MIT Press, 2003.

Pasricha, Anjana. *Climate Change Impacts India's Tea Growing Region.* Voice of America News, January 4, 2011. http://www.voanews.com/english/news/asia/south/Climate-Change-Impacts-Indias-Tea-Growing-Region-112860129.html.

Sargsyan, Gevorg, Mikul Bhatia, Sudeshna Ghosh Banerjee, Krishnan Raghunathan, and Ruchi Soni. *Unleashing the Potential of Renewable Energy in India.* New York: South Asia Energy Unit, Sustainable Development Department, World Bank, 2010.

Union of Concerned Scientists. "Each Country's Share of CO2 Emissions," 2010, http://www.ucsusa.org/global_warming/science_and_impacts/science/each-countrys-share-of-co2.html.

Environmental Justice and Climate Change in the United States

Although not often explicitly linked, there are clear relationships between environmental justice and climate change within the United States. The research presented has shown that environmental injustices occur as a result of industrial pollution. The facilities discussed contribute to high greenhouse gas emissions and toxins that contribute to social degradation, specifically in areas of low socioeconomic status and minority groups, such as African Americans, women, children, and the elderly. The research examined has also shown that climate change is a specific consequence of greenhouse gas emissions and toxins released from concentrated animal-feeding operations (CAFOs) and coal-burning plants, which leads to environmental degradation. Therefore, there are several possibilities for research opportunities to examine how these and other industrial polluters link climate change and environmental justice as a joint issue within the United States.

Climate change occurs in response to the accumulation of greenhouse gas emissions in the atmosphere. The greenhouse gases that are the greatest contributors of climate change are carbon dioxide (CO_2), nitrous oxide (N_2O), methane (CH_4), tropospheric ozone (O_3), and chlorofluorocarbons (CFCs). Carbon dioxide is the most concentrated greenhouse gas, with methane and nitrous oxide also acting as key forcing contributors. Human activities have accelerated greenhouse gas emissions through the burning of fossil fuels, changes in industrial processes, and land usages. The effects of climate change are characterized by the increase in catastrophic storms (i.e., strong thunderstorms, hurricanes, and tornadoes) and also through extreme temperature shifts, resulting in colder winters, hotter summers, flooding, and severe droughts. Conversely, major contributing forces of climate change, such as industry polluters, are situated within certain communities in the United States and contribute to air toxins, polluted water supplies, and higher rates of adverse health conditions. Environmentally unhealthy conditions, such as toxic waste dumps and coal-burning facilities located in neighborhoods where individuals live, work, or play, spurred an environmental justice movement in the United States in the early 1980s. Resistance against environmental injustices has been documented earlier than the 1980s, but it was not until then that it developed as a political movement. Originally the movement was used to address environmental injustices in predominantly African American neighborhoods. Originally termed "environmental racism," the movement has evolved in the United States to include all groups depressed of environmental rights. While this includes all individuals, the groups that are most vulnerable tend to be poor minority groups with limited political clout and limited access and abilities for involvement in environmental decision making within their communities. This is important in the United States, because currently the United States ranks highest among nations as an emitter of carbon dioxide, methane, and nitrous oxide. Climate change therefore is felt in the United States through the degradation it poses on the environment and humans.

Toxic Waste Dump: Warren County, North Carolina

Toxic landfills contribute to atmospheric greenhouse gases largely from methane emissions but also carbon dioxide emissions. These landfills not only emit greenhouse gases that are forces of climate change but also contribute negatively to human health through airborne toxic water contaminants. In 1982, Warren County, North Carolina, was selected by the state as the area to house a hazardous waste landfill. The Warren County PCB Landfill was created to contain soils that were contaminated by the spraying of polychlorinated biphenyl (PCB) containing oil along the roadsides of multiple counties in North Carolina. The Ward Transformers Company illegally sprayed 31,000 gallons of PCBs in 14 counties in 1973. At that time, Warren County was a rural community that consisted of predominantly poor African Americans and did not contain a city council or mayor. The county was among 6 in North Carolina's rural Southern Black Belt region. The Southern Black Belt region got its name because southern rural African Americans comprise the majority of the population, dating back to ancestral ties from early agriculture and slavery. Due to the risks associated with toxic dumps and groundwater contamination, community members from Warren County organized and protested, which attracted national support and attention. Trucks that attempted to carry multiple loads of toxic waste to the dump were met with resistance, resulting in the arrest of more than 500 protestors. This particular event created a political movement that led to studies and reports that linked social justice and environmental issues. Thus, the environmental justice movement was created in the United States and addressed environmental inequalities at the state and federal levels.

Concentrated Animal-Feeding Operations

CAFOs are intensive industrial agriculture operations that house thousands of livestock, typically consisting of cattle, pigs, or chickens. Large agricultural corporations commonly own CAFOs. The corporations typically have little to no physical contact with the actual operation and instead use contract farmers comprised mainly of migrant workers and nonwhites to oversee the facilities. CAFOs are a large contributor to climate change in the United States both directly and indirectly. They directly contribute to climate change through the emission of methane from concentrated livestock manure and indirectly through carbon dioxide and nitrous oxide emissions produced from intensive agricultural production of corn and soy that is necessary for animal feed. Greenhouse gases produced from livestock and crop production are commonplace in agriculture, but the intensification of production through CAFOs has contributed to the tripling of emissions over a relatively short period of time, making CAFOs causal to significant environmental degradation. CAFOs are concentrated within specific regions within the United States, depending on the type of livestock, and are predominately located in low-income nonwhite neighborhoods. Hog production, for example, is largely concentrated in North Carolina and also mostly within the state's counties that make up the rural Southern Black Belt. The article "Environmental Injustice in North Carolina's Hog Industry" states that North Carolina ranks second among U.S. states in industrial hog production, where the operations are mostly concentrated in the lower coastal region. Constructing CAFOs in this region of North Carolina is particularly risky due to its topographic makeup of high water tables and low floodplains. According to the article, the large amount of hog manure produced from hog CAFOs poses risks to the safety of the drinking water and is responsible for the release of airborne toxins and noxious odors. This study examined the locality and effects of

Hot Spot

New Jersey and Long Island

Rising waters are nibbling at the coastlines of New Jersey and Long Island. Some beachfront vacation homes on Long Island have been raised on stilts as tides have risen. Hoboken subway stations also have been reinforced with tide guards. Vivian Gornitz, a sea-rise specialist at NASA's Goddard Institute of Space Studies in New York City, said that the amount of land lost to the sea could accelerate to between two and five times the present rate by the end of the 21st century. She said that seas in the area could rise between 4 and 12 inches by 2020, 7 inches to 2 feet by 2050, and 9.5 inches to almost 4 feet by 2080.

Jim Titus, project manager for sea-level rise at the U.S. Environmental Protection Agency, said that such a rise could imperil all beaches in the area. "The data doesn't lie," he said. "The sea is rising. The shore is eroding because the sea is rising. There's no doubt about that." At Long Beach Island, said Titus, sea levels already have risen a foot in 70 years. Two days a month, high tides at full moon (called spring tides) often cause flooding in the area. Homes are now built on stilts, and garbage cans are anchored to prevent them from floating away on the tides. Local officials said that future storms could flood transportation infrastructure, including the Newark-Liberty International Airport, which lies less than 10 feet above present sea levels.

Bruce E. Johansen

2,514 hog CAFOs in North Carolina and their relationships to populations based on racial and economic backgrounds. The findings were conclusive that the hog operations were disproportionately placed in areas of high poverty where primarily nonwhite persons resided. These results evoke considerations of environmental justice issues not only due to the environmental degradation and bacterial contamination of the drinking water but also to the future economic and social stagnation it causes to a politically powerless and economically depressed community. The emission of greenhouse gas pollutants from CAFOs is also an environmental public health concern, linked to the disruption of proper respiratory and brain function. While this study was able to link the occurrences of environmental degradation in high-poverty nonwhite communities to hog CAFOs and prove environmental injustices in

North Carolina, further research is necessary to link CAFO pollution directly to environmental justice and climate change.

Coal-Burning Power Plants

The United States is greatly dependent on coal, mainly for electricity. The chemical composition of coal may pose environmental and health risks during the incineration process. When burned, coal releases carbon dioxide into the atmosphere. Carbon dioxide is the largest greenhouse gas contributor to climate change, and emitting large quantities is detrimental to environmental health. Coal-fired power plants in the United States currently account for approximately 3 million tons of CO_2 emitted per year. Burning coal also produces atmospheric chemicals such as nitrogen and sulfur oxides, radioactive materials such as thorium and uranium, and mercury emissions. These by-products are harmful to and seriously jeopardize human health. Coal-fired power plants also produce coal ash as a waste product after the coal is burned. Coal ash may contain toxic chemicals such as nickel, zinc, chlorine, barium, aluminum, antimony, beryllium, cobalt, boron, manganese, molybdemum, thallium, vanadium, and arsenic. The inhalation, ingestion, and consumption of these metals has been linked to severe public health issues, such as certain cancers, neurological defects, nervous system impairments, and birth defects. Coal-burning power plants are present in every state except Idaho, Vermont, and Rhode Island, with the highest polluters heavily concentrated in Pennsylvania, Georgia, Indiana, Texas, North Carolina, West Virginia, Alabama, Ohio, Kentucky, New Mexico, Florida, and Wyoming. The Environmental Protection Agency (EPA) has implemented air pollution regulations to offset atmospheric greenhouse gas emissions, using scrubbers to help catch atmospheric toxins and coal ash. The toxic waste captured in scrubbers has been found to leach into rivers, lakes, and groundwater with little cost to or regulation for the facilities. Also, the requirements of emissions regulations are not implemented on plants built before 1977, giving older plants the freedom to determine their emission outputs. In Massachusetts, for example, the older plants that are not regulated produce three to four times more atmospheric emissions than newly regulated plants. The environmental degradation caused by coal-fired power plants makes this specific industrial operation a major concern of environmental justice in communities. In the article "Unequal Exposure to Ecological Hazards: Environmental Injustices in the Commonwealth of Massachusetts," the authors examine 368 communities in Massachusetts that suffer environmental degradation from facilities such as coal-fired power plants, linking socioeconomic status and race to the geographic dispersal of these operations.

The study used the five dirtiest power plants in the state based on emission rates and found that four of the five are located in predominantly low-income neighborhoods. Of the four plants that were disproportionately positioned in communities with higher rates of poverty, only one plant was in a community of both low income and high minority population. The individuals in these communities, such as pregnant women, children, and the elderly, are most at risk for diseases caused from the inhalation of air pollutants, airborne mercury emissions, and groundwater contamination. These diseases include various types of cancers, reproductive disorders, and neurological defects. The study concluded that while the majority of polluting facilities, such as coal-burning power plants, are predominantly located in areas of lower wealth and higher minority populations, environmental justice issues also disproportionately affect working-class communities and may be gender or age specific. Further research is necessary to directly link the environmental and social impacts of coal-burning power plants to climate change.

The background and case studies presented regarding environmental justice in the United States demonstrate how and why the movement originated and what it looks like today. Environmental justice has evolved to include all individuals, but as the studies demonstrate, issues of race, class, and gender are deeply rooted within the movement. The industries that were examined, such as toxic waste dumps, CAFOs, and coal-fired power plants, contribute to greenhouse gas emissions that directly impact climate change. Therefore, research linking environmental justice and climate change in the United States is necessary to drive future environmental policy that will address both issues together rather than separately.

Kristen Casper LaRusse

Further Reading

Agyeman, J., R. D. Bullard, and B. Evans. "Introduction." In *Just Sustainabilities: Development in an Unequal World,* edited by J. Agyeman, R. Bullard, and B. Evans, 1–16. London: Earthscan Publications, 2003.

Agyeman, U. *Sustainable Communities and the Challenge of Environmental Justice.* New York and London: New York University Press, 2005.

Candy, J. C. "Environmental Justice Cited in Fight against Power Plant near Hillburn, N.Y." The Record, November 19, 2001, http://ezproxy.chatham.edu:2048/login?url=http://search.ebscohost.com/login/aspx?direct=true&db=nfh&AN=2W63417687359&site=ehost-live.

Chalvatzaki, E., and M. Lazaridis. "Estimation of Greenhouse Gas Emissions from Landfills: Application to the Akrotiri Landfill Site (Chania, Greece)."

Global NEST Journal 12(1) (2010): 108–116. http://www.gnest.org/journal/Vol12_no1/108-116_681_Lazaridis_12-1.pdf.

"Coal Ash: The Toxic Threat to Our Health and Environment." http://www.psr.org/assets/pdfs/coal-ash.pdf.

"Coal and Climate Change Facts." http://pewclimate.org/global-warming-basics/coalfacts.cfm.

Cutter, S. L. "Race, Class and Environmental Justice." *Progress in Human Geography* 19(1) (1995): 111–122. http://geogrpaphy.ssc.uwo.ca/faculty/baxterj/readings/Cutter_environmental_justice_PIHG_1995.pdf.

Duhigg, C. "Cleansing the Air at the Expense of Waterways." *New York Times,* October 12, 2009, http://www.nytimes.com/2009/10/13/us/13water.html.

Faber, D. R., and E. J. Krieg. "Unequal Exposure to Ecological Hazards: Environmental Injustices in the Commonwealth of Massachusetts." *Environmental Justice* 110(2) (2002): 277–287. http://www.ncbi/nlm.nih.gov/pmc/articles/PMC1241174/pdf/ehp110s-000277.pdf.

"50 Dirtiest U.S. Power Plants Named." Environmental News Service, July 26, 2007, http://www.ens-newswire.com/ens/jul2007/2007-07-26-05.asp.

Freedman, A. "New York Times Publishes a Searing Drought Story, but Completely Misses the Climate Change Angle." Web log message, July 12, 2011, http://www.climatecentral.org/blogs/new-york-times-writes-a-searing-drought-story-but-misses-the-climate-change-angle.

Gabbard, A. "Coal Combustion: Nuclear Resource or Danger." *ORNL Review* 26 (1993), http://www.ornl.gov/info/ornlreview/rev26–34/text/colmain.html.

Hoerner, J. A., and N. Robinson. "A Climate of Change: African Americans, Global Warming, and a Just Climate Policy for the U.S. Environmental Justice and Climate Change Initiative." 2008, www.ejcc.org.

Jones, R. E., and S. A. Rainey. "Examining Linkages between Race, Environmental Concern, Health and Justice in a Highly Polluted Community of Color." *Journal of Black Studies* 36(4) (2006): 473–496. doi:10.1177/0021934705280411.

Keys, E., and W. J. McConnell. "Global Change and the Intensification of Agriculture in the Tropics." *Global Environmental Change* 15 (2005): 320–337. doi:10.1016/j.gloenvcha.2005.04.004.

Lappe, A. *Diet for a Hot Planet: The Climate Crisis at the End of Your Fork and What You Can Do about It.* New York: Bloomsbury, 2010.

Lawrence, L. L., and N. McDonald. *Environmental Justice: A Growing Movement.* New York: NY Affordable Reliable Electricity Alliance, 2007, http://www.area-alliance.org/documents/environmentaljustice.pdf.

Martinez-Alier, J. "Mining Conflicts, Environmental Justice and Valuation." In *Just Sustainabilities: Development in an Unequal World,* edited by J. Agyeman, R. Bullard, and B. Evans, 201–228. London: Earthscan Publications, 2003.

Wing, S., D. Cole, and G. Grant. "Environmental Injustice in North Carolina's Hog Industry." *Environmental Health Prospectives* 108(3) (2000): 225–231.

Environmental Movement: 1980–2000

If the 1970s mark the decade in which the more traditional notion of conservation was replaced with a more fully conceptualized idea of an "environment," then the 1980s was the decade in which the politics of this concept began to take shape and its truly global nature first became apparent. Movements in the 1980s and 1990s were bolstered by public opinion that had become entrenched in favor of environmental progress but were also forced to respond to problematic political trends such as deregulation, hegemonic liberalism and its attendant reliance on market solutions, and the increasingly global nature of environmental crises.

Two major events set the stage for 1980s environmentalism. First, scientists began to develop increasingly solid evidence that climate change was occurring and began to identify the central causes, primarily in chlorofluorocarbons (CFCs), methane, and other emissions—collectively categorized under the name "greenhouse gases." The ozone layer that these gases were said to be depleting became a household name, testament to the widespread success with which environmental groups were able to call attention to the problem. Alarm over the destruction of ozone protections and the specter of global warming—rather than cooling, as had previously been feared—led to the establishment of the National Climate Program Act in the United States, a fleeting and poorly funded but symbolically important step. Second, global energy crises provided a critical window for environmental groups to cohere around a central theme of renewable energy. Calling energy "the greatest challenge our country will face during our lifetimes," President Jimmy Carter famously had 32 solar panels installed on the White House roof in 1977, a symbolic gesture no doubt but a leader's way of marking and calling attention to an impending political, economic, and environmental specter. In 1979 an energy crisis, exacerbated by escalating oil prices and scarcity during the 1979 Iranian Revolution, set the stage for environmental groups to link various threads of crisis and concern, which they did with some degrees of success. The taking of 52 American hostages by Iranian militants facilitated the election of Ronald Reagan as president of the United States and drove home for Americans the relationship between energy politics and national security—a connection that some environmentalists attempted to exploit throughout the 1980s and 1990s but never successfully mainstreamed into the fabric of the environmental movement's rhetorical strategy.

One crucial legacy of the 1970s was the congealing and entrenching of a public consensus around the importance of environmental protection, much of which was the result of grassroots organization rather than governmental intervention. Whereas the main result of these developments in the 1970s was a series of legislative and regulatory victories, the antienvironmentalism of the Reagan administration in the 1980s forced environmentalists to change their posture. In particular, the Reagan administration's antiregulatory posture degraded American environmentalists' trust in the government's will to continue the work of the 1970s and enforce key legislation. National projects established during the 1970s such as the Clean Water Act were dismantled and left unenforced as the Environmental Protection Agency (EPA) gave way to a public preference for local action—a move in deference to states' rights adherents but that resulted in little more than inaction. The result for those partaking in the environmental movement was a decisive move away from the promised land of regulation toward other modes of engagement, including an uptick in litigation to force the Reagan administration to act.

The attack on environmental regulation waged by the Reagan administration had the paradoxical effect of arousing an American environmental movement that had grown complacent in the late 1970s. The early Reagan years served as crucial recruitment tools as groups utilized Reagan's high-profile antienvironmentalist and probusiness appointments; James Watt's appointment as secretary of the interior and Anne Burford as head of the Environmental Protection Agency served as rallying cries for a nation that in the 1970s had begun to take environmental matters seriously. As a result, during this time the ranks of more mainstream environmental groups swelled. The Sierra Club, for example, witnessed a threefold increase in membership, and the Wilderness Society grew sixfold. Groups from Greenpeace to Friends of the Earth to the World Wildlife Fund experienced similar booms in membership.

Early on, Reagan-era policies spawned a reorganization of environmental organizations and priorities, clarifying for American environmentalists critical differences between the long-term strategic prospects for decentralized nonstate grassroots political movements and those of cooperative efforts with government, much of which would prove toothless in a deregulated anti-environmental political climate. On the one hand was the rise of so-called third-wave environmentalism in which environmental groups such as the Environmental Defense Fund negotiated with governmental agencies and corporations to develop environmentally sound market mechanisms. This movement, which arose in part as a rejection of more radical forms of environmentalism (especially deep ecology), sought to marginalize grassroots activists and work within existing—especially anthropocentric and technologically driven—frameworks rather than through radical reforms. On the other hand, this trend also spawned a series of radical movements toward deep ecology and an explicit rejection of reform environmentalism that arose as a concession to the market strategies of the Reagan years.

The early days of the Reagan administration thus inspired the congealing and mass mobilization of citizen-activist–fueled nongovernmental organizations (NGOs). Among the most radical of these new groups was Earth First! Formed in 1979 and rallying around its slogan "No compromise in the defense of Mother Earth," Earth First! advanced an ecocentric approach inspired in large part by Edward Abbey's novel *The Monkey Wrench Gang* by advocating direct engagement with loggers, polluters, and other groups seen as destroying the environment. These groups, which were the early progenitors of later groups such as the Earth Liberation Front, turned away from negotiation and consensus building toward strategies of ecotage and other forms of direct action. Groups such as Greenpeace, which was formed in the early 1970s but rose to prominence in the 1980s, made nuclear waste and the near extinction of beloved species such as whales its signature issues. Greenpeace utilized high-profile celebrities who appeared regularly on MTV and other networks to raise popular awareness, especially among younger people.

The 1980s also saw the birth of the so-called environmental justice movement, in large part a response to the claim that toxic dumping and other environmental disasters were increasingly foisted upon vulnerable communities, especially communities of poverty and color. The environmental justice movement ushered in a dispreference for large centralized groups and a turn toward grassroots efforts organized in networks, many of which engaged in the redress of local and regional environmental crises. The watershed event for the environmental justice movement occurred in 1982 when community leaders in Shocco Township, North Carolina, mobilized against the

EPA-approved Warren County PCB Landfill, which was slated to store about 60,000 tons of contaminated soil. Throughout the 1980s and 1990s, various states established councils and working groups to address environmental justice issues, particularly concerning the disproportionate burdens placed on disempowered populations with regard to siting issues. The movement would achieve theoretical coherence with the statement titled "Principles of Environmental Justice" issued at the First National People of Color Environmental Leadership Summit, held in Washington, D.C., in October 1991. By the late 1990s, groups such as the Basal Action Network and the Global Anti-Incinerator Alliance would mark the enduring—and increasingly transnational—basis of the global environmental justice movement as well as its increased concern with the developing world.

In the mid to late 1980s, global organizations also began to take shape to specifically investigate and address warnings about the specter of climate change. In 1987, 140 nations, including the United States, signed the Montreal Protocol on Substances That Deplete the Ozone Layer, in which signers pledged to reduce greenhouse gases such as CFCs. Significantly, in 1988 the United Nations Intergovernmental Panel on Climate Change (IPCC) was established; in 1990 the IPCC would release its first findings that Earth had warmed by half a degree Celsius within a century. The IPCC urged strong action and sounded the first major alarm of the crisis of the next two decades.

In a major indication of the success of 1980s environmental movement, by 1990 environmental issues—and climate change in particular—were no longer considered marginal but instead were becoming increasingly mainstreamed into a wide range of political thinking and organizing. Public opinion had militated against Reagan's early attempts to undo 1970s environmental gains to such an extent that his vice president, George H. W. Bush, ran for election under the banner of "environmental president" and renewed at least the nation's rhetorical commitment to environmental protection, restaffing the EPA with scientists and bureaucrats concerned about environmental protection rather than antiregulators and deregulators. Like the attention given to Love Canal, Bhopal, and other consciousness-raising events, President Bush used the awareness generated by the *Exxon Valdez* oil spill of 1989 to encourage Congress to pass amendments to the Clean Air Act, which it did in 1990 to address urban smog, acid rain, and various types of toxic emissions. Though Bush had declared that the act would "make the 1990's the era for clean air," environmentalists were frustrated at successful attempts by business interests to stymie the bill's implementation. In particular, environmentalists watched as the heightened environmental promises of Bush's first two years fell by the wayside amid economic recession. According to

Carl Pope, associate executive director of the Sierra Club, "In his commitments during the 1988 campaign, he was good on some issues, mediocre on others, and very weak on a few. . . . [W]hat we did get in the first two years was environmental moderation. He did not treat the environment as an issue he would throw away to the right wing of the Republican party and the least responsible sectors of American business. . . . But what we have seen in the last six months is an almost compete reversion to the policies pursued by the Reagan Administration."

The election of President Bill Clinton saw, in 1992, the convergence of an international movement, marked by a major international conference, the United Nations Conference on Environment and Development (UNCED), known as the Earth Summit, held in Rio de Janeiro. Among the many frameworks established at Rio was an agreement by 154 nations to reduce greenhouse gas emissions to 1990 levels by 2000, a goal that would not be met. More generally, however, the 1990s were marked by a recognition of the increasingly global nature of the environmental crisis. This demanded an increase in the transnational organization of environmental movements, to which many organizations responded, as well as a general rethinking of the role that states could play. The 1990s saw the environmental movement turn decidedly away from state-based approaches as global NGOs and transnational environmental activist organizations became increasingly key. No longer were the more national or even regional concerns of the American conservation or 1970s environmental movement or the the localized—often "Not in My Backyard" (NIMBY)—politics of much of the 1980s environmental justice movement sufficient. At the same time, 1990s environmental discourse became increasingly organized around the language of sustainable development, which exchanged earlier talk of population limits, resource scarcity, and ecocentricity for the idea that a symbiotic vision of economic and environmental life was possible. This discourse came to drive much of work being done at the level of international diplomacy—with Rio marking the height of its pervasiveness—and established the need for environmental organizations to rethink their relationship with market-based environmental strategies.

The language of sustainable development would be tested in the 1990s by a new level of crisis rhetoric that arose in response to climate change, propelled by new and strong scientific evidence. With these changes came a renewed activity by more radical environmental groups, many of which were inspired by deep ecology and were skeptical of the market-based approaches that had become de rigueur in the 1990s. Earth First! became increasingly active, but its attempt to seek wider recognition and membership—as well as its general respect for private property—led to the formation of a splinter group, the Earth Liberation Front (ELF). ELF, driven in part by commitments to anarchist philosophy and decentralized organizational techniques, came to see illegal and sometimes violent activities as increasingly necessary to resist and push back against the nexus formed by various industries, states, and international institutions. In a direct response to neoliberal market environmentalism, ELF sees itself as engaged in ecodefense, aiming directly at profits derived at the expense of Earth.

The Rio conference set the stage for the most important development of global environmental cooperation in the 1990s, even if frustrating to environmentalists: the development of the Kyoto Protocol in 1997. The Kyoto Protocol "sets binding targets for 37 industrialized countries and the European community for reducing greenhouse gas emissions," to be implemented by 2012. Yet post-Kyoto politics were characterized by a new brand of apocalypticism—not of the world but (at least in the United States) of the economy, again exacerbating differences in market ideology among environmental groups. The contemporary manifestation of this compulsory neoliberalism has had a critical impact on climate change arguments. With the rising of waters and the destruction of ecosystems seemingly too large to grasp, environmental groups have begun to underscore not only the health impacts but also the rising costs that are likely to follow from rising temperatures. Later in the decade, however, a challenge to this neoliberal consensus began to take shape. In large part due to trade developments such as the North American Free Trade Agreement (NAFTA), environmental movements began to operate under larger umbrellas. At first in the late 1990s these groups partook in a new antiglobalization movement, which peaked in explosive protests in Seattle at the meetings of the World Trade Organization. Seattle was a critical turning point for its ability to merge environmental activism with other groups that were often ideologically apposite, including protectionist labor unions and libertarians, signaling the mainstreaming of environmental politics into other political issue platforms.

Post-Kyoto implementation meetings have been met with persistent protest by environmental groups, many of which are frustrated by both feet-dragging nations and an inability and unwillingness on the part of many developed nations to take much-needed next steps in the Kyoto framework. These protests continue to benefit handily from the organizational inspiration drawn from the 1999 Seattle protests and a sense that neoliberal strategies such as cap and trade and carbon sequestration disadvantage developing nations and fail to address the real problems of climate change, which require fundamental reconfigurations of economic policy.

The vantage point of the early 21st century forces a decidedly mixed analysis of the environmental movement in the 1980s and 1990s. Public opinion remained firm in its recognition of the importance of dealing with environmental problems in general and global climate change in particular. Yet actual commitment to the sacrifices and policy challenges have proven elusive, marked most extremely in the United States by President George W. Bush's withdrawal of support from the Kyoto process in 2001. Despite high public concern about the environment in the abstract, most Americans agree with or have been convinced by Bush's position that Kyoto would "harm our economy and hurt our workers." In short, mainstream environmental groups proved unable to turn widespread public support into large-scale governmental action—it would not be until later, in the first decade of the 21st century, that a green discourse that promised that environmental protection and progress could be made compatible with economic growth would appear.

More drastically, the movement is faced with a daunting challenge of combating naysayers who maintain, against all evidence to the contrary, that global climate change is something addressable through technology, diplomacy, and policy. Tellingly, a December 2010 Rasmussen poll found that "Most U.S. voters continue to be concerned about global warming but still are more inclined to think it's caused by planetary trends rather than human activity."

Global climate change is an issue that demands global cooperation; however, it has also empowered action at the local levels all over the world. Not only are lowlands such as Bangladesh in the climate change crosshairs, but expensive beachfront property is among the most vulnerable to erosion, flooding, and the adverse effects of changes in extreme weather as a result of rising sea temperatures. Climate change, in other words, marks the first truly global environmental disaster that resists the redistribution of ultimate costs and risks to disempowered classes.

But many of the successes that the environmental movement had between 1980 and 2000 depended upon tangible progress that could be linked directly to quality of life issues, such as cleaner air or water, health, and beautification projects. The necessary steps to address climate change are unlikely to yield such outcomes. In fact, the tipping points that may put climate change beyond the realm of redress will likely serve as the too-late markers of inaction—thus making climate change a particularly elusive political question. This poses significant challenges for environmentalists, who must convince a skeptical public to look beyond their own backyards, to think beyond not only their neighborhoods but also their nations. The probusiness approach of the Reagan years and the neoliberal strategies of Kyoto and the Clinton years will likely come to blows with the realization that the stakes of climate change are such that they require a de-prioritization of provincial economic interests. A large part of the green movement is built around the belief that technological fixes will be sufficient, in part because such efforts believe that ultimate solutions to climate change can—or should—come from strategies that provide business opportunities rather than calling upon businesses to accept lower profits in the name of global survival. As in the 1980s, these debates pit probusiness neoliberal models against more radical environmental organizations.

At the same time, environmentalists in the 21st century must cautiously defend the role that science plays in guiding responses to climate change, aware that short-term business interests will create friction against attempts to deal with long-term global problems. Science, in short, is not enough in a political environment that has made many people suspicious of ideological agendas in the scientific community. Climate scientists learned this lesson in stark terms with the accusation that Nobel Prize–winning researchers for the ICPP had cherry-picked evidence in their 2007 report that found that "warming of the climate system is unequivocal." Though the reaction to the revelations by climate change skeptics was overblown, it underscored a fundamental challenge facing environmental advocates in maintaining unassailable scientific standards. Such standards are critical, considering that 21st-century environmental movements face not only the daunting task of addressing climate change but also keeping at bay a vociferous cadre of critics ready to cease upon even the slightest crack in the otherwise solid scientific consensus about the facts of and challenges posed by climate change. The quite daunting challenges of organizing a global environmental movement that can respond to the global challenges of this century will require rethinking the very nature of environmentalism itself, which now operates on a scale that few global problems—with the possible exception of nuclear holocaust—have posed.

Daniel Skinner

Further Reading

Dryzek John S., and David Schlosberg. *Debating the Earth: The Environmental Politics Reader.* New York: Oxford University Press, 2005.

Rawcliffe, Peter. *Environmental Pressure Groups in Transition.* Manchester, UK: Manchester University Press, 1998.

Szaz, Andrew. *EcoPopulism: Toxic Waste and the Movement for Environmental Justice.* Minneapolis: University of Minnesota Press, 1994.

Torgerson, Douglas. "Farewell to the Green Movement? Political Action and the Green Public Sphere." *Environmental Politics* 11(1) (2000): 133–145.

Wapner, Paul. "Politics beyond the State: Environmental Activism and World Civic Politics." *World Politics* 47 (1995): 311–340.

Environmental Thought in the 19th Century

In the early 1800s, many Americans believed that their nation was turning a corner from being a settler society to becoming a more civilized nation, to rival those of Europe. In trying to stimulate such development, many Americans made extensive comparisons between the United States and the long-standing European societies. The young American nation compared unfavorably in many categories, especially arts and culture. However, in its natural wonders, there could be no disputing the majesty of the United States. For this reason, some Americans sought new ways to highlight the natural splendor that distinguished the United States from Europe. They came to believe that even though the United States had little history in comparison with European nations, it could instead offer natural history. By getting to know the continent's nature better, argued early naturalists, the United States could gain its identity.

In this story of intellectual history, the idea of nature changed for many Americans as their nation's economic standing changed over the course of the 19th century. Some historians argue that economic advancement enhanced Americans' ability to find nonutlitarian—idealistic—priorities regarding natural resources. For instance, the ability to see aesthetic beauty in the natural world grew as Americans relied less on it for their everyday survival. At the very least, though, this intellectual progression marks the establishment of a critical foundation for later environmental thought.

Imaging America's Natural History

For a few scientists, chronicling North America's everyday nature as well as its natural history became an effort of both science and patriotism. Charles Wilson Peale worked with the Philosophical Society of Philadelphia to initiate this process in 1784 when he established the first natural history museum in the United States. His efforts to preserve and catalog the species of North America were shared by Thomas Jefferson. Referred to as "natural history," this effort to know the continent through the creatures living on it spurred at least one of the young nation's first unified federal undertakings: Peale's effort to excavate a mastodon skeleton from New York state in the late 1700s.

Natural History and National Meaning

When the skeleton was excavated by Peale from a Hudson River Valley farm in 1801, he could not yet verify what type of creature it was. After studying it in Philadelphia, he identified the skeleton as that of a mastodon. In Europe and elsewhere, the mastodon had been the focus of a debate about the prehuman past as well as about the concept of extinction. Due to this heightened level of interest, Peale's mastodon discovery became important world news. The bones came to serve as an international puzzle for scientific-minded people everywhere.

The ability to excavate and reassemble the skeleton also became an important symbol for the stability of the young United States. For many Americans, the animal's symbolic meaning far outweighed its scientific significance as evidence of extinct species or a prehuman past. "Indeed," writes historian Paul Semonin, "while Lewis and Clark were exploring the western wilderness, Peale had remounted his skeleton with its tusks pointing downward to magnify its ferocity." Most historians view this as a representative moment of scientific naïveté. And yet, Semonin suggests that this understandable lapse instead demonstrates that the mastodon was "the nation's first prehistoric monster" used by the nation's founders as a "a symbol of dominance in the first decades of the new republic."

Also a product of the zeal for natural history was the Lewis and Clark expedition. It is important to note that President Thomas Jefferson had Meriwether Lewis travel to Philadelphia to receive advice from Peale, the nation's leading naturalist. In particular, Jefferson hoped not that the expedition would create more bones; instead, he actually hoped that the explorers would find a living mastodon.

Symbolizing the United States

If they had found such a beast roaming the American West, America's national emblem may have turned out very differently. As it was, most interested Americans focused on fowl to serve as a national emblem. However, there was a bit of debate over which bird was best.

The bald eagle (Haliaeetus leucocephalus) was chosen because of its long life, great strength, and majestic looks and also because it was believed to exist on this continent only. The national bird of the United States is the only eagle unique to North America. The common name dates from a time when "bald" meant "white," not hairless. The bald eagle ranges over most of the North American continent, from the northern parts of Alaska and Canada to northern Mexico.

As a national symbol, the eagle eventually was placed on the backs of coins. In addition, the Great Seal of the United States depicts the outstretched eagle with a shield covering his breast. On the shield one finds 13 perpendicular red and white stripes surmounted by a blue field with the same number of stars. The eagle clutches a bundle of 13 arrows in his left talon, an olive branch in his right, and finally a scroll in his beak inscribed "E Pluribus Unum."

There were some dissenters to this imagery and selection. Benjamin Franklin wrote that the eagle was "a bird of bad moral character, he does not get his living honestly, you may have seen him perched on some dead tree, where, too lazy to fish for himself, he watches the labor of the fishing-hawk, and when that diligent bird has at length taken a fish, and is bearing it to its nest for the support of his mate and young ones, the bald eagle pursues him and takes it." Franklin favored the humble turkey as a symbol of America.

Other symbols of strength and permanence emanated from North America's natural elements. Jefferson decided that one of the preeminent examples was the Natural Bridge near his home in Virginia. Historian Charles Miller hypothesizes that Jefferson likely saw the bridge for the first time in 1767 and then became its first American owner in 1774 when his family purchased it with the adjoining 150 acres. Jefferson proceeded to enlist artists to paint the bridge and erected a small cabin for tourists and guests. The Natural Bridge is just one example that from the earliest days of the 1800s, natural wonders often embodied many of the ideals that Americans wished to present to the world.

Creating an American Natural History

This passion for natural history inspired young naturalists to record for posterity what they found. John James Audubon began using his painting talents to preserve each species of bird that he could find (and kill) in North America. His collection *The Birds of North America* first appeared in 1824.

The hunter-artist Audubon represents well the mixed motives of most collectors who were impressed with the natural wonders of North America. Audubon, writes biographer Richard Rhodes, "engaged birds with the intensity (and sometimes the ferocity) of a hunter because hunting was the cultural frame out of which his encounter with birds emerged. In early 19th-century America, when wild game was still extensively harvested for food, observation for hunting had not yet disconnected from observation for scientific knowledge." Audubon observed American fowl extensively, but he also killed samples (at least six and as many as hundreds) of each species. The catch was skinned and stuffed with frayed rope. Audubon then posed each sample in positions he had observed in the wild. He used the samples as puppets or manikins to create the illusion that the bird sample was still alive, and then he would paint it. Audubon wrote that "By means of threads I raised or lowered a head, wing, or a tail and by fastening the threads securely I had something like life before me."

Audubon's efforts in painting were mirrored in the form of the written word by the Bartrams. The American tradition in nature writing grew out of the efforts of John Bartram and his third son, William. Keeping journals during their extensive travels throughout the southeastern United States, the Bartrams gave many American and European readers their first understanding of the details of American nature. Most of their trips took place in the late 1700s; however, the writings that they published inspired the writers who followed during the 1800s.

Although the details of his exploration brought new understanding and appreciation to his readers, the real value of the Bartram accounts derived from the basic aesthetic appreciation of the nature with which the father and son approached the landscapes of North America. John is given credit, for instance, for being the first naturalist to use the term "sublime" to describe nature. Charged with the meaning of romantic nature, sublimity valued nature for completely nonutilitarian reasons. He also wrote about the great virtues of native peoples whom he contacted—particularly their relationship with and appreciation of natural surroundings. In each of these cases, the Bartrams countered the accepted approach of most of American culture. John's description of a mountain thunderstorm in Rabun County, Georgia, near the North Carolina border, provides a good example:

> It was now after noon; I approached a charming vale, amidst sublimely high forests, awful shades! darkness gathers around, far distant thunder rolls over the trembling hills; the black clouds with august majesty and power, moves slowly forwards, shading regions of towering hills, and threatening all the destructions of a thunderstorm; all around is now still as death, not a whisper is heard, but a total inactivity and silence seems to pervade the earth. . . .

> The face of the earth is obscured by the deluges descending from the firmament, and I am deafened by the din of thunder; the tempestuous scene damps my spirits, and my horse sinks under me at the tremendous peals, as I hasten for the plain.

Accounts such as this one formed the foundation of American Romanticism, which took shape as a multimedia effort in the early to mid-1800s. The first realm for the aesthetic appreciation of nature was on the canvas of oil painting.

A National Commitment to Natural History

Inviting the nation to the opening of his museum, Charles Wilson Peale wrote:

> Mr. Peale respectfully informs the Public, that having formed a design to establish a MUSEUM, for a collection, arrangement and preservation of the objects of natural history and things useful and curious, in June 1785, he began to collect subjects, and to preserve and arrange them in Linnaean method. . . .
>
> . . . [T]he museum having advanced to be an object of attention to some individuals . . . he is therefor the more earnestly set on enlarging the collection with a greater variety of birds, beasts, fishes, insects, reptiles, vegetables, minerals, shells, fossils, medals, old coins. . . .
>
> With sentiments of gratitude, Mr. Peale thanks the friends of the Museum, who have beneficially added to his collection a number of precious curiosities, from many parts of the world;—from Africa, from Indies, from China, from the Islands of the great Pacific Ocean, and from different parts of America.

In order to establish his museum, Peale relied heavily on the help of his sons: Rubens, Franklin, Titian II, Rembrandt, and Raphaelle. Together, the Peales accepted donations of trophy animals shot all over the world from many Americans, including George Washington. Other American collectors donated insects, shells, and plants collected internationally. Finally, Meriwether Lewis and William Clark presented Peale with many specimens taken during their exploration of the American continent, including a prong-horned antelope (the only known specimen of its kind), Lewis's woodpecker, Clark's crow, a western tanager, and a large California condor.

By April 1799, Peale listed his holdings as including more than 100 quadrupeds, 700 birds, 150 amphibians, and thousands of insects, fishes, minerals, and fossils. Peale also began to collect and catalog various specimens of unknown creatures and biological oddities. One of his earliest unidentified specimens was a lizard from Louisiana presented by President Jefferson. There were also a Ripley-Believe-It-Or-Not feel to specimens such as cows with additional heads and tails.

The commitment to this record of the nation's natural history represented an important watershed to the United States. Through it the young nation discerned itself from every other nation through the celebration of its unique and bountiful natural resources.

Romantics and Artists Claim That Nature's Value Derives from Beauty

Ideas of beauty vary significantly between different cultures and are rarely consistent even within specific societies. Although it contrasted with much of the aesthetic taste appreciated by European sensibilities, many American thinkers and artists of the early 1800s made one theme a primary portion of their emerging tastes: nature. Grouped with the aesthetic movement known as Romanticism, this aesthetic turned typical views of nature on their head. Based in biblical teachings as well as in the experience of struggling against nature for survival, most of Western thought before 1800 emphasized the need to civilize and alter nature. In revisionist thinking, nature could be beautiful without being civilized, settled, or used by human influence.

The Hudson River School

As wealthy Americans longed for their young nation to form a unique cultural tradition that would stand out among the nations of the world, it seems unsurprising that artists would focus on the natural wonders of North America. Known as the Hudson River School, the first internationally recognized genre of art to be initiated in the United States grew between the 1820s and the late 19th century. Initially, their paintings were organized around scenes of the Hudson River Valley and the adjoining mountains of New York and Vermont. Eventually, as a view of nature and not a region was identified as the primary organizational device for the genre, Hudson River School artists would paint natural wonders from all over the world.

Hudson River painters conveyed their natural wonders with a fairly clear and unapologetic ideology. They painted with an almost religious reverence for the magnificence of the American wilderness. Their effort to re-create the unique beauty of the American landscape for the public can be viewed as one of the first expressions of the American desire for preservation. For the Hudson River painters, of course, instead of national parks or policies, they set aside locations and moments in time within the boundaries of a frame. In this view, the artists could make nature appear aesthetically wonderful, even exaggerating the existing beauty of a site.

Unlike the orderly landscapes of European artists, Thomas Cole and the artists who followed him created vast awe-inspiring scenes that were designed to convey threatening characteristics of the American wilderness. Cole hoped to use his influence to encourage more people to love, enjoy, and protect nature. His work inspired a new class of wealthy American and international tourists to the areas he painted. In fact, the irony is that by inspiring visitation, Cole's work increased development.

These larger societal changes created a border between developed and undeveloped parts of America that also inspired art. Maybe the best example is George Innes's *Lackawana Valley.* This canvas, which was completed in 1855, provides a dramatic visual demonstration of the machine and garden paradigm. Unapologetically depicting stumps in the foreground, the scene centers on a train and one of the technological wonders of the rail age, a round house. Overall, though, the scene highlights the vibrant colors of a natural environment that seems to be accepting such changes.

Thomas Cole's *Oxbow*

Another painting that demonstrated the vicious dichotomy of America's view of nature was Thomas Cole's painting of 1836 titled *The Oxbow* (the Connecticut River near Northampton). In the painting, the tension between wilderness and garden, savagery and civilization, is recorded visually as European conventions of landscape painting are employed to comment on the state of the physical place of America. The savagery of the storm clouds over the wilderness retreats from the advancing cultivated landscape of civilization. Art historian Barbara Novak writes that "Cole used the storm-ravaged tree, a palimpsest of time associations, almost as a signature. This natural picturesque is of a totally different order from what we might call the man-made or unnatural picturesque, the cut stump of the wood-chopper, utilitarian man feeding his ten thousand fires."

Cole seemed to express the concern of a growing number of 19th-century Americans: Was wilderness being lost? His paintings dramatically left no doubt that the answer was affirmative.

Transcendentalism

Early in the 19th century, the aesthetic appreciation for nature had no intellectual foundation. After the 1820s, though, writers and intellectuals began knitting together ideas and influences from other parts of the world with sensibilities such as those of visual beauty expressed by Coles. The literary and intellectual movement that grew out of this increased interest in nature is referred to as transcendentalism. This realm of belief became a portion of American Romanticism, ultimately combining spirituality and religion.

Writers and reformers including Ralph Waldo Emerson, Henry David Thoreau, Margaret Fuller, and Amos Bronson Alcott developed this line of thinking in New England between 1830 and 1850. Their actions helped to transform transcendentalism—at least partly—into an intellectual protest movement; however, it continued to carry with it a new appreciation for nature. Most often, transcendentalists connected to the ideas of philosophical idealism that derived from German thought, either directly or through the British writers Samuel Taylor Coleridge and Thomas Carlyle.

Emerson emerged as the intellectual leader of this group when he connected Romanticism with Unitarianism. By 1825, Unitarianism had many followers in Massachusetts, where they openly attacked the orthodoxy of the Puritans who dominated New England. In place of Puritan thinking, the Unitarians offered a liberal theology that stressed the human capability for good. Four years after resigning as pastor at Boston's Second Church, Emerson published *Nature* in 1836. He directly challenged the materialism of the age, and his writing was adopted as the centerpiece of transcendentalism.

In the Boston area, the Transcendental Club began to meet in order to refine and disseminate the ideas that Emerson had voiced in his writing. This group of intellectuals also created the famous Brook Farm experiment in communal living (1840–1847) in West Roxbury, Massachusetts. Young Thoreau became active with the club and began working with its publication, the *Dial.* Thoreau's writing emphasized the role of nature in Americans' lives. He published his greatest work, *Walden,* in 1854. This book was Thoreau's account of transcendentalism's ideal existence of simplicity, independence, and proximity with nature. In *Walden,* Thoreau extended Emerson's ideas of replacing the religion of early 19th-century America with the divine spirit out of nature. In this paradigm, the natural surroundings took on spiritual significance. More than ever, Thoreau created a model of transcendentalist thought connected to nature.

Thoreau's message from Walden Pond urged Americans to escape from mechanical and commercial civilization in order to be immersed in nature, even if only for a short time. Although few Americans in 1850 either read *Walden* or immediately came to see nature differently, Thoreau and other transcendentalists laid the foundation for a new way of viewing the natural environment. No longer simply raw material for industrial development, nature possessed aesthetic or even spiritual value.

Writers, poets, and artists, through the efforts of transcendentalists, argued for America to be nature's nation. And the symbolic nature of the United States was not necessarily the manicured beauty of the manicured French and British gardens but instead was the raw wilderness not found in Europe. In April 1851, Thoreau lectured at Concord Lyceum in Massachusetts. After beginning by saying that he "wished to speak a word for nature," he answered proponents of developments and civilization. Finally, he shared a timeless insight when he stated that "In Wildness is the preservation of the world."

With such a statement, Thoreau forged a connection between the intellectual approach of transcendentalism to wilderness and on to American ideals of democracy, independence, and beauty. The attraction of nature would eventually also include an interest in primitivism—one interpretation of Thoreau's "wildness." As society became more industrialized, developed, and urban, a contrary impulse attracted some Americans to seek innocence in raw nature.

Ironically, this interest in primitivism also occurred when the civilizing forces of the United States were mobilized against the primitive native occupants of North America. Clearly, in literature and in art this interest in primitivism influenced a new appreciation and romantization of native culture. Whereas many Americans saw the native peoples as a national problem and a nuisance, some Romantic viewers began to see them as a people uniquely and admiringly in tune with the natural environment.

Wild nature thus became a source of national pride as the root of character traits for a unique national identity. This imagery was most notable in paintings, which through their reconfigured realities allowed artists to create a new creation myth for America that emphasized the primacy of the white settlers. Artists of this genre granted a privileged role for an American elite and ennobled the white discovery and settlement of the wilderness by evoking images of classical heroes. Nature was also given a reinterpreted role: rather than presenting nature as an obstacle to the establishment of a civilization, American authors and painters alike upheld nature as the source of the animating spirit behind the American character. The following are excerpts from two of the genre's most famous authors:

Ralph Waldo Emerson, Essay VI, *Nature*

There are days which occur in this climate, at almost any season of the year, wherein the world reaches its perfection, when the air, the heavenly bodies, and the earth, make a harmony, as if nature would indulge her offspring; when, in these bleak upper sides of the planet, nothing is to desire that we have heard of the happiest latitudes, and we bask in the shining hours of Florida and Cuba; when everything that has life gives sign of satisfaction, and the cattle that lie on the ground seem to have great and tranquil thoughts. These halcyons may be looked for with a little more assurance in that pure October weather, which we distinguish by the name of the Indian Summer. The day, immeasurably long, sleeps over the broad hills and warm wide fields. To have lived through all its sunny hours, seems longevity enough. The solitary places do not seem quite lonely. At the gates of the forest, the surprised man of the world is forced to leave his city estimates of great and small, wise and foolish. . . .

These enchantments are medicinal, they sober and heal us. These are plain pleasures, kindly and native to us. . . . Nature is always consistent, though she feigns to contravene her own laws. She keeps her laws, and seems to transcend them. She arms and equips an animal to find its place and living in the earth, and, at the same time, she arms and equips another animal to destroy it.

Henry David Thoreau, *Walden,* Chapter 16

After a still winter night I awoke with the impression that some question had been put to me, which I had been endeavoring in vain to answer in my sleep, as what—how—when—where? But there was dawning Nature, in whom all creatures live, looking in at my broad windows with serene and satisfied face, and no question on her lips. I awoke to an answered question, to Nature and daylight. The snow lying deep on the earth dotted with young pines, and the very slope of the hill on which my house is placed, seemed to say, Forward! Nature puts no question and answers none which we mortals ask. She has long ago taken her resolution. "O Prince, our eyes contemplate with admiration and transmit to the soul the wonderful and varied spectacle of this universe. The night veils without doubt a part of this glorious creation; but day comes to reveal to us this great work, which extends from earth even into the plains of the ether."

William Henry Jackson and Romantic Landscape Photography

During the mid-1800s, the new technology of photography became a new way of expressing the romantic beauty of the American landscape. The efforts of these early photographers became an important portion of the growing American interest in western lands and their eventual use as national parks. William Henry Jackson picked up where the romantic thinkers and painters left off and created actual views that spoke to the same passionate beauty in natural forms.

Jackson formed his sensibility beginning at the age of 10 when he received his first formal artistic training. His emphasis was drawing, and his first job came in 1858 when he was hired as a retoucher for a photographic studio in Troy, New York. In this job, his duty was to enhance or warm up black and white portraits by tinting them with watercolors and India ink. While working at this job, Jackson learned to use cameras and the darkroom techniques. He began to apply his artistic sensibility to this new technology. In 1869 he moved to Omaha, Nebraska, and established his own photographic studio. From this base, Jackson made the first subject of his photography American Indians from the nearby Omaha reservation and the construction of the Union Pacific Railroad.

He established a reputation with these early images and soon was contacted by Dr. Ferdinand Hayden, who was organizing an expedition that would explore the geologic wonders along the Yellowstone River in Wyoming Territory. Hayden realized that a photographer would provide a valuable record of the sights of the exotic western landscape. In addition, Jackson had already gotten quite capable of managing the complexities of early photography on the move.

The visual record that Jackson created provided important verification to legends of geysers and waterfalls in the territory. Jackson's images helped to arouse public interest and fueled the push for Congress to officially designate Yellowstone National Park in 1872. Although the well-known painter Thomas Moran also created images based on the travels of the Hayden expedition, Jackson's images struck even more Americans as genuine and unembellished. In fact, during the expedition, Moran often helped Jackson locate the best views and then used the images afterward to compose his paintings. Because no member of Congress had seen Yellowstone, Hayden and his colleagues brought Jackson's photos, along with Moran's watercolors, to Capitol Hill. A year later, in 1873, the Department of the Interior compiled 37 of Jackson's photographs into a portfolio, which was presented to Congress in an effort to lobby funds for future expeditions to the West.

Jackson and Hayden worked with the Department of the Interior to release these and later images as stereo views, which increased their general appeal. These views were made with a special double camera that used two horizontal lenses. Each lens recorded the image as seen by each human eye. The double-prints were then placed on a card that could be viewed in a device that allowed the viewers' eyes to combine the images into one scene. More than any other image, stereoviews allowed Americans to place themselves into the scene. The stereoviews figured significantly in Yellowstone National Park's growing popularity. For some, this was their first and only view of Yellowstone.

Jackson and Hayden joined forces for the U.S. Geological Survey for almost another decade. Their subject was almost entirely the landscape of the western United States. After his work at the Geological Survey, Jackson continued to work in the West, opening a studio in Denver, Colorado. He traveled widely to photograph railroad construction to mining towns in the Rockies. Many of his images were viewed by the millions of visitors to the World Columbian Exposition in Chicago in 1893.

Through the efforts of these writers and painters, among others, a new paradigm became part of American culture. In this new mind-set, nature was granted worth in its own right, particularly for its aesthetic beauty. Although the majority of Americans maintained a utilitarian view of nature, the intellectual construction of what would develop into a conservation ethic in the later 1800s had begun.

Bringing America's Nature Aesthetic to Life in Parks

Although numerous intellectuals and artists conceived of the Romantic in writing and on canvas, a few also began to do so on the landscape. The first examples of such planning were part aesthetic creation and part utilitarian. Disposal of the dead had become a serious problem for growing urban areas in the early 1800s. Thus, the need for new cemeteries became one of the first applications of the Romantic view of nature.

Mount Auburn Cemetery

The landscape that began the rural cemetery movement, Mount Auburn Cemetery opened in Cambridge, Massachusetts, in 1831. Unlike previous places of interment, Mount Auburn prioritized the natural beauty of the surroundings, not just the utilitarian storage of dead bodies. By doing so, it marked an important change in American community planning as well as in Americans' relationship with nature.

With few parks or public areas, the American landscape of the early 1800s still was organized entirely by the everyday needs of residents. There was little interest in the frivolity of aesthetic beauty. And what was considered beautiful by Americans was rarely natural. Mount Auburn began to change this by presenting a real landscape that contained ideas of beauty seen in paintings, particularly those of Romantic artists.

In Mount Auburn, tombstone markers were not lined up tightly in order to bury as many people as possible. Instead, the landscape was designed as a unit by Andrew Jackson Downing. He wove the markers in with the hills of the site and interspersed them with miniature species of exotic shrubs and trees. He used winding paths and benches to create private areas for repose or mourning. However, simultaneously winding wagon trails allowed visitors relaxing for the day to interact with natural beauty. The patterns and choices of his design became the typical characteristics of the rural cemetery movement, which would see similar areas created in many cities by the 1850s.

At Mount Auburn, Americans learned that the landscape could be art. A primary component of this canvas, of course, was the natural details that Americans had seen as horrible just a few decades prior. Following the success of rural cemeteries, Americans also became interested in public parks. Each of these developments helped to spur the growth of landscape architecture in the United States.

The Rural Cemetery Movement

Prior to 1830, most communities considered cemeteries to be necessary for disposal of the dead but of little aesthetic importance. Normally, urban cemeteries were found adjacent to churches—in churchyards, as the area was often called—or in central common areas of a town or city. Clearly, from the first European settlement in New England, Americans carried on the European tradition of burying the dead amid the living. By the turn of the century, many of these churchyards had become overcrowded. Many residents believed that this created a genuine health threat to the living. This was particularly problematic in cities that had come to use their cemeteries as multipurpose areas for common activities, including playing, relaxing, and mourning. As a result, cemeteries became one of the first planned landscapes in the United States.

In the young republic of the United States, the rural cemetery movement was inspired by Romantic perceptions of nature, art, national identity, and the melancholy theme of death. It drew upon innovations in burial ground design in England and France, particularly the Père Lachaise Cemetery in Paris, established in 1804 and developed according to an 1815 plan.

Between 1830 and 1855, new ideas for interring the dead were grouped under the terms "rural" or "garden" cemetery. Nearly a complete contrast from its predecessor, the rural cemetery was most concerned with aesthetics. Its appearance was designed to provide peaceful surroundings by using the beauty of nature, including ornamental shrubs and trees and romantic layouts such as winding carriage trails and ornamental monuments. Even though the rural cemetery movement represented a social change in American life, it also marked an important moment in the American relationship with nature.

Whereas 19th-century community cemeteries typically were organized and operated by voluntary associations that sold individual plots to be marked and maintained by private owners according to individual taste, the memorial park was comprehensively designed and managed by full-time professionals. Often these cemetery parks were operated as a business venture, prioritizing economy of price and natural beauty. In other cases, nonprofit corporations ran the cemetery. Regardless, the natural beauty of cemetery sites continued to be enhanced through landscaping, while often incorporating picturesque hills with an overall design of rolling terrain. The cemetery clearly took on an overall design through initial planning and ongoing maintenance.

The rural cemetery movement clearly grew out of European trends in gardening and landscape design. However, the cemetery form acted as a conveyor of this aesthetic into the lives of many Americans who had not yet considered the time, expense, and effort needed to sculpt their home landscapes with such beauty. The English roots, of course, grew from traditions of the 1700s that had seeped out of the tastes of English nobility who had urged their gardeners to use classical landscape paintings as their models. The greatest names in English garden designers of this era included Lancelot "Capability" Brown, William Kent, Sir Uvedale Price, Humphrey Repton, and John Claudius Loudon. In an effort to imitate natural forms, these designers built gracefully curving pathways and streams into their rolling land forms. They created a picturesque model of design by combining these elements with contrast and variation by massing trees and plants and incorporating ornamental features. By the end of the 18th century, the picturesque landscape was a fairly concrete and well-known form, including open meadows of irregular outline, uneven stands of trees, naturalistic lakes, accents of specimen plants, and here and there incidental objects such as an antique statue or urn on a pedestal to lend interest and variety to the scene.

Concentrated around urban areas, these changes in cemetery layout emanated from the Boston area. Ultimately, though, rural cemeteries sprang up throughout the United States. Typically, these cemeteries resembled what would later be called parks. Their construction therefore required open tracts of land. For this and aesthetic reasons, the new cemeteries were most often sited on the outskirts of cities and towns. Their location also marked an important step in suburbanization, when human communities began to spread out of urban areas. Well-to-do visitors treated the rural cemeteries less as a place for somber reflection and more for outdoor leisure in beautiful surroundings. A new appreciation of nature began. People began to see nature as something to be enjoyed as well as tamed.

Mount Auburn was followed by the formation of Laurel Hill Cemetery in Philadelphia in 1836; Green Mount in Baltimore in 1838; Green-Wood Cemetery in Brooklyn and Mount Hope Cemetery in Rochester, New York, in 1839; and ultimately many others. Later in the 19th century, the design format would change to that of the perpetual care lawn cemeteries or memorial parks of the 20th century. The lawn plan system shifted from the rural design by de-emphasizing monuments in favor of unbroken lawn scenery or open space.

In the rural design, however, open space would never be a priority. Balance and overall design grew from hilly, wooded sites, even if ground needed to be moved in order to create the appropriate scene. Such settings, it was thought, stirred an appreciation of nature and a sense of the continuity of life. By their example, the popular new cemeteries started a movement for urban parks that was encouraged by the writings of Andrew Jackson Downing and the pioneering work of other advocates of picturesque landscaping.

Andrew Jackson Downing

At Mount Auburn and many other sites, Andrew Jackson Downing introduced a new tastefully designed nature to the American public. A landscaper, Downing wrote widely about his ideas and philosophies of gardening. Ultimately he created guidebooks and design manuals that made him a definer of taste—similar to the contemporary persona of Martha Stewart. Downing's influence took off in 1846 when he founded the magazine the *Horticulturist* to diseminate his ideas to upper-class consumers. In addition to stylistic concerns, the magazine spread Downing's interest in scientific agriculture. The primary readers, of course, were the very wealthy who had leisure time to consider such things. By presenting them with factual information, though, Downing helped to create a class of upper-class gentlemen farmers and Victorian women most appreciative of beauty in the landscape.

Where, though, were country farmers to live? Most wealthy Americans lived in cities, and very few had vacation homes. Picking up on English traditions, Downing sought to interest his readership in a new type of countryside. Collaborating with Alexander Jackson Davis in 1842, Downing published *Cottage Residences,* which was a pattern book of country houses. In his drawings, the homes picked up the architecture of the English countryside and blended in aspects of Romantic landscape design seen at Mount Auburn and elsewhere. Sounding similar to Jefferson decades before, Downing argued the worth of living close to the land in rural areas in which one could interact with nature. Escape from the urban confusion interested many Americans who were wealthy enough to consider such a move. Soon, Downing had legions of followers.

Downing's priorities and his interest in helping Americans to interact with nature contributed to the American park movement. Ultimately the spirit of his country homes would inspire the first American suburbs.

Although Downing's home designs were directed to the wealthy, he also hoped to create places that would be enjoyed by all classes of society. This desire led him to begin advocating large inner-city parks. Downing saw a civilizing aspect of open spaces and wanted to bring one to nearby New York City. Finally, after many years, New York City set aside land for Central Park. Downing and his partner Calvert Vaux devised preliminary plans for the park.

At roughly the same time, Downing was asked to design the public grounds in Washington, D.C. This included the Mall and land around the White House. These stunning designs contrast markedly with the monumental landscape that would eventually be built in the nation's capitol. Downing used his Romantic sensibilities to intermingle landscape features with the monuments. Unfortunately for his Central Park and Washington plans, in 1852 Downing died in a steamboat accident on the Hudson River. He had, however, spurred a version of American taste with the landscape as a vital component.

Planning Beauty in Landscapes

With the rapid growth of urban centers later in the 19th century, landscape design and city planning merged in the work of Frederick Law Olmsted, the country's leading designer of urban parks. Olmsted and his partners were influential in reviving planning on a grand scale in the parkways they created to connect units of municipal park systems. Although Olmsted was more closely tied to the naturalistic style of landscape planning, his firm's work with Daniel H. Burnham in laying out grounds for the World's Columbian Exposition of 1893 in Chicago conformed to the classical principles of strong axial organization and bilateral symmetry.

Evolving over a century, the tradition of design that had begun with rural cemeteries demonstrated the value of planning. By fusing Romantic ideas of the picturesque with the engineering of landscape planning, many of the parks became symbols of the possible middle ground that could take shape, with nature at its core.

During the 20th century, the work of John C. Olmsted and Frederick Law Olmsted Jr., successors of the elder Olmsted and principals of the Olmsted Brothers firm, expanded these ideas throughout the country.

Olmsted Helps to Define the American Movement for Parks

Central Park in New York City became an emblem for a new era in the perception of the natural world. Many communities sought to plan natural spaces into their central areas. Also, Americans began to reconsider what should be done with areas designated as parks belonging to the entire nation—national parks. In each case, a single American thinker revolutionized the nation's spaces and ultimately its expectations for its natural environment.

Although the passion for nature in one's everyday life and the preservation interest in national parks each existed before 1900, they did not intersect and evolve into the modern parks movement until the 1910s. The growing interest in the natural environment did not immediately alter the places in which Americans chose to live. In fact, the American landscape urbanized at an increasingly rapid rate in the early 1900s.

With the population concentration in more urban regions, the impulse to preserve accessible areas of nature became even more imperative. These impulses drew a direct relation to those driving the design of Central Park, the nation's first planned park, and its designer, Frederick Law Olmsted. Olmsted urged Americans to appreciate the psychological and restorative power of nature. His plans for additional urban parks and the early suburbs brought nature nearer to the lives of most Americans. Olmsted also worked to inspire a national set of parks that would celebrate the nation's symbolic appreciation of its natural resources.

After making his name in New York City, Olmsted moved his practice to Brookline, Massachusetts, in 1883. He had begun work on a park system for the city of Boston, and eventually he focused much of his time on the area known as the Emerald Necklace. With his reputation reaching an international scale, Olmsted received one of the greatest assignments of his career: the grounds of the 1893 World's Fair in Chicago. In this remarkable site, Olmsted created a monumental landscape that served as the setting for the works of the world's greatest living architects. Although the entire landscape (including its buildings) would not be allowed to last beyond the term of the temporary fair, Olmsted's design created a spectacle that enhanced his renown.

In each of his designs of this era, Olmsted sought to advance a shared sense of community among all of its members. In Olmsted's mind, landscape architecture provided a critical opportunity for the natural environment to shape healthy and productive ways of American life. Particularly in congested urban areas, Olmsted believed that parks offered an antidote to stress and artificiality that would help prevent mental decay. In greenswards and ornamental trees, he sought to spread calmness and democracy. He believed that if properly designed, landscapes might enhance American ideals. His concept of democratic recreation was perfect for park spaces that sought to appeal beyond local needs.

It follows therefore that Olmsted took a pioneering role by defining the form of the emerging national park system. He had written the initial report to establish Yosemite National Park in 1865. Drawing on his Central Park experience, Olmsted viewed the valley's preservation as the creation of a work of art. He argued in terms of the psychosociological theory honed in the Central Park campaign that "It is a scientific fact that the occasional contemplation of natural scenes of an impressive character, particularly if this contemplation occurs in connection with relief from ordinary cares, change of air and change of habits, is favorable to the health and vigor of men and especially to the health and vigor of their intellect." Without such recreation, in situations "where men and women are habitually pressed by their business and household cares," they are susceptible to "a class of disorders" that include such forms of "mental disability" as "softening of the brain, paralysis, palsy, monomania, or insanity."

Through his work, Olmsted had helped to transform American taste. Similar to Gifford Pinchot and others, Olmsted's work also altered the nation's concept of nature. The idea of parks as well as the growing number of natural areas set aside and placed under federal jurisdiction increased legislators' interest in formalizing the government's role in conservation. The effort to connect the growing interest in nature with the idea of national parks gained energy during the 1890s and early 1900s, and Congress voted to create additional parks in Sequoia, Yosemite (to which California returned Yosemite Valley), Mount Rainier, Crater Lake, and Glacier.

During these same years, western railroads helped spur tourism to the new parks by building large hotels and rail access. Simultaneously, Congress added other types of sites to the national collection. Prehistoric Indian ruins and artifacts were first preserved by Congress at Arizona's Casa Grande Ruin in 1889. In 1906 Congress added Mesa Verde National Park and passed the Antiquities Act that authorized presidents to set aside "historic and prehistoric structures, and other objects of historic or scientific interest," in federal custody as national monuments.

Behind many of these initiatives was President Theodore Roosevelt. He used the act to proclaim 18 national monuments, including El Morro, New Mexico, the site of prehistoric petroglyphs and historic inscriptions, as well as natural features such as Arizona's Petrified Forest and the Grand Canyon. Congress later converted many of these natural monuments into national parks. Although these new federal sites reflected the changing interest in preserving nature and history, there remained no unified ethic to tie together these sites. This ethic emerged in 1916 with the passage of the National Park Service Organic Act, which established a separate federal agency to oversee the parks.

Forcing the American Tradition of National Parks

Do Americans need nature in their lives? Romantic writers and painters said yes. However, their's was a wondrous, overwhelming, and sublime nature. Other intellectuals took this impulse and used it to create a model of a usable natural form that became known as parks. Still, many Americans viewed the idea as excessive waste.

Although other nations had established parks and planned natural areas for leisure use, the United States coined the model of setting specific areas aside from development for no reason related to religion or historic importance. This cultural tradition became known as preservation. The name given to these federally owned treasures became "national park." In fact, this term was used before anyone even knew what it meant and against the wishes of some Americans who feared the dangerous precedent of taking land out of private hands and locking it away. Not only did many Americans disagree with setting land aside, but many also argued that the federal government had much more important things to do than administer and care for such parks.

The idea for national parks was a blend of intellectual traditions, including ideas of Romanticism particularly as they took physical form in the work of artists and photographers. The exact moment of origin for the idea of national parks, though, remains the stuff of legend. In national park folklore, it is said that the idea originated in September 1870 among the members of the Washburn-Doane Expedition (a largely amateur party organized to investigate tales of scenic wonders in the area).

The legendary origin myth proceeds like this: During an evening campfire discussion near Madison Junction, where the Firehole and Gibbon Rivers join to form the Madison River in present-day Yellowstone National Park, the explorers recalled the natural spectacles of the day. Americans should see these wonders, all the campers agreed. But how long could they last before developers, realizing what great profits could be made, exploited the natural attractions? The concerns of the group convinced them that action needed to be taken to protect unique sites such as Yellowstone. Everyone around the fire agreed that Yellowstone's awe-inspiring geysers, waterfalls, and canyons should be preserved as a public park. In short, it was Yellowstone's oddity that fueled the designation of the world's first national park.

Convincing politicians to enforce this designation was something else entirely. Although painters and photographers could use their skills to portray the physical beauty of areas such as Yellowstone to the American public, 19th-century Americans needed to know something basic about the areas to be set aside from development. In an era of rapid development and growth, 19th-century Americans were unable to envision the luxury of leaving natural resources unused. For this reason, before they could consider setting Yellowstone aside they needed to first establish that it was "worthless." In presenting the idea of preservation of the Yellowstone area to Congress, proponents of the park needed to at once describe its beauty and also establish that the area contained no resources of value.

The description would have been convincing regardless of its author. However, Hayden's reputation was such that his report proved very influential. In order to convince Congress of the need to create the park, Hayden convinced followers that they needed to establish the park's uselessness for all but scenic enjoyment. George Edmunds of Vermont opened the brief debate by declaring that Yellowstone was "so far elevated above the sea" that it could not "be used for private occupation at all." He therefore assured his colleagues that they did "no harm to the material interests of the people in endeavoring to preserve" the region.

Critics countered that the only rebuttal of significance came from Senator Cornelius Cole of California, who stated that "I have grave doubts about the propriety of passing this bill." Although he was convinced of there being "very little timber on this tract of land," he could not believe that it was off-limits to grazing and agriculture. He and other critics argued that preservationists overstated the threat to the area. For instance, he argued, what harm would come to the geysers and natural curiosities if they fell into private control?

Throughout the Rocky Mountains, argued Cole, were many areas that would make a splendid public park; however, Yellowstone was a place "where persons can and would go and settle and improve and cultivate the grounds, if there be ground fit for cultivation." Further guarantees by Senator Edmunds that Yellowstone was "north of latitude forty" and "over seven thousand feet above the level of the sea" failed in the least to quiet Cole's objections. "Ground of a greater height than that has been cultivated and occupied," he retorted before asking, "But if it cannot be occupied and cultivated, why should we make a public park of it? If it cannot be occupied by man, why protect it from occupation? I see no reason in that." The argument proved moot. In just a matter of a few years, evidence of Yellowstone's vulnerability to development appeared.

This proposal moved through high political circles and was approved within a year and a half. In 1872 Yellowstone National Park was established, making it the world's first national park. The actual congressional act included that Yellowstone "is hereby reserved and withdrawn from settlement, occupancy, or sale under the laws of the United States, and dedicated and set apart as a public park or pleasuring-ground for the benefit and enjoyment of the people; and all persons who shall locate or settle upon or occupy the same, or any part thereof, except as hereinafter provided, shall be considered trespassers and removed therefrom."

The Act also sought to establish basic guidelines for the use and management of the new park:

> SECTION 2. That said public park shall be under the exclusive control of the Secretary of the Interior, whose duty it shall be, as soon as practicable, to make and publish such rules and regulations as he may deem necessary or proper for the care and management of the same. Such regulations shall provide for the preservation, from injury or spoliation, of all timber, mineral deposits, natural curiosities, or wonders within said park, and their retention in their natural conditions. The secretary may in his discretion, grant leases for building purposes for terms not exceeding 10 years, of small parcels or ground; at such places in said park as shall require the erection of buildings for the accommodation of visitors; all of the proceeds of said leases, and all other revenues that may be derived from any source connected with said park, to be expended under his direction in the management of the same, and the construction of roads and bridle-paths therein. He shall provide against the wanton destruction of the fish and game found within said park, and against their capture or destruction for the purposes of merchandise or profit. He shall also cause all persons trespassing upon the same after the passage of this act to be removed therefrom, and generally shall be authorized to take all such measures as shall be necessary or proper to fully carry out the objects

> and purposes of this act. (Forty-Second Congress, Session 2, chap. 24, March 1, 1872)

Even though it had created the park, Congress did not approve funding for Yellowstone until 1877, and even that was insufficient to manage and protect the reserve. In 1884, additional problems came about from the proposal to construct an access railroad across the northeast corner of the park. Proponents argued that the railroad was the only way to remove gold-bearing ores from Cooke City, just east of the park, to the Northern Pacific Railway at Gardiner Gateway, Yellowstone's northern entrance. Although Congress turned down the plan, writes Sellars, "the project was denied more because of what the mines lacked rather than what the tracks would have threatened." The Cooke City mines actually never lived up to expectations. Sellers reports that "In truth, Dr. Ferdinand V. Hayden had been vindicated; his assessment in 1871 that few of Yellowstone's volcanic formations contained precious metals was correct." Still, the willingness of many parties to consider the railway demonstrated that Yellowstone's status still hinged on its worthlessness.

At Yellowstone and soon at Yosemite, Americans learned that a new ethic could be applied to land. In the case of Yellowstone, the actions of those seated around the fire have allowed Americans nearly 150 years later to see at least a version of what confronted them. The Hayden explorers attributed the name "Yellowstone" to the Native Americans living in the area at the time of the Hayden Expedition. Because of the high yellow cliffs surrounding the waterway that passed through this region, they referred to it as the Yellowstone River. Eventually this geological oddity also was used to name the nation's first national park.

Today, the park spans 2.2 million acres. Although most of the acreage is contained in northwestern Wyoming, the park also reaches into Montana and Idaho. Initially the odd geological landforms within the park, including geysers and fissures in the earth, attracted the imaginations of many Americans. After initiating Americans' conception of national parks, Yellowstone's role as a symbol for all parks has never diminished. Today, Yellowstone continues to serve as an active battleground as Americans strive to define the meaning of preservation and wilderness.

The other great symbol of the American movement for national parks can be found in northern California; however, its origin story differs significantly from that of Yellowstone. Originally referred to as America's Switzerland, the Yosemite Valley possessed great mountains, waterfalls, and forests that, similar to Yellowstone, struck preservationists as unique oddities. The great girth and height of the Giant Sequoias, for instance, made the Mariposa

Grove area one of the nation's most fabled landscapes by the 1870s. President Abraham Lincoln signed the bill to set aside the Yosemite Valley and the Mariposa Grove during the American Civil War in 1864. This bill, however, did not designate the area as a national park; instead, it granted the land to the State of California as a public trust.

One of the earliest observers of Yosemite was America's leading proponent of parks, Frederick Law Olmsted. Prior to the establishment of the National Park Service in 1916, planners such as Olmsted were the leading authorities on the American idea of parks and park development. In an 1865 report, he cautioned that the great features of the park might be exploited by development if government restrictions were not placed on the park. He writes:

> It was during one of the darkest hours, before Sherman had begun the march upon Atlanta or Grant his terrible movement through the Wilderness, when the paintings of Bierstadt and the photographs of Watkins, both productions of the War time, had given to the people on the Atlantic some idea of the sublimity of the Yo Semite, and of the stateliness of the neighboring Sequoia grove, that consideration was first given to the danger that such scenes might become private property and through the false taste, the caprice or requirements of some industrial speculation of their holders, their value to posterity be injured. To secure them against this danger Congress passed an act providing that the premises should be segregated from the general domain of the public lands, and devoted forever to popular resort and recreation, under the administration of a Board of Commissioners, to serve without pecuniary compensation, to be appointed by the Executive of the State of California.
>
> By no statement of the elements of the scenery can any idea of that scenery be given, any more than a true impression can be conveyed of a human face by a measured account of its features. It is conceivable that any one or all of the cliffs of the Yo Semite might be changed in form and color, without lessening the enjoyment which is now obtained from the scenery. Nor is this enjoyment any more essentially derived from its meadows, its trees, streams, least of all can it be attributed to the cascades. These, indeed, are scarcely to be named among the elements of the scenery. They are mere incidents, of far less consequence any day of the summer than the imperceptible humidity of the atmosphere and the soil. The chasm remains when they are dry, and the scenery may be, and often is, more effective, by reason of some temporary condition of the air, of clouds, of moonlight, or of sunlight through mist or smoke, in

the season when the cascades attract the least attention, than when their volume of water is largest and their roar like constant thunder.

There are falls of water elsewhere finer, there are more stupendous rocks, more beetling cliffs, there are deeper and more awful chasms, there may be as beautiful streams, as lovely meadows, there are larger trees. It is in no scene or scenes the charm consists, but in the miles of scenery where cliffs of awful height and rocks of vast magnitude and of varied and exquisite coloring, are banked and fringed and draped and shadowed by the tender foliage of noble and lovely trees and hushes, reflected from the most placid pools, and associated with the most tranquil meadows, the most playful streams, and every variety of soft and peaceful pastoral beauty.

This union of the deepest pest sublimity with the deepest beauty of nature, not in one feature or another, not in one part or one scene or another, not any landscape that can be framed by itself, but all around and wherever the visitor goes, constitutes the Yo Semite the greatest glory of nature.

Although Olmsted, similar to Romantics, believed that Americans needed to interact with nature in order to keep from becoming overcivilized, he also sought to exploit and develop this interest of the wealthy into a bona fide American tradition of landscape design.

Dominating the design of urban parks for a generation, Olmsted was uniquely suited to establish how a federally sponsored system of larger park areas would need to be administered and constructed. In addition to planning the roads and grounds of Yosemite, Olmsted wrote widely on humans' need to maintain a connection with nature in an era of increasing industry and urbanity. Later in his report on the Yosemite Valley, Olmsted wrote:

It is a scientific fact that the occasional contemplation of natural scenes of an impressive character, particularly if this contemplation occurs in connection with the relief from ordinary cares, change of air and change of habits, is favorable to the health and vigor of men and especially to the health and vigor of their intellect beyond any other conditions which can be offered them, that it not only gives pleasure for the time being but increases the subsequent capacity for happiness and the means of securing happiness.

Olmsted's forecast proved correct by the 1870s, when a tourist landscape of roads, hotels, cabins, teamed with incursions by cattle and hogs, threatened to overtake Yosemite's majesty. The park was moved to federal authority in 1890, which eventually improved the misuses. Together, Yellowstone and Yosemite defined an American original in the human relationship with nature: the national park.

In the wilderness setting and with a backdrop of the vast, dramatic landscape of the western frontier, the origin of the national park idea seemed fitting and noble. However, the reality of America's national parks is not quite this simple. The effort to establish the first parks forced Americans to consider the idea of preservation, but no one defined what exactly that meant. Future debates were needed to establish the logic of the American national parks.

Conclusion: Marsh Spurs Consideration of Industrialization

In each Romanticized approach to nature, little criticism was leveled at 19th-century life. Primarily, Romantics celebrated nature while not detracting from wealthy consumers. However, as nature's beauty was seen worthy of celebration and reverence, some observers were growing increasingly unwilling to overlook the abuses wrought on it by the insensitive. Leading this vocal criticism, George Perkins Marsh emerged in the 1860s as one of the only spokesmen for a scientific orientation who was willing to take on the culture of industrialization.

With the growth of cities in the United States during the 19th century, there was a dramatic increase in industry, and as industry grew, the natural environment was adversely impacted in immediately visible ways. For example, the machinery of many factories was fueled by coal that caused smokestacks to belch black smoke into the air, and industrial by-products flowed into the waterways, leaving them polluted. The impact of these industries did not go unnoticed to young Marsh, who in a letter to the botanist Asa Gray in 1849 wrote the following:

> I spent my early life almost literally in the woods; a large portion of the territory of Vermont was, within my recollection, covered with natural forests; and having been personally engaged to a considerable extent in clearing lands, and manufacturing, and dealing in lumber, I have had occasion both to observe and to feel the effects resulting from an injudicious system of managing woodlands and the products of the forest.

A trained geographer, Marsh still had influences that could instill in him a Romantic view of the natural world. For instance, his cousin James Marsh, a philosopher and the president of the University of Vermont, helped to redefine transcendentalism. Instead of the idealism of much of New England transcendentalism, with its interest in conservation or a primitivism, Marsh's view of transcendentalism advocated taming wilderness. He advocated for practical informed decisions and increased command over nature. Concerning human use of natural resources, he felt that it was important to weigh the results and act accordingly.

Seeing the damage to the natural environment occur right before their eyes, some people became alarmed and began to search for ways to create a balance between industrial progress and the preservation of natural resources. These very early conservationists included George Perkins Marsh, who wrote *Man and Nature.* Marsh argued that the growth of industry was upsetting the natural balance of nature. The scale and scope of this action overwhelmed knowledgeable observers such as Vermont statesman James Marsh. While acknowledging the need for human use of the natural environment, George Marsh used his 1864 book *Man and Nature* to take Americans to task for their misuse and mismanagement of their national bounty. Marsh writes:

> Nature, left undisturbed, so fashions her territory as to give it almost unchanging permanence of form, outline, and proportion, except when shattered by geologic convulsions. . . . In countries untrodden by man, the proportions and relative positions of land and water . . . are subject to change only from geological influences so slow in their operation that the geographical conditions may be regarded as constant and immutable. Man has too long forgotten that the earth was given to him for usufruct alone, not for consumption, still less for profligate waste. . . . But she has left it within the power of man irreparably to derange the combinations of inorganic matter and of organic life. . . . [Man] is everywhere a disturbing agent. Wherever he plants his foot, the harmonies of nature are turned to discords. . . . [O]f all organic beings, man alone is to be regarded as essentially a destructive power.

To reach his conclusions in *Man and Nature,* Marsh drew from his observations as a youth in Vermont as well as those from travels in the Middle East. The philosophies expressed above, such as referring to humans as "disturbing agents," contradicted the conventional ideas of the time.

In geography, for instance, the work of scholars including Arnold Guyot and Carl Ritter argued that the physical aspects of Earth were entirely the result of natural phenomena, mountains, rivers, and oceans. To suggest that humans could disrupt and ultimately manipulate these forms and patterns was profound. Marsh was the first to describe the interdependence of environmental and social relationships. Lowenthal writes that "Like Darwin's *Origin of Species,* Marsh's *Man and Nature* marked the inception of a truly modern way of looking at the world. Marsh's ominous warnings inspired reforestation, watershed management, soil conservation, and nature protection in his day and ours."

In addition to constructing this intellectual framework for future generations, Marsh used various occupations to influence approaches to land use, including lawyer, newspaper editor, sheep farmer, mill owner, lecturer, politician, and diplomat. As a congressman in Washington (1843–1849), Marsh helped to found and guide the Smithsonian Institution. He served as U.S. minister to Turkey for five years, where he aided revolutionary refugees and advocated for religious freedom, and he spent the last 21 years of his life (1861–1882) as U.S. minister to the new United Kingdom of Italy.

One of the lasting influences of Marsh's thought was to celebrate and eventually to preserve the remaining unspoiled places. Years of living in the Middle East afforded him time to travel throughout Egypt and part of Arabia. On one of these journeys he developed an obsession for the camel and was convinced that the animal might thrive in the American deserts. In addition to transportation, Marsh thought that the camel could prove useful in wars in the Southwest. Inspired by a lecture that Marsh delivered at the Smithsonian upon his return to the States, Congress ordered 74 camels from the Middle East to be shipped to Texas in 1856. The experiment failed, mostly because of the onset of the American Civil War and the unfamiliarity with the ways of the camel on the part of the army's equestrian division.

Regardless, Marsh brought an alternative paradigm to ideas of development and land use. His ideas galvanized a growing sentiment in the 19th century that wanton growth was not the only model for national development and that the federal government would serve as an important tool in reining in uncontrolled development.

Brian C. Black

Further Reading

Carr, Ethan. *Wilderness by Design.* Lincoln: University of Nebraska Press, 1988.

Hales, Peter B. *William Henry Jackson and the Transformation of the American Landscape.* Philadelphia: Temple University Press, 1988.

Lowenthal, David. *George Perkins Marsh, Prophet of Conservation.* Seattle: University of Washington Press, 2000.

Marsh, G. P. *The Earth as Modified by Human Action: Man and Nature.* 1874; reprint, New York: Scribner, Armstrong, 1976.

Nash, Roderick. *Wilderness and the American Mind.* New Haven, CT: Yale University Press, 1982.

Novak, Barbara. *Nature and Culture.* New York: Oxford University Press, 1980.

Opie, John. *Nature's Nation.* New York: Harcourt Brace, 1998.

Roper, Laura Wood. *FLO: A Biography of Frederick Olmsted.* Baltimore: John Hopkins University Press, 1973.

Runte, Alfred. *National Parks: The American Experience.* 3rd ed. Lincoln: University of Nebraska Press, 1997.

Schuyler, David. *Apostle of Taste: Andrew Jackson Downing, 1815–1852.* Baltimore: Johns Hopkins University Press, 1996.

Sellars, Richard West. *Preserving Nature in the National Parks: A History.* New Haven, CT: Yale University Press, 1997.

Semonin, Paul. *American Monster: How the Nation's First Prehistoric Creature Became a Symbol of National Identity.* New York: New York University Press, 2000.

European Union Climate Policy

Introduction

The European Union (EU) helped pioneer elements of climate policy. Internationally, negotiations toward the two keystone United Nations (UN) climate treaties, the UN Framework Convention on Climate Change (UNFCCC) in 1992 and the Kyoto Protocol in 1997, were strongly supported by EU countries. At a regional level, the EU has implemented several climate policy packages originating within member states and the European Parliament.

For example, the European Climate Change Programme (ECCP) was established in June 2000 to help the European Commission (EC) identify and take the most environmentally effective and most cost-effective policies and measures on cutting EU greenhouse gas (GHG) emissions to achieve its target under the Kyoto Protocol. Each of the member states has put in place its own domestic actions that either build on the ECCP measures or complement them.

The first ECCP produced one of the most significant EU policy responses to climate change, the European Union Emissions Trading System (EU ETS), aiming to achieve emission reductions cost-effectively. Launched in 2005, the EU ETS became the world's first large-scale multinational GHG emission trading mechanism. This program currently covers some 11,000 power stations and industrial plants across 30 participating countries, including the present 27 member states (EU-27) plus 3 non-EU states. It has been the engine of the global carbon market, running at steadily increased trading volumes with significant credit values, and moreover has provided a market for Clean Development Mechanism (CDM) credits generated in developing countries.

Beyond the ETS, in 2007 EU leaders endorsed an integrated approach to ambitious climate and energy goals of limiting EU GHG emissions by at least 20 percent by 2020 from 1990 levels and achieving a target of 20 percent of total EU primary energy use through renewable energy. To achieve these objectives, a climate and energy package was proposed by the EC and became law in June 2009. The EU has recently offered an increase of the emissions reduction target to 30 percent by 2020 on condition that other major emitting countries in both the developed and developing worlds make reasonable and fair commitments under future global climate agreements.

Mitigation under the EU ETS

The ETS, the EU's primary policy tool for reducing GHGs, was conceived to cut industrial carbon emissions across the participating member states, ostensibly to fulfill their Kyoto obligations but more directly to achieve EU goals. At present, the scheme covers only energy and industrial sectors, including electric power, oil refineries, coke ovens, metal ore and steel, cement kilns, glass, ceramics, and paper and pulp, and the approximately 11,000 installations are collectively responsible for close to half of the EU's emissions of carbon dioxide (CO_2) and 40 percent of its total GHG emissions.

Market Mechanism

Emissions trading, also known as cap and trade, is a market-based policy instrument that provides flexibility to companies in choosing how to meet their emission targets. A cap, or a limit of the total amount of certain GHGs that can be emitted, is set to factories, power plants, and other industrial installations in the scheme. Based on the cap, companies receive emission allowances, which can be surrendered to comply with their individual targets. The emission allowances allocated and primarily traded under the EU ETS are called European Union Allowances (EUAs). One EUA unit equals one tonne of CO_2 equivalent. For compliance, each company must turn in a number of allowances that covers the amount of its GHG emissions as monitored and reported during each trading year; otherwise, heavy fines are imposed.

Hot Spot

Europe

Europe, heretofore usually a temperate place with a climate shaped by the ocean, has undergone notable extremes in recent years, including heat waves, droughts, deluges, and melting glaciers in the Alps. Such extremes in the Alps could become routine fare there, according to a 2004 report by the European Environment Agency. The report said that rising temperatures could eliminate three-quarters of Alpine glaciers by 2050 and bring repeats of Europe's mammoth floods of 2002 and the heat wave of 2003. Global warming has been evident for years, but the problem is becoming acute, Jacqueline McGlade, executive director of the Copenhagen-based agency, told the Associated Press. "What is new is the speed of change," she said. Ice melt, for example, reduced the mass of Alpine glaciers by 10 percent in 2003 alone, the report said.

Mikhail Koslov and Natalia G. Berlina analyzed records from a Lapland reserve on the Kola Peninsula of Russia during 1930 and 1998. The researchers found "a decline in the length of the snow-free and ice-free periods by 15 to 20 days due to both delayed spring and advanced autumn/winter." Emissions of sulfur dioxide from a nearby industrial plant may have contributed to the cooling, which was associated with a snowfall increase of more than 40 percent during the same period.

Bruce E. Johansen

Fines are designed to be much higher than the price of carbon credits transacted in the market. A participating firm thus can choose the installation(s) at which it reduces emissions most cost-effectively to meet its overall quota, sell allowances to those with higher mitigation costs if it holds extra, or purchase allowances to make up the difference. In this way, all parties are supposed to benefit from trading, and the overall environmental target is achieved at the lowest cost.

Regulatory Framework

Regulation of the ETS is decentralized to the government of each member state but under guidance and coordination of the EC. Member states have some discretion to operate their national registry over the affected companies, in terms of allocating allowances, tracking emissions and transactions, and monitoring, reporting, verification, and other enforcement procedures. The overall cap of the EU ETS is determined as the sum of 30 separate state decisions concerning the total number of EUAs that each member state could distribute to affected installations within its jurisdiction. Each government proposes a national allocation plan (NAP) containing the quantity of EUAs, a list of installations, and a distribution plan, which is subject to review and approval by the EC according to procedures and criteria specified in the EU Emissions Trading Directive. Each member state also operates its own registry to record the creation, transfer, and surrender of allowances; however, a high degree of uniformity is required through a central registry, called the Community Independent Transaction Log (CITL), that provides publicly accessible information of affected installations across all national registries.

Evolution of the EU ETS

Origins The EU had taken action on climate change before the UNFCCC was ratified at the Earth Summit in Rio de Janeiro in June 1992. Prior to the Rio conference, the EU had attempted to establish a carbon tax across all member states; however, the proposition failed in 1997 due to intense opposition from member states that were concerned about sovereignty. Despite this hurdle, the EU participated intently in Kyoto negotiations following the UNFCCC, producing the Kyoto Protocol, which was signed in 1997 and committed the 15 pre-2004 EU member states (EU-15) jointly to an 8 percent GHG emissions reduction below 1990 levels by 2012. This collective commitment was translated into differentiated national emission targets for each of the EU-15, and these are now binding under EU law. To enable the EU to fulfill its Kyoto obligation, the EU ETS developed swiftly: it was politically agreed by the EU Council of Environment Ministers in 2002, and its authorizing legislation Emissions Trading Directive was formally adopted and entered into force the next year. With the accession of 10 new EU members in May 2004 from Central and Eastern Europe, the EU ETS started operating on January 1, 2005, including the 25 initial member states, before the Kyoto Protocol went into effect on February 16.

Launch Phase I (2005–2007) Having been enacted before the Kyoto Protocol became legally binding in international and EU law, the EU ETS launched on January 1, 2005; it thus would have become operational even if the Kyoto Protocol had not entered into force one month later. The initial phase was expected to provide the experience and establish the infrastructure to ensure success in the mitigation period when the EU Kyoto obligation began. Phase I started with covering only carbon dioxide and a limited number of sectors. Available data suggest a 2 percent decline in carbon emissions (40 to 100 $MtCO_2e$ annually) attributable to the ETS during this trial period. Three non-EU members—Norway, Iceland, and Liechtenstein—joined the scheme in 2007.

Operation Phase II (2008–2012) Presently, EU ETS is operating in its second trading period, with the participation of EU-27 (2 new EU accession states but not Annex I countries—Cyprus and Malta—joined the scheme in Phase II) plus three non-EU countries. Phase II runs successively for five years from 2008 to 2012, which corresponds to the first commitment period of the Kyoto Protocol. Gaining significant experience from Phase I, a functioning market for carbon credits has developed quickly. Phase II is in place, equipped with the cap and trade infrastructure of market institutions, registries, monitoring, reporting, and verification, and a significant portion of European industry is getting used to taking into account the price of CO_2 emissions as a factor of their regular production decisions. Compared to the first trial period, Phase II involves tighter overall caps, increases the portion of auction activities to replace free allocation, expands to other GHGs beyond CO_2 (primarily nitric oxide), and adds additional sectors under coverage. Starting from 2012, aviation will be included into the ETS: missions from all domestic and international flights from or to anywhere in the world that arrive at or depart from a EU airport will be covered in the trading scheme.

Outlook Phase III (2013–2020) The third trading period of the EU ETS will begin in 2013 with an extension to new sectors (e.g., aluminum) and new GHGs. In prospect, a harmonized regulatory framework and market surveillance system reform has been proposed by which an overall EU-wide cap would be determined centrally, with allowances then allocated to member states; moving from allowance distribution to mandatory auctioning, at least 50 percent of allowances would be auctioned in 2013 compared to around 3 percent in Phase II (EC 2011a); a unified legal accounting and taxation framework would be established across the EU; and more restricted limits would be imposed on using offsets. Also, Phase III corresponds to the implementation phase of the Effort Sharing Decision (ESD) on the overall 20 percent EU emission reduction, which binds differentiated annual targets to each member state.

Flexibility in the ETS

One flexible mechanism toward fulfilling companies' obligations under the EU ETS is called banking and borrowing. If the actual emissions are lower than the allowances received, besides selling the surplus allowances on the market, companies can bank them to cover future emissions; by the same token, they can borrow a future year allowance to meet a current year target. There is no restriction on banking or borrowing of allowances within each trading period, which allows the annually issued allowances to cover emissions in any year within the trading period. Among phases, neither banking nor borrowing was allowed between Phase I and Phase II; unlimited interperiod banking, but not borrowing, will be allowed for the following trading periods.

Another flexible mechanism links the EU ETS to two project-based carbon markets that were set up under the Kyoto Protocol. One is the CDM whereby industries that attempt to meet Kyoto targets can earn Certified Emission Reduction (CER) credits by implementing emission reduction projects in developing countries, and the other is the Joint Implementation (JI) whereby industries can implement emission reduction projects in other industrialized countries to receive Emission Reduction Units (ERUs). EU legislation provides for the EU ETS operators to use JI/CDM credits up to 50 percent of the EU-wide reductions below 2005 levels between 2008 and 2010, but a concrete limit is determined in the NAP, which varies among member states and in some cases by sectors within a member state. Two types of JI/CDM credits—nuclear and temporary forest (afforestation and reforestation)—cannot be used at present.

EU ETS in the Global Carbon Market

The EU ETS has been the engine of the global carbon market. EUA transactions in 2009 reached US$118.5 billion (€88.7 billion), compared to a total trading value of US$4.3 billion in other Annex I region allowance markets, making the EU ETS the largest existing carbon market. It inevitably has significant connections with other Annex I players. In 2009, evidence strongly indicated that U.S. funds and trading companies participated substantially in the EU ETS, which represented 10 to 15 percent of trade volume on the London European Climate Exchange, and private players from Japan were also active although in lower proportions. Although no firm used the project-based credits (such as those from the CDM) to offset their emissions until Phase II, EU ETS companies have been the primary private participants dominating the demand side, with a share of over 80 percent on the CDM markets.

Issues and Controversies

Allocation During Phase I and II, most carbon allowances were allocated for free. The revised Emission Trading Directive envisions a fundamental change of allocation method as of 2013: ambitious benchmarking will apply to industrial and heating sectors, and auctioning will be the rule for electricity production. Rules for allowance distribution in Phase III will be fully harmonized across all EU member states, and benchmarks are to be developed on a product-by-product basis, which reflects the average performance of the 10 percent most efficient installations in the EU producing that product. As the power sector will be totally subject to auctioning, at least half the allowances are expected to be auctioned from Phase III.

Carbon Price Volatility During the first trading period, the price of EUAs tripled in the first six months but collapsed by half in one week of 2006, when it began the decline toward zero over the next 12 months. The sharp fall in EUA price led to a controversy about overallocation. Lacking extensive baseline emission data and without accurate sector definitions, projecting the caps in Phase I is full of uncertainties, and therefore the reduction goals were modest: the expected overall reduction for the EU-25 was only 1 to 2 percentage points. As a result, too many allowances were distributed to member states, precipitating an EUA price collapse when the extent of the overallocation was discovered. The volatility was exacerbated by the prohibition of banking between Phase I and II. Even with banking added, the issue of overallocation and the associated EUA price volatility, however, did not disappear in Phase II. The price drop in Phase II had begun in the second half of 2008 as the financial crisis intensified, and by February 2009 EUA prices had plummeted to €8, versus €30 nine months earlier. Due to the economic slowdown, output declined in industrial and power sectors in Europe, and companies were therefore holding more allowances than they needed for compliance. These companies sold excess EUAs on the market, thereby exerting downward pressure on the EUA price. In the trading market, the carbon price should signal reasonable expectations between supply and demand of carbon credits; high price volatility therefore may discourage investments in low-carbon technologies or emission reduction projects.

Mitigation beyond the EU ETS

Effort-Sharing Decision

GHG emissions from sectors that currently are not included in the EU ETS such as transport, buildings, agriculture, and waste will be cut by 10 percent EU-wide from 2005 levels by 2020, under the so-called Effort Sharing Decision (ESD). Member states, according to their relative wealth, are expected to contribute to this effort differently: richest nations will limit GHG emissions by 20 percent, while poorer countries have lower or even negative reduction targets up to –20 percent during the period 2013–2020, bound by annual national targets. At the community level, strong monitoring activities will be taken to ensure that member states fulfill their obligations, and flexibility mechanisms will help to increase the cost-effectiveness for them to comply.

Energy and Low-Carbon Technologies

The EU Climate and Energy Package set up emission reduction goals but also seeks to transform Europe into a low-carbon economy and increase its energy security. EU-wide energy efforts include mandating increased use of renewable energy sources, such as wind, solar, hydro, and biomass, and of renewable transport fuels, such as biofuels; adopting measures on energy performance of buildings and a wide array of equipment and household appliances; and further promoting emission reductions in buildings with energy labeling systems to inform consumers. Technology policy in the EU has focused on carbon capture and storage (CCS) to trap and store CO_2 emitted by power stations and other large installations, with substantial funding allocated to CCS demonstration. Also, the development of new technologies in energy efficiency and renewable energies will make an important contribution to moving the EU toward a low-carbon economy. The EU has established an institutional and legal framework to ensure safe deployment and to facilitate commercialization of innovative technologies.

Transportation

As the second-largest–emitting sector after energy, transportation contributes around a quarter of EU GHG emissions, more than two-thirds of which are from road transport alone, accounting for about one-fifth of the EU's total CO_2 emissions. Transport emissions have increased 36 percent since 1990, while other sectors have generally lowered their levels. In addition, aviation will be covered in the EU ETS in 2012. The EU has so far put a range of policies in place, aiming to lower emissions from the transport sector. Adopted in 2007, a comprehensive strategy set up a goal of limiting average CO_2 emissions from new cars to 120 grams per kilometers by 2012; it also contains various measures that have been or are to be implemented to help achieve the objective. Concerning fuel quality, EU legislation establishes a Low Carbon Fuel Standard, calling for a 10 percent reduction in GHG intensity of the fuels used in vehicles, which will be calculated in a life-cycle basis.

Forests

The ECCP set up a working group on forest-related sinks in 2002 that has proposed promising measures that can increase the contribution of forests to the mitigation of climate change. Internationally, the EC has put great efforts in negotiations in terms of reducing tropical deforestation, making recommendations to develop a Global Forest Carbon Mechanism that provides financial rewards for developing countries taking deforestation reduction actions.

Adaptation

In a recent report, the European Environment Agency (EEA) summarized the climate change impacts in Europe, stating that the vulnerability to climate change varies significantly across regions and sectors in Europe, making adaptation a context- and location-specific challenge. Being aware of the need to adapt to climate change, some European countries have implemented or adopted adaptation strategies, and some others have started developing or preparing these. National adaptation strategies were developed for each of 32 EEA member countries, and the EU seeks to enable sharing and communication of lessons learned among member states. Currently there is perceived to be inadequate knowledge sharing on climate change impacts, vulnerability, and adaptation EU-wide. To address the issue, a European Clearinghouse (EC) is being created for sharing and maintaining information. Within this clearinghouse, a comprehensive adaptation strategy is under development and is expected to be in place by 2013. Moreover, the EC has published or is in the process of publishing several frameworks for its adaptation measures and policies: a green paper on adapting to climate change in Europe, options for EU action in 2007, a white paper on adaptation to climate change in 2009, and a communication on mainstreaming adaptation and mitigation in 2011. A communication on disaster risk prevention was adopted in 2009 that aims to integrate policies and instruments related to disaster risk assessment, forecasting, prevention, preparedness, and recovery and also calls for better sharing of data in the context of the EU civil protection mechanism. The connection between disaster risk reduction and climate change adaptation will be further explored in the coming years.

Yiyun Cui and Nathan Hultman

Further Reading

Convery, Frank. "Origins and Development of the EU ETS." *Environmental Resource Economics* 43 (2009): 391–412.

EC. "Auctioning," 2011, http://ec.europa.eu/clima/policies/ets/auctioning_en.htm.

EC. "Benchmarks for Free Allocation," 2011, http://ec.europa.eu/clima/policies/ets/benchmarking_en.htm.

EC. "Deforestation: Forests and the Planet's Biodiversity are Disappearing," 2010, http://ec.europa.eu/clima/policies/forests/deforestation_en.htm.

EC. "Effort Sharing Decision," 2010, http://ec.europa.eu/clima/policies/effort/index en.htm.

EC. "Emissions Trading System," 2010, http://ec.europa.eu/clima/policies/ets/index_en.htm.

EC. "The EU Climate and Energy Package," 2010, http://ec.europa.eu/clima/policies/package/index_en.htm.

EC. "First European Climate Change Programme," 2010, http://ec.europa.eu/clima/policies/eccp/first_en.htm.

EC. "Forests and Climate Change," 2010, http://ec.europa.eu/clima/policies/forests/index_en.htm.

EC. "Fuel Quality," 2011, http://ec.europa.eu/clima/policies/transport/fuel/index_en.htm.

EC. "Legal Framework," 2010, http://ec.europa.eu/clima/policies/effort/framework_en.htm.

EC. "Linking the EU ETS to other Emissions Trading Systems and Incentives for International Credits," 2010, http://ec.europa.eu/clima/policies/ets/linking_en.htm.

EC. "Low Carbon Technologies," 2010, http://ec.europa.eu/clima/policies/lowcarbon/index_en.htm.

EC. "Reducing Emissions from the Aviation Sector," 2011, http://ec.europa.eu/clima/policies/transport/aviation/index_en.htm.

EC. "Reducing Emissions from Transport," 2011, http://ec.europa.eu/clima/policies/transport/index_en.htm.

EC. "Road Transport: Reducing CO2 Emissions from Light-Duty Vehicles," 2010, http://ec.europa.eu/clima/policies/transport/vehicles/index_en.htm.

EC. "What Is the EU Doing on Climate Change?" 2010, http://ec.europa.eu/clima/policies/brief/eu/index_en.htm.

EEA. *Adapting to Climate Change: SOER 2010 Thematic Assessment,* November 28, 2010, http://www.eea.europa.eu/soer/europe/adapting-to-climate-change.

EEA. *Application of the Emissions Trading Directive by EU Member States, Reporting Year 2008.* Luxembourg: Office for Official Publications of the European Communities, 2008.

EEA. "Climate Change Policies," 2010, http://www.eea.europa.eu/themes/climate/policy-context.

EEA. *Greenhouse Gas Emission Trends and Projections in Europe 2009: Tracking Progress towards Kyoto Targets.* Copenhagen: European Environment Agency, 2009.

EEA. "Mitigating Climate Change: SOER 2010 Thematic Assessment." November 28, 2010, http://www.eea.europa.eu/soer/europe/mitigating-climate-change.

EEA. "National Adaptation Strategies," 2011, http://www.eea.europa.eu/themes/climate/national-adaptation-strategies.

EEA. "Tracking Progress towards Kyoto Targets and 2020 Targets in Europe." October 12, 2010, http://www.eea.europa.eu/publications/progress-towards-kyoto.

Ellerman, Denny, and Paul Jaskow. *The European Union Emissions Trading System in Perspective.* Arlington, VA: Pew Center on Global Climate Change, 2008.

Kossoy, Alexandre, and Philippe Ambrosi. *State and Trend of the Carbon Market, 2010.* Annual report. Washington, DC: World Bank, 2010.

Pew Center. *The European Union Emission Trading Scheme (EU ETS): Insight and Opportunities.* White paper. Washington, DC: Pew Center, 2005.

UNFCCC. "CDM in Numbers," 2010, http://cdm.unfccc.int/Statistics/Registration/RegisteredProjAnnex1PartiesPieChart.html.

Extinctions: Early Earth History

Extinction is a fact of life for individual organisms, species, and other taxonomic groups. Indeed, estimates of the number of species that have gone extinct since the beginning of life on Earth range above 99 percent. Though most of these die-offs resulted from the constant trickle of species loss caused by regional environmental changes and Darwinian competition (background extinctions), some accompanied the five catastrophic mass extinctions (global biodiversity losses) in the geological record: end-Ordovician, ≈ 439 million years ago; end-Devonian, ≈ 364 million years ago; end-Permian, ≈ 251 million years ago; end-Triassic, ≈ 199–214 million years ago; and end-Cretaceous, ≈ 65 million years ago (the demise of the dinosaurs).

As late as 1800, however, if you had queried a naturalist or any other educated Westerner about whether an organism had or could go extinct, more likely than not they would have contended that the "Creator" would not allow a broken link in his "great chain of being," the hierarchical order of nature laid out in Genesis that granted mankind—the ultimate creation—dominion over the natural world. Devoted amateur scientist and sometime politician Thomas Jefferson, for example, argued in his *Notes on the State of Virginia* (1787) that "such is the economy of nature, that no instance can be produced of her having permitted any one race of her animals to become extinct; of her having formed any link in her great work so weak as to be broken." Thus, when Jefferson and his 18th-century contemporaries sought to explain the fossil remains of animals and plants frequently discovered in the Old and New Worlds, they argued that these creatures might still be found somewhere in the as-yet-unexplored corners of the planet. In fact, Jefferson instructed Meriwether Lewis and William Clark to keep a sharp lookout on their cross-continent expedition (1804–1806) for information about or living examples of animals "deemed rare or extinct," including it seems mammoths *(Mammuthius primigenius*) and the giant ground sloth (*Megatherium*).

Jefferson himself adhered to the no-extinction theory almost until his death in 1826. But by that time most of his contemporaries had changed course, largely in response to the wide dissemination of the work of French naturalist Georges Cuvier (1769–1832), whose comparisons of the anatomies of extant quadrupeds such as African and Indian elephants (he also determined that these were separate species) with the fossil remains of former inhabitants of the Paris Basin, such as the mammoth, convinced him that though the latter might have been part of the same genus as modern elephants, it was morphologically so distinct from its contemporary cousins that it was indeed a separate species, one that had not only died out in France but for the same reasons (Cuvier postulated a natural cataclysm of some sort) had most likely died out all over the globe. In a series of groundbreaking papers starting in the 1790s and then in multiple publications over the next decades—including his best-known works, *Recherches sur les ossemens fossiles de quadrupèdes* (1812) and *La Règne Animal* (1817)—Cuvier analyzed the anatomy of several other extinct quadrupeds, largely creating the field of vertebrate paleontology in the process. Meanwhile, in cooperation with Alexandre Brongiart (1770–1847) and Alcide d'Orbigny (1802–1857), Cuvier mapped out the fossil-bearing strata of the Paris region, showing that entire systems of terrestrial and marine flora and fauna (biota), 27 in all, had succeeded each other in the long history of life on Earth.

Theoretically, then, strata could be dated, at least in relative terms, by their fossils, and the system could be applied all over the world. In fact, not long after Cuvier's *Descriptions géologiques des environ de Paris* came out in 1811, the British canal surveyor and devoted amateur paleontologist William Smith (1769–1839) released a similar study for Great Britain, *A Delineation of the Strata of England and Wales with part of Scotland* (1815). Over the next decades almost feverish work by British, European and American paleontologists filled out much of the geologic record, including the first discoveries in English Jurassic strata of giant reptile fossils named "*Dinosauria*" (fearful reptiles) by British geologist Richard Owen (1804–1892). Not only did these efforts permanently lay to rest any lingering doubts about the reality of extinction—after all, no one had seen the giant *Megalosaurus* roaming the British countryside—but they also showed that Earth was a very old place formerly populated by strange and fascinating life forms.

Broad agreement on the stratigraphic column (the divisions of fossil strata into geological periods, such as Jurassic, Triassic, etc.) and extinction, however, did not mean that scientists were in accord about the pattern and process of either background or mass extinctions. Cuvier argued, for instance, that since flora and fauna living in one location for eons must be well adapted to that particular ecological niche, only a natural catastrophe (for example, the flooding of the Paris Basin by saltwater) could explain the wholesale extinctions illustrated by abrupt transitions between fossil strata in the Paris region and elsewhere. Other catastrophists of a more religious bent, such as the influential British geologist William Buckland (1784–1856), claimed that these seemingly abrupt shifts were actually evidence of a giant deluge such as Noah's flood from Genesis. In a famous 1837 address to a meeting of the Swiss Society of Natural History, the naturalist Louis Agassiz proposed another catastrophist possibility: that much of Northern Europe had once been covered by a giant ice sheet. The deep cold that made this ice age possible also killed off the supposedly tropical animals such as hyenas, lions, and tigers as well as mammoths and mastodons roaming as far north as Scandinavia and Siberia.

Yet the catastrophists had a difficult time explaining where succeeding biota had come from. Cuvier sidestepped the issue by asserting that the new flora and fauna simply migrated from elsewhere, while those of the biblical persuasion were stuck arguing that the six days of creation in Genesis were more figurative than literal, that is, God might have intervened several times to destroy life, after which he created the world anew.

An opposing group, the uniformitarians, argued instead, according to Donald Grayson, that the "extinction of species is a predictable, natural, and ongoing phenomenon, that can be expected to occur slowly through the ages" rather than by sudden convulsions. Based on his own field observations in Great Britain, France and Italy (later North America and elsewhere), British geologist Charles Lyell (1797–1875), for example, argued in his seminal *Principles of Geology* (1830–1833) that contrary to catastrophist theories, the "present was key to the past," and therefore the steady-state forces driving geological change and extinction in the present had been constant through time, and by extension, studying current evidence of change provided the key to understanding the geological record and extinction. Equally as important and in direct contrast to catastrophist theories of complete biotic extinction and new creation, Lyell argued that there were many species that had persisted through several geological ages, some even down to the present era. In fact, he sought to rework the definition of stratigraphic succession by focusing on the number and percentage of marine mollusks that continued into the succeeding phase of the fossil record.

Lyell had little to say, however, about why certain organisms persisted through time while many others simply (or not so simply) disappeared, and he had no explanation—climate change excepted—for the absence of mammals in the earliest rock strata. Lyell's correspondent and friend Charles Darwin (1809–1882) did, however, and in *Origin of the Species* (1859) Darwin argued for natural selection as the mechanism of species change and disappearance. Those animals and plants not well adapted to the continuous natural competition for resources and mates, or those incapable of adjusting to environmental changes and the limitations of the resource base died out in the struggle (and eventually went extinct), to be replaced by those better suited that would in turn pass the advantage on to their offspring. Repeated over the long history of Earth, Darwinian natural selection and descent with modification could be made to explain the variety of current life on Earth and in the fossil record as well as change (and to a lesser extent persistence) over time. It was, in short, the ideal companion piece to Lyellian uniformitarianism, though Lyell himself was something of a reluctant convert to Darwin's theories, since they did not posit a progression of species in the geological record culminating in the emergence of man and since Lyell believed that the fossil record was too fractured to provide evidence of the missing links between evolving life forms.

Over the next century and a half, many of those missing links were unearthed, starting with the discovery of the transitional reptile-bird archaeopteryx in 1861. Meanwhile, the tree of life has taken on much more definite form, particularly with the development of sophisticated absolute dating techniques (i.e., radiometric dating, assessing the age of a fossil by measuring the rate of decay of one or more radioactive isotopes of the rock in which it was found) and assessment (molecular structures, DNA comparisons where possible). With the important exception of the concept of punctuated equilibrium, however, none of these endeavors seriously challenged the dominant descent through modification or uniformitarian paradigms to explain gradual change and/or extinction in the fossil record. In fact, those mechanisms are still part of the standard explanatory framework for background extinction in Earth's history.

Then in 1982, University of Chicago paleontologists Jack Sepkoski and David Raup published a synoptic paper on marine fossil data in *Science* that indicated that even though "background extinction rates appear to have declined" since the explosion in sea life (the only kind there was at the time) of the Cambrian era (540–505 million years ago), there were four clear episodes (and possibly a fifth) of "mass extinction" in the marine fossil record. Though the first point fit comfortably within the uniformitarian paradigm, as it was "consistent with the prediction that the optimization of fitness should increase through evolutionary time," the second point most definitely did not. In fact, it seemed to land paleontology back to the catastrophist threshold. What else could explain these massive die-offs—and massive they were, with 95 percent of marine species disappearing at the Cretaceous-Paleogene (K-T) boundary (about 65 million years ago) alone and more than 80 percent of all terrestrial and marine life lost at the Permian-Triassic (P-T) boundary (around 250 million years ago)—except for some cataclysmic event that broke through the normally very slow evolution, spread, specialization, and eventual extinction of species?

It turned out that an earlier (1980) article in *Science* had already provided a partial answer to the question implied in the Sepkoski and Raup piece (at least for the K-T extinction event), and it proved to be much more controversial, for as the title "Extraterrestrial cause for the Cretaceous-Tertiary Extinction" implied, lead author and Berkeley geologist Luis Alvarez was arguing that the dinosaurs died out because of something from outer space. The culprit in this case was a roughly 10-kilometers-wide asteroid that, according to Alvarez's hypothesis, slammed into Earth sometime near the end of the Cretaceous era, setting off a chain of events that led to the extinction of not just dinosaurs but also the long-surviving ammonites (marine invertebrate cephalopods related to modern squid and octopuses). What initially led Alvarez to this controversial conclusion was his observation of a layer of iridium-enriched soil at a K-T fossil boundary site in Italy, an "anomaly" (iridium is a very rare element in Earth's crust but is more common in asteroids and comets) that turned up in the same strata in other widely separated sites (initially Denmark and then New Zealand and elsewhere), indicating a global phenomenon.

The sheer novelty of Alvarez's theory meant that he and his coauthors (paleontologist Helen Michael, nuclear chemist Frank Asaro, and Alvarez's father and Nobel laureate Walter Alvarez) had to endure a storm of skepticism until confirming evidence of an asteroid impact turned up in 1986 in the form of shocked quartz deposits (only found outside the laboratory at nuclear test sites and asteroid impact locations) at a K-T boundary site in the U.S. West. Several years later another team of scientists discovered a 180-kilometers-wide crater off the coast of the Yucatán Peninsula in Mexico, confirming that something very large had indeed impacted Earth near the end of the Cretaceous period some 65 million years ago.

Finding the proximate cause for the K-T mass extinction did not, however, explain the kill mechanism. According to the Alvarez article, besides the deaths resulting from the shock wave, heat pulse, and explosions from the asteroid impact itself, the impact's aftermath raised a giant cloud of dust that blocked sunlight from reaching Earth, causing global temperatures to plummet and thus arresting photosynthesis and primary productivity at the bottom of the food web. In other words, climate change, in the form of a nuclear winter (not Alvarez's term) of cold and starvation, killed off the dinosaurs and thousands of other species at the end-Cretaceous boundary, not the asteroid itself.

But not everyone agreed with this scenario. O'Keefe and Aherns argued in a 1989 article in *Nature* that instead of a nuclear winter, the Alvarez asteroid actually instigated a long period of intense global warming resulting in the K-T extinction event. Because the asteroid hit deposits of marine limestone and gypsum, Ahrens and O'Keefe theorized, it would have released massive amounts of CO_2, creating a heat blanket effect that might have made much of the continental landmasses (and the continents were then approaching their current position) hostile to life in any form.

Still others, however, debate the importance of the Alvarez asteroid impact as either the chief proximate cause or kill mechanism for the K-T mass extinction, opting instead for the global impact of massive volcanic basalt flows (the Deccan Traps in India), which also coincided with the end-Cretaceous extinction event. Indeed, in the absence of convincing proof of other bolide impacts (asteroids, meteors, and so on), volcanism has become the prime suspect in initiating several of the mass extinctions but particularly the largest of them all, the Permian-Triassic, the extinction event that came closest to obliterating life on Earth 251 million years ago.

The suspect in this case might have been the Siberian Traps, a million-year-long volcanic eruption that spewed out enough basalt to cover 2 million square miles. The volcanic ash and sulfur aerosols emanating from the Siberian Traps could have blocked enough sunlight to induce a long period of glaciation and sea-level drop (water locked up in ice), while the SO_2 would have produced so much acid rain that it killed off the phytoplankton at the foundation of the marine food web. Then the longer-lasting CO_2 released by the Siberian eruptions, in concert with the release of methane-rich gas hydrates on the ocean floor (remember the drop in sea-level), might have combined to create a many-thousand-year-long period of global warming and rapid sea-level rise from melting glaciers, killing off almost everything that survived the initial global cooling.

Global cooling figures prominently in explanations for both the end-Ordovician (439 million years ago) and end-Devonian (364 million years ago) extinction events. Earth's first ice age quite possibly precipitated the end-Ordovician event, with widespread glaciation occurring on parts of the globe now in the tropics (in the Sahara and the Amazon) but then on a large continent (Gondwana) positioned over the much colder South Pole. As in the next two glacial epochs (Early Carboniferous to Early Permian and our own), the end-Ordovician glacial expansion locked up much of the world's water, leading to a severe and maybe sudden drop in sea level, which in turn led to the extinction of many marine species from habitat loss and the cooling itself.

Though glaciation and sea-level drop played central roles in the end-Devonian mass extinction as well, one of the triggering mechanisms for global cooling in this case might have been the evolution of plants and their spread across much of Earth's surface in the mild Devonian climate. The new plant life locked up CO_2, thus removing significant volumes of a global warming greenhouse gas (GHG) from the atmosphere, exacerbating cooling trends. Meanwhile, according to Paul Wignall, both the end-Ordovician and the end-Devonian saw "a near-total shut-down of marine primary productivity" and a consequent "holocaust in the water column" as the "single-celled photosynthesizing algae" disappeared, and with them the base of the marine food web, with disastrous effects on dependent sea life.

Scholars continue to debate both the proximate causes and kill mechanisms for each of the five mass extinctions. But what has become clear after more than a century of research is that extinction itself (whether background or mass) seems to strike first and hardest those taxa with, according to Paul Taylor, "large body size, dietary, thermal and other kinds of ecological specialism," fewer offspring, and "slow growth rate" as well as "restricted geographical range and low population density." On the other hand, generalist species that reproduce quickly and easily tend to persevere even through mass extinctions, particularly if they have wide global dispersal. Indeed, some microorganisms such as the cyanobacteria have been around since the Pre-Cambrian era (3,690 million–590 million years ago) in much the same form as now. If we are in fact entering what some scientists call the sixth mass extinction, one almost entirely driven by *Homo sapiens,* then it might be comforting to note that some life forms made it through the narrow bottlenecks of the previous five, enough to repopulate Earth and rebuild biodiversity and stability, albeit in a timescale measured in the millions of years.

Jacob Jones

Further Reading

Alvarez, Luis, Walter Alvarez, Frank Asaro, and Helen V. Michael. "Extraterrestrial Cause for the Cretaceous-Tertiary Extinction: Experimental Results and Theoretical Interpretation." *Science* 208(4448) (June 6, 1980): 1095–1108.

Benton, Michael J. *When Life Nearly Died: The Greatest Mass Extinction of All Time.* London: Thames and Hudson, 2003.

Bowler, Peter J. *The Norton History of the Environmental Sciences.* New York: Norton, 1992.

Boylan, Patrick J. "Lyell and the Dilemma of Quaternary Glaciation." In *Lyell: The Past Is Key to the Present,* edited by D. J. Blundell and A. C. Scott, 145–159. London: Geological Society of London Special Publications 143, 1998.

Grayson, Donald K. "Nineteenth-Century Explanations of Pleistocene Extinctions: A Review and Analysis." In *Quaternary Extinctions: A Prehistoric Revolution,* edited by Paul S. Martin and Richard G. Klein, 5–41. Tucson: University of Arizona Press, 1984.

Hallam, Anthony, and Paul B. Wignall. *Mass Extinctions and Their Aftermath.* Oxford: Oxford University Press, 1997.

Jablonski, David. "Extinctions in the Fossil Record." *Philosophical Transactions of the Royal Society of London: Biological Sciences* 344(1307) (April 29, 1994): 11–17.

Raup, David M., and J. John Sepkoski Jr. "Mass Extinctions in the Marine Fossil Record." *Science* 215(4539) (March 19, 1982): 1501–1503.

Rowland, Stephen M. "Thomas Jefferson, Extinction, and the Evolving View of Earth History in the Late Eighteenth and Early Nineteenth Century." In *The Revolution in Geology from the Renaissance to the Enlightenment:*

Geological Society of America Memoir 203, edited by G. D. Rosenberg, 225–246. Boulder, CO: Geological Society of America, 2009.

Rudwick, Martin J. S. *Georges Cuvier, Fossil Bones, and Geological Catastrophes.* Chicago: University of Chicago Press, 1997.

Schopf, J. William. "Extinctions in Life's Earliest History." In *Extinctions in the History of Life,* edited by Paul D. Taylor, 34–60. Cambridge: Cambridge University Press, 2004.

Taylor, Paul D. "Extinction and the Fossil Record." In *Extinctions in the History of Life,* edited by Paul D. Taylor, 1–34. Cambridge: Cambridge University Press, 2004.

Wignall, Paul B. "Causes of Mass Extinctions." In *Extinctions in the History of Life,* edited by Paul D. Taylor, 119–150. Cambridge: Cambridge University Press, 2004.

Extreme Weather

At the outset, it should be stressed that weather and climate are distinct concepts. Weather refers to the temperature, air pressure, precipitation, and other atmospheric conditions relating to a particular time and place. Climate is the consistent pattern of weather for a particular region over a relatively long period of time. Characterizing the climate of a region of the globe requires a large amount of weather data covering decades.

Extreme weather therefore includes both predictable and unpredictable events. The extremes of high and low temperatures, wind, and precipitation for any given region during an average year are relatively predictable. These include periods of rain and drought associated with monsoons. Unpredictable meteorological conditions in a particular season or year include record-breaking storms, floods, or droughts, events that are too rare to be included as part of a normal climate pattern. However, this perception may be misleading, since weather rhythms sometimes occur over relatively long cycles.

Researchers who are interested in past climate must rely on indirect methods of gathering data, including studies of the rings of ancient fossilized trees, geologic sedimentary deposits, and layers of glacial ice. In its roughly 4.5 billion-year history, Earth has experienced four major ice ages, that is, periods with extensive glaciation. Corresponding roughly with the Quaternary geological period, the fourth ice age started about 2.6 million years ago and has oscillated numerous times between periods of relative cold (glacials) and warmth (interglacials). With the end of the previous glacial period about

Hot Spot

Australia

Pollution, drought, salinization, and habitat loss due to overdevelopment and brushfires are some of Australia's most pressing environmental problems. Most Australians live in fertile coastal communities that surround the harsh outback, known locally as the bush—the continent's vast interior desert. This has limited the amount of land available for human habitation, which has prompted real estate developers to build houses on land that once served as habitat for koalas and kangaroos, among other animals. The country's growing population and high consumption continue to put stresses on Australia's environment. Land degradation, deforestation, loss of biodiversity, and air and water pollution (especially in Australia's large coastal cities) continue to plague the country.

Australia's worst environmental catastrophe in recent memory is the country's multiyear drought, which lasted from approximately 2002 to 2009. The lack of rainfall sparked huge brushfires throughout the country and dried up many of its major rivers. And in April 2007, Prime Minister John Howard announced that the government would turn off the supply of irrigation water to the Murray-Darling Basin in southeastern Australia, which yields 40 percent of the nation's agricultural produce, unless a significant rainfall occured soon. The crisis may be the first global-warming–driven disaster ever to strike a developed nation.

11,600 years ago, human beings have benefited from the relatively warm and stable Holocene epoch.

Because accurate weather records have only been kept since the late 19th century, researchers do not have detailed data about day-to-day weather patterns from the more distant past. The earliest accounts of weather (often by sailors and farmers) are piecemeal and go back only a few thousand years to the beginning of written language. Therefore, evidence for the single largest storm, the greatest amount of rain, or the longest drought is irrevocably lost.

Guessing which regions of Earth are likely to experience periods of extreme cold or heat on a regular basis is sometimes a matter of common sense. The closer a landmass is to the poles, the cooler is its average temperature (because the Sun's rays reach the surface at a glancing angle) and the more likely it is to be hit with a cold snap. The closer to the equator, the warmer the place is and the more likely it is to be hit with a heat wave. Proximity to large bodies of water tends to have a moderating impact on extremes of temperature, but this is likely to increase the amount of precipitation.

Several locations vie for the title of the coldest or the hottest places on the planet, depending on whether the figures are the annual average temperature, a seasonal average temperature, or the record lowest or highest recorded temperature or if they include factors taking account of wind chill or humidity. The coldest place on Earth, unsurprisingly, is Antarctica, which holds the records for coldest natural temperature recorded—minus 128.6 degrees F—as well as seasonal and average low temperatures. The North Pole, where only one layer of polar ice covers the warmer ocean currents, is warmer than the Arctic Circle landmasses of northern Russia, Canada, Alaska, Greenland, and Scandinavia. The coldest average winters in the United States occur in Alaska, North Dakota, Minnesota, and northern New England. One of the warmest places on the surface of the world is Al Aziziyah in Libya, with a recorded temperature of 136 degrees F. Other hot spots include Indouf, Algeria; Death Valley, California, in the United States; and the Danakil Depression (part of Africa's Great Rift Valley), which seems to be the most consistently hot place on Earth with an average annual temperature of over 100 degrees F.

The surface temperature of Earth's oceans plays a significant role in recurring extremes of weather, since it is a crucial determinate in a number of special ocean circulations. Warm water in the Atlantic Ocean helps generate the African Monsoons, and the warm Indian Ocean helps create the storms that bring needed seasonal rains to Pakistan and India. Such variations in the temperature of surface water have a direct relationship to the severity of storms and the paths that they take. In the central and eastern Pacific Ocean, warmer than usual water temperatures result in a pattern of heavy rains in South America, California, and East Africa with corresponding droughts in Australia, India, and Southeast Asia. This effect, known as El Niño, occurs roughly every five years. When cooler than normal ocean temperatures occur, a complementary pattern of rainfall called La Niña follows, bringing cool and wet weather to India and Southeast Asia and droughts to California and South America. Together, El Niño and La Niña are often referred to as the El Niño–Southern Oscillation (ENSO).

Sometimes combinations of factors, including variations of air currents in the upper atmosphere, ocean currents, or even cosmological or planetary events such as sunspots and the eruption of volcanoes, result in unusual weather events. Such extreme events can be characterized in a number of ways: the longest duration, the most precipitation, the highest winds, the

Our climate system has been severely disrupted and many countries are paying a high and only growing price because of it.

. . . [I]n my own country . . . over the past 50 years we went from being hit by one extreme hydrometeorological event per decade in the 1960s and 1970s, to nine in the last ten years.

The Tropical Depression 12E that struck El Salvador two months ago was the last in this terrible and destructive sequence. For ten days, it dumped up to 1.5 meters of rain. Among other effects, we experienced landslides, flooding of one-tenth of our territory, and huge losses of infrastructure and agriculture.

While in comparison to previous events, we did manage to reduce the number of lives lost due to a fairly strengthened civil protection system; 40 people still died. Moreover, we could not avoid the economic losses that totaled $840 million—4% of our GDP.

Our country received a little over $6 million in emergency assistance, and my government is grateful for the solidarity of the United Nations, our regional banks, Norway, Taiwan, Switzerland, Canada, USA, Korea, Spain, Japan, and Brazil among others.

We were especially moved by Guatemala, who was also affected by Tropical Depression 12E, but nevertheless decided to share with us some of the emergency relief that their country had received.

It is precisely this spirit of solidarity and shared responsibility that we have to bring to our negotiations here in Durban. Do not make the mistake of thinking that there are currently other crises that are more important than the disruption of our climate system.

Herman Rosa Chavez, Environment and Natural Resources minister, El Salvador, 2011

---

most damage caused, or the most lives lost. A particularly cataclysmic event stemmed from the explosion of Mount Tambora in Sumbawa, Indoensia, in 1815. The volcano's injection of ash and sulfate aerosol into the atmosphere produced significant cooling for the global climate, which disrupted agriculture and (directly or indirectly) led to tens of thousands of fatalities during 1815–1818.

The largest storm in recorded weather history occurred in the Pacific Ocean on October 5, 1979. With sustained winds of 190 miles per hour and a diameter of over 1,300 miles, Typhoon Tip never passed over land while at

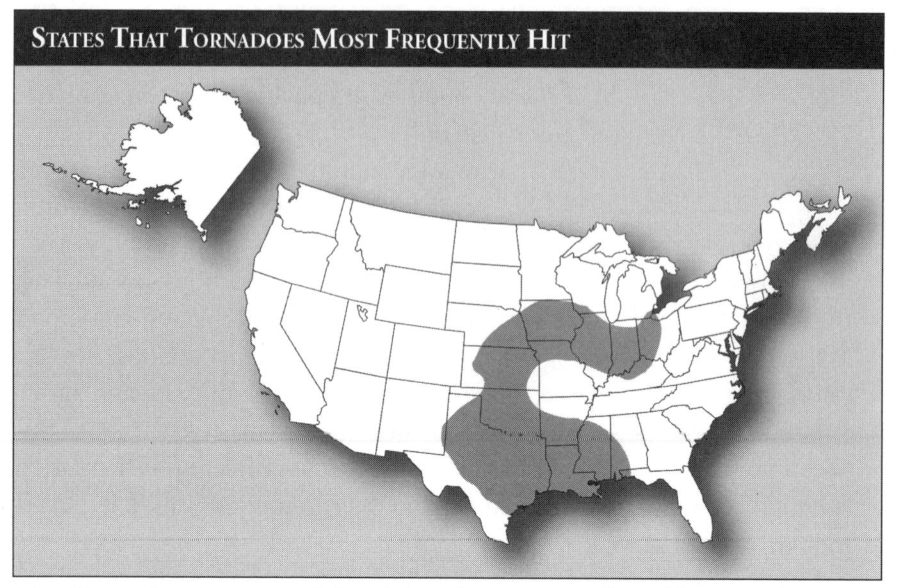

Map illustrating the areas of the United States most affected by tornadoes. The area is known as Tornado Alley. In coming years, extreme weather is expected to occur even more frequently in these areas. (ABC-CLIO)

full strength, thus causing minimal damage. On the other hand, the so-called Tri-State Tornado of 1925 cut a path of destruction 219 miles long across the states of Illinois and Indiana. No other U.S. tornado has matched it for size, speed, or death toll.

Some of the most devastating examples of extreme weather have occurred when storms have caused large-scale flooding in low-lying coastal areas. It is probable that the account of the flood in the Bible is based on some ancient inundation, perhaps as some scholars have suggested along the coast of the Black Sea. A storm surge in the year 1228 has been estimated to have caused the deaths of 100,000 people in the Low Countries of Belgium and Holland, while the Great Hurricane of October 10–16, 1780, is the deadliest Atlantic hurricane on record, drowning more than 20,000 people in the Lesser Antilles. On November 12, 1970, the Bhola Cyclone swept the coast of eastern Pakistan, wiping out entire villages and killing up to half a million people.

Since the 1980s when anthropogenic global warming became a major scientific and political concern, there has been an ongoing debate about the possible connection between global warming and extreme weather. It is clear that Earth's land and ocean surface temperatures have increased during the last three decades. Trends regarding the frequency and severity of storms are less certain, but many analyses suggest an increase. Nevertheless, even if we

Greenpeace activists swim into the ocean pretending to be world leaders and delegates attending the UN Climate Change Conference in Cancun, Mexico, Friday, December 10, 2010. The protest was set inside the sea to send a message that while delegates and world leaders keep discussing environmental issues without arriving at agreements, the world is being flooded. (Israel LealAP/Wide World Photos)

accept such analyses of temperature and storm trends, making a causal link between them is difficult.

One of the first researchers do so was atmospheric scientist Kerry Emanuel of MIT. During the summer of 2005 (shortly before Hurricane Katrina struck America's southern coast), Emanuel published an article in *Nature* that strongly linked increasing storm strength to increasing ocean temperatures. A number of scientists challenged Emanuel's ambitious study. Scientists at the U.S. National Oceanic and Atmospheric Administration tended to believe that the increase in storm activity could be attributed to changes in ocean circulation that were unrelated to global warming. Meteorologist William Gray of Colorado State University made especially influential challenges based on his reputation as an expert on hurricanes. Gray's objections highlighted a difference of style between different climate scientists. Whereas Emanuel claimed a causal connection between global warming and increased hurricane strength based on extensive theoretical analysis and computer modeling, scientists such as Gray tend to distrust modeling and to restrict their work to collecting and analyzing reliable data. They searched for correlations between data but were hesitant to impute causation.

The relative casual importance of global warming and ENSO on extreme weather is a related subject of debate, as is the debate over the possible future effect of warming on ENSO. Most general circulation computer models indicate that ENSO is sensitive to warming trends. However, the models do not yet agree on whether warming will lead to more El Niño–like or more La Niña–like conditions. Such uncertainty about ENSO increases uncertainty about extreme weather. Gray has suggested that increased El Niño activity leads to decreased Atlantic hurricane activity due to conditions of higher vertical wind shear. Therefore, if global warming leads to more El Niño–like conditions, then this might bring fewer and not more Atlantic hurricanes.

As of the present writing, the causal connection between global warming and extreme weather remains uncertain. Nevertheless, the consensus of the climate science community is that the connection is a serious enough possibility to warrant further scientific study and consideration when making public policy. The fourth Intergovernmental Panel on Climate Change report (of 2007) considers it "more likely than not" that the two are linked and that it is "likely" that extreme weather will increase during the 21st century. The changing global climate pattern will have significant effects on many regions. For example, North American cities will probably experience more frequent and severe heat waves, and coastal communities will be subject to greater flooding and storm activity. Europe will be somewhat more adversely affected and experience greater differences between its regions, in terms of seasonal variation, amount of precipitation, and extreme weather. Countries in Southern Europe will probably be subject to higher temperatures and reduced water availability and therefore worsening crop productivity. The many coastal areas of Africa and Asia will likely be subject to more extensive and frequent flooding, while many inland regions will be subject to droughts.

Helen York and Gary J. Weisel

Further Reading

Burt, Christopher C. *Extreme Weather: A Guide and Record Book.* New York: Norton, 2004.

Collier, Michael, and Robert H. Webb. *Floods, Droughts, and Climate Change.* Tucson: University of Arizona Press, 2002.

IPCC. *Climate Change 2007: Synthesis Report; Contribution of Working Groups I, II, and III to the Fourth Assessment Report of the Intergovernmental Panel on Climate Change.* Edited by R. K. Pachauri and A. Reisinger. Geneva, Switzerland: IPCC, 2007.

Laskin, David. *Braving the Elements: The Stormy History of American Weather.* New York: Doubleday, 1996.

Mooney, Chris. *Storm World: Hurricanes, Politics, and the Battle over Global Warming.* New York: Houghton Mifflin Harcourt, 2007.

Smith, Rupert. *The Hottest Place on Earth.* Lion Television Productions, Broadcast on BBC One, UK, March 19, 2009.

F

Feedback Mechanisms

In what amounted to near unanimity on the issue, the scientists of the International Panel on Climate Change (IPCC) declared in the 2007 fourth assessment report (AR4) that evidence of global warming in the last half century was now "unequivocal" and that the observed increase in anthropogenic greenhouse gas (GHG) concentrations was "very likely" (greater than 90 percent probability) the cause for that warming trend. Certainty about the impact of GHG forcings—the initial drivers of a climate shift—on the planet's temperature gradient, however, does not mean that the IPCC is equally certain about the path ahead in the 21st century and beyond. One source of that uncertainty is the question of how human societies will respond to the challenge of curbing GHG emissions. But even if we manage to substantially decrease those discharges in the next few years, GHGs already in the atmosphere will continue to warm the planet for decades with possibly dire consequences, not necessarily as a result of the forcing itself but instead from the knock-on effects of feedbacks, which can amplify (positive feedback) or dampen (negative feedback) the influence of the initial forcing.

Some studies suggest that the negative feedbacks might balance out the positives in the long term if GHG emissions and global CO_2 levels are kept below a certain point. It is just possible, in other words, that the equable climate that mankind has enjoyed over the last 10,000–12,000 years (since the waning of the last ice age) might remain relatively stable but in a significantly altered state. On the other hand, paleoclimatologists have found ample proof over the last decades that at several points in the past, Earth swung abruptly between climate regimes, particularly in the shift between glacial-interglacial epochs. Scientists do not yet know with certainty what pushed the climate past these tipping points, though atmospheric carbon dioxide and other gas concentrations along with their attendant feedbacks were clearly part of the process. Meanwhile, some of the general circulation models (GCMs) that scientists use to help predict the future of Earth's climate also suggest that sudden swings might result even from a level of GHG forcing not much above the current numbers. Thus, until scientists better understand the dynamics of the major climate feedbacks—water vapor, clouds, ice-albedo, methane release, and so on—they cannot be sure where or when the tipping points might appear and thus whether Earth's climate systems might not experience another abrupt and calamitous shift in the not-too-distant future.

Curiosity about the possible magnifying effects of feedbacks came early on in what we now call climate change research. By 1861, for example, Irish British physicist John Tyndall had not only proved that carbon dioxide and water vapor were the main heat-trapping gases in Earth's atmosphere (rather than its chief components, oxygen and nitrogen) but also that water vapor blocked far more outgoing radiation than CO_2. To Tyndall, however, such an effect had only positive implications, since water vapor served as a "blanket, more necessary to the vegetable life of England than clothing is to man."

Building on the findings of Tyndall and others, Swedish physical chemist Svante Arrhenius sought to prove that shifts in atmospheric CO_2, combined with water vapor and other feedbacks, were sufficient to explain the onset and retreat of glacial eras. Using a simplified energy budget model (keeping track of incoming and outgoing energy), Arrhenius calculated that a halving of atmospheric CO_2 would result in a 4°C average temperature decline at some latitudes. With two positive feedbacks, one due to changes in atmospheric water vapor and the other due to changes in Earth's overall surface reflectivity, this cooling trend might have initiated ice ages in the past. As cooling took place, Earth's atmosphere held less water vapor, which resulted in further cooling. In addition, more ice and snow formed (especially at the poles), leading to higher reflectivity, or albedo, and therefore to further cooling. Conversely, Arrhenius found that a doubling of atmospheric CO_2 (from natural or anthropogenic sources such as burning coal) combined with feedbacks from water vapor and surface reflectivity would be enough to raise global temperatures by 5–6°C. Thus, as he famously concluded in his landmark 1896 paper on the subject "On the Influence of Carbonic Acid in the Air upon the Temperature of the Ground," "if the quantity of carbonic acid increases in geometric progression, the augmentation of temperature will increase nearly in arithmetic progression."

It was Arrhenius's contemporary, American geologist Thomas Chrowder Chamberlin, who best explained the wide temperature differentials between glacial and interglacial periods as primarily a function of the volume of water vapor in the atmosphere. Like Arrhenius and others, Chamberlin agreed that a rise in atmospheric CO_2 might well lead to a higher global mean surface temperature but not enough to explain the 10–12°C bump of the interglacial periods. That kind of temperature swing, Chamberlin argued, did not come from CO_2 warming alone or even from Scottish geologist James Croll's contemporary theory that Earth's orbital variations around the sun could account for the initiation and recession of glacial epochs. Instead, Chamberlin asserted, only the strong feedback linkage between CO_2 and H_2O could account for the glacial-interglacial temperature gradients. "Water vapor," Chamberlin wrote, "is dependent on temperature for its amount." When CO_2 is added to the atmosphere, its radiative effects will raise global temperatures by itself. But more important, "it calls into function a certain amount of water vapor which further absorbs heat, raises the temperature and calls forth more vapor, thus building a dynamic pyramid on its apex" sufficient to trigger and sustain an interglacial warming period.

After a gap of several decades in which water vapor research languished, German meteorologist and geophysicist Fritz Möller published a paper in a 1963 issue of the *Journal of Geophysical Research* that argued that a doubling of atmospheric CO_2 could bring a temperature increase of 10°C or higher as a result of runaway water vapor feedback. A sophisticated GCM constructed by Syukuro Manabe and Richard Wetherald of Princeton's NOAA (National Oceanic and Atmospheric Administration) Geophysical Fluid Dynamics Lab, however, indicated that a doubling of CO_2 content in the atmosphere (given a fixed relative humidity) would result in a temperature increase of about 2°C, a number that has remained relatively consistent in GCM projections up to the present. Nonetheless, the key assumption of Möller's calculations (as it had been for Chamberlin's)—that, according to Spencer Weart, the "water vapor content of the atmosphere should increase with increasing temperature" (since warmer air holds more moisture)—has remained constant as well, and the most recent studies indicate that water vapor feedback alone could double the effect of CO_2 forcing.

Möller's was not the only runaway feedback model to appear at midcentury. In 1956, Maurice "Doc" Ewing and William Donn of the Lamont Geological Observatory in New York, argued that the zig-zag pattern of Pleistocene glacial-interglacial periods began when, according to Weart, "the North Pole wandered into the Arctic Ocean basin," setting off a series of cooling feedbacks that resulted in heavy snow and ice accumulation in the northern latitudes and the onset of an ice age. That freezing, however, set the stage for equally rapid deglaciation, as a frozen Arctic Ocean did not provide sufficient moisture to maintain the snowfall necessary to sustain the continental glaciers, and thus, according to Weart, "the sheets began to dwindle, the seas rose, warm currents spilled back into the Arctic Ocean," and the ice melted. Ewing and Donn's arguments were based on so many linked assumptions that the model did not hold up under scrutiny, though the popular press covered the arguments enough to scare many into thinking that a new ice age might well be imminent.

Nonetheless, the prize for midcentury models of climate calamity from feedback loops did not go to Ewing and Donn but rather to Soviet geophysicist and climatologist Mikhail Budyko and his American counterpart William D. Sellers. In the 1960s, Budyko developed a simple heat-balance model of Earth that suggested that if current global average temperatures increased by only a small amount, then all of the world's remaining ice (some of which was still left over from the last glacial era) would disappear, taking its high reflectivity (high albedo) with it and leaving the bare earth and the oceans to absorb even more solar radiation, magnifying the initial heating trend to the point where the planet might return to the warm wet conditions of the Jurassic era (200 million–146 million years ago), albeit without the dinosaurs. Conversely, if average temperatures fell by only a small amount, Budyko's equations suggested that the world might be plunged into such a severe ice age that the entire planet could become ice-covered, even at the tropics.

Meanwhile, working separately at the University of Arizona, Sellers arrived at similar conclusions using different calculations. His model predicted that a mere 2 percent decline in incoming solar energy could increase snow and ice cover to such an extent that much of that radiation would be reflected back into space, resulting in the same "snowball Earth" that Budyko predicted.

Though critics lambasted the simplicity of these models, arguing that such equations could not match the complexity of the real world's climate systems, physical evidence continued to confirm that Earth's climate had shifted abruptly in the past, probably as a result of seemingly small changes in solar radiation (the Milankovitch cycles) magnified by ice-albedo and other feedbacks. For example, both deep ocean sediment cores and ice cores taken from Greenland and Antarctica in the 1970s and 1980s all confirmed that the transitions between glacial and interglacial periods had sometimes been abrupt indeed, with only years or decades separating the two climate states. In the meantime, evidence sent back by Mars satellites showed that the red planet once had abundant water flow but was now covered in ice. Carl Sagan and others argued that the shift occurred due to a runaway ice-albedo effect. Finally, evidence began to accumulate that glaciation might well have extended as far as the tropics in the distant geological past. In a 2010 paper in *Science,* lead author and Harvard geologist Francis A. Macdonald claimed that an iron-rich layer of glacial sediment in the remote Canadian Yukon (a region of the continent carried from the tropics by continental drift) indicated that what "climate modeling has long predicted" turns out to be true, that "if sea ice were to develop within 30 degrees latitude of the equator, the whole ocean would rapidly freeze over" due to the "high albedo of ice."

Even if, as seems increasingly likely, our planet experienced several snowball Earth periods in the deep past of the geologic record, then these are more a testament to the strength of the ice-albedo feedback itself rather than a forecast of what will happen in our own near future, even in the IPCC's worst-case scenario. Given the current warming trend, it is likely that the water vapor and albedo feedbacks will be strongly positive and will more than double the impact of GHG forcings. But there are also feedbacks that might partially constrain these trends (for example, clouds). Clouds cover about 60 percent of Earth's surface, the IPCC's AR4 pointed out, and they "are responsible for up to two-thirds of the planetary albedo, which is about 30%" overall. But such simple numbers hide the complexity of the cloud feedback parameters, as cloud albedo can be altered by cloud shape, height, thickness, and physical properties among other factors. In fact, just by altering the cloud parameters, all of the main GCMs, as well as pioneering intermodel comparisons by atmospheric scientist Robert D. Cess in 1989 and by others since, have yielded a wide range of potential climatic outcomes (1.9°C to 5.4°C of warming) from a doubling of atmospheric CO_2 since the preindustrial era (the standard reference point for all GCMs and a number we are steadily approaching).

These varied results do not necessarily mean, however, that increased cloud cover, type, and composition as a result of global warming will work as an overall negative feedback sufficient to offset global warming (or return Earth to an imaginary preindustrial equilibrium state). True, at least one recent study, that by Rebecca Lindsey, suggests that the increased cloud cover in the Arctic Ocean Basin, for example, might compensate for the albedo feedback from melting snow and ice. On the other hand, a 2010 *Science* paper by Texas A&M atmospheric scientist Andrew Dressler contends that on a global basis, increased cloud cover from global warming will trap more heat than it reflects, resulting in a positive feedback. "Based on my results," Dressler concludes, "I think the chances that clouds will save us from dramatic climate change are pretty low."

Meanwhile, the climate-system deck may still hold several wildcards, the most ominous of which might be methane (CH_4). Though its atmospheric half-life is less than 10 years, versus the century-long persistence rate of carbon dioxide, methane is a much more potent GHG than CO_2. Moreover, the volume of CH_4 in the atmosphere has more than doubled in the last 250 years, and recent trends suggest that the rate of release is accelerating as the snow, ice, and permafrost layers of the northern latitudes (including the Arctic) begin to melt in response to global warming. As the permafrost melts, the frozen peat bogs and organic detritus previously locked up in the deeply frozen soil begin to degrade, releasing tremendous amounts of methane. In 2005, as discussed by Ian Sample, researchers announced that a vast section of Siberia, an area "the size of France and Germany combined," was starting to thaw out, and this area alone could pump enough methane out to "double the levels of the gas" in the atmosphere and lead to a "10% to 25% increase in global warming" as more methane caused more warming, more melting, and more methane release. But the main concern of feedback scientists is the potential release of methane from clathrates (gas hydrates) previously locked up in the subsea permafrost under the now-warming Arctic Ocean and connecting seas. Though stable under near-freezing water and high pressure, if water temperatures on the Arctic Sea shelf warm sufficiently, then the methane in these hydrates might release rapidly. The "release of even a fraction of the methane stored in the shelf," the National Science Foundation warns, "could trigger abrupt climate warming."

Despite these potentially calamitous feedbacks, however, scientists cannot yet predict with certainty global or even regional balances between positive and negative feedback responses to global warming. Thus, research continues on all fronts, from the water vapor, cloud, ice-albedo, and methane feedbacks noted above to the as yet not fully known knock-on effects from changes in ocean circulation and carbon cycles (oceanic and terrestrial/biotic) that might result from the current warming trends.

Jacob Jones

Further Reading

Budyko, Mikhail I. "The Effect of Solar Radiation Variations on the Climate of the Earth." *Tellus* 5 (1969): 611–619.

Cess, Robert D., et al. "Interpretation of Cloud-Climate Feedback as Produced by 14 Atmospheric General Circulation Models." *Science* 245 (1989): 513–516.

Fleming, James Rodger. *Historical Perspectives on Climate Change.* New York: Oxford University Press, 1998.

Held, Isaac M., and Brian J. Soden. "Water Vapor Feedback and Global Warming." *Annual Review of Energy and the Environment* 25 (November 2000): 441–475.

Hoffman, Paul F., and Daniel P. Schrag. "The snowball Earth Hypothesis: Testing the Limits of Global Change." *Terra Nova* 14 (2002): 129–155.

Kueppers, Lara M., Margaret Torn, and John Hart. *Quantifying Ecosystem Feedbacks to Climate Change: Observational Needs and Priorities.* Report to the Office of Biological and Environmental Research and the Office of Science, U.S. Department of Energy. May 10, 2007, draft. http://faculty.ucmerced.edu/lkueppers/pdf/Feedbacks%20Report%20pq_10May07.pdf.

Le Treut, Hervé, Richard Somerville, et al. "Historical Overview of Climate Change Science." In *Climate Change 2007: The Physical Science Basis; Contribution of Working Group I to the Fourth Assessment Report of the Intergovernmental Panel on Climate Change,* 93–127. Cambridge. Cambridge University Press, 2007.

Lindsey, Rebecca. "Arctic Reflection: Clouds Replace Snow and Ice as Solar Reflector." NASA, Earth Observatory, January 31, 2007. http://earthobservatory.nasa.gov/Features/ArcticReflector/.

Manabe, Syukuro, and Richard T. Wetherald. "Thermal Equilibrium of the Atmosphere with a Given Distribution of Relative Humidity." *Journal of Atmospheric Science* 24 (1967): 241–259.

McGuire, A. David, F. S. Chapin III, John E. Walsh, and Christian Wirth. "Integrated Regional Changes in Arctic Climate Feedbacks: Implications for the Global Climate System." *Annual Review of Environmental Resources* 31 (2006): 61–91.

Möller, Fritz. "On the Influence of Changes in the Concentration in Air on the Radiation Balance of the Earth's Surface and on the Climate." *Journal of Geophysical Research* 68 (July 1, 1963): 3877–3886.

National Science Foundation. "Methane Releases from Arctic Shelf May Be Much Larger and Faster Than Anticipated." Press release 10-036, March 4, 2010.

Overpeck, Jonathan T., and Julia E. Cole. "Abrupt Change in Earth's Climate System." *Annual Review of Environmental Resources* 31 (2006): 1–31.

Randall, D. A., R. A. Wood, et al. "Climate Models and Their Evaluation." In *Climate Change 2007: The Physical Science Basis; Contribution of Working Group I to the Fourth Assessment Report of the Intergovernmental Panel on Climate Change.* Cambridge: Cambridge University Press, 2007.

Sample, Ian. "Why the News about Warming Is Worse Than We Thought: Feedback." *Guardian,* February 3, 2007.

Sellers, W. D. "A Global Climatic Model Based on the Energy Balance of the Earth-Atmosphere System." *Journal of Applied Meteorology* 8 (1969): 392–400.

"Snowball Earth: New Evidence Hints at Global Glaciation 716.5 Million Years Ago." *Science Daily,* March 5, 2010.

Weart, Spencer. "Simple Models of Climate Change." American Institute of Physics, The Discovery of Global Warming, 2003–2007, http://www.aip.org/history/climate/simple.htm.

Food and Nutrition Security

Climate change and variability will significantly affect food and nutrition security. Droughts and water scarcity resulting from climate change will reduce overall food security and dietary diversity, leading to increased protein-energy and/or micronutrient malnutrition. Infectious disease, which interacts with malnutrition in a vicious circle, will also proliferate. Increased flooding is likely to result in a greater number of people exposed to diarrheal and other infectious diseases, reducing nutrient absorption capacity and compromising immunity.

According to the fourth assessment report (AR4) of the Intergovernmental Panel on Climate Change (IPCC), climate change is most likely to adversely affect South Asia and sub-Saharan Africa, which already are the most vulnerable to food insecurity and malnutrition.[1] Both have had substantial amounts of agricultural land compromised or lost due to desertification, soil salinization, and other climate change–related trends. These negative effects are projected to continue and perhaps accelerate. For example, Bangladesh, which has suffered a permanent loss of land due to persistent and extreme flooding, may experience salinity intrusion of up to 60 kilometers within the next century.

Impacts on Food Security

Climate change will affect all four dimensions of food security: availability, access, stability, and utilization. Food availability is determined by domestic production, import capacity, existence of stocks, and food aid. Access depends on poverty levels, household purchasing power, food prices, the existence of transportation and market infrastructure, and distribution systems. Supply stability depends on weather, price fluctuations, man-made disasters, and other political and economic factors. The safe consumption and utilization of food depends on care and feeding practices, food safety and quality, and access to clean water, health, and sanitation and is the dimension most closely related to human nutrition.

Hot Spot

Britain

As cold-water fish abandon waters near Britain, species that usually live in southerly waters have taken their places. Sightings of warm-water fish have been plentiful since about 1990 in British coastal waters and have received considerable publicity in local newspapers. Warm-water species have been turning up regularly off the coasts of Devon and Cornwall for more than a decade; some scientists believe that this is a clear indicator of global warming. In the late 1980s, southern species such as sunfish and torpedo rays began to appear; by the late 1990s, such visitors were no longer regarded as oddities.

A series of small cold-water marine animals such as copepods were replaced by their warm-water cousins. "As fish are very dependent on the temperature of the water, it is sensible to link these changes with changes in water temperature. They would be consistent with predictions of climate change," said Douglas Herdson of the National Marine Aquarium.

Bruce E. Johansen

There are many pathways through which climate change affects food security. By reducing yields, climate change diminishes availability. Additional less direct pathways involve the demographic and socioeconomic repercussions of climate change. Direct pathways include:

- Impacts of severe weather events, gradual temperature increases, and water scarcity on plant and animal physiology;
- Increased crop and livestock diseases and pest infestations; and
- Reduced forestry, wild fisheries, and aquaculture production.

Indirect pathways include:

- Demographic shifts due to disease and migration;
- Decreased labor supply due to disease and migration;
- Political unrest due to competition for diminished resources; and
- Reduced reach of government and social services due to political unrest.

AR4 concludes that 200 million–600 million more people will suffer food insecurity in 2080 than would in a climate change–free world.

Availability

AR4 suggests that moderate increases in temperature (1–3°C), along with associated CO_2 increase and rainfall changes, can have small beneficial impacts on major rain-fed crops (e.g., maize, wheat, and rice) in mid- to high-latitude regions. However, in seasonally dry and tropical regions, even slight warming (1–2°C) reduces yield. Further warming has increasingly negative impacts in all regions. Overall, tropical and subtropical developing countries, many of which have poor land and water resources, are thus facing lower production potential due to climate change. In Africa and elsewhere, trade in cereals, livestock, and forestry products will likely increase, with most developing countries becoming increasingly dependent on imports. Exports of temperate zone food products to tropical countries will grow.

Sea-level rise as a result of climate change poses a major threat to food availability in coastlands, lagoons, and mangrove forests. Coastal areas are also threatened by temperature increases leading to changes in marine fishery distribution and community interactions. Regional changes in fish species distribution and productivity are expected due to continued warming. Climate change also poses a risk to inland fisheries and aquaculture, which are important to African and Asian food and nutrition security.

Stability and Access

Increased frequency of extreme weather events as well as climate change–related pests and diseases will affect supply stability and access. The purchasing power of smallholders and others with agricultural or fishing livelihoods may be reduced. Decreased crop yields, lower forest productivity, changes in aquatic populations, and increased costs all raise budget constraints.

Climate change and variability also contribute to food price volatility. Temperature increases of more than 3°C could increase prices by up to 40 percent. This has especially important implications in Africa, which imports 25–50 percent of its food.

Food Utilization: Care and Feeding

Climate change exacerbates many of the socioeconomic and environmental variables that can affect care and feeding practices, such as breastfeeding during the first six months of life and complementary feeding thereafter. These include increased competition for natural resources, loss of biodiversity, HIV/AIDS and other pandemics, and violent conflict, all of which affect women's

Extreme flooding, such as this experienced in Bangladesh during the monsoon season, destroys agricultural potential and hampers normal economic development. (PhotoDisc, Inc.)

time to care for their infants and young children. These stresses reduce resilience in the agricultural and related sectors, with associated welfare effects.

Food Safety and Quality

Rising temperatures increase biotoxin, chemical, and food contaminant prevalence. Aflatoxin-contaminated feed can transfer toxins to humans through milk and meat. PCB and dioxin contamination of soil and feed is associated with climate change–related extreme events, such as floods. Warmer seas may contribute to increased cases of vibrio-related outbreaks associated with shellfish consumption as well as increased levels of methyl mercury in fish and marine mammals. These increase 3–5 percent for each 1°C rise in water temperature.

Access to Clean Water

More than 2 billion people live in dry regions and suffer disproportionately from malnutrition related to contaminated or insufficient water. Other risks include exposure to waterborne diseases such as cholera and exposure to chemicals and vector-borne diseases (e.g., malaria). By 2080, climate change is projected to increase the number of water-scarce people from 1.1 billion to 3.2 billion.[2] Any increase in infectious disease has direct implications for food utilization, nutrient absorption, and malnutrition.

Health and Sanitation

Increased flooding, precipitation, rising temperatures, and other aspects of climate change are projected to increase the diarrheal disease burden in low-income regions by 2–5 percent by 2020. Higher temperatures are associated with increased episodes of diarrheal disease in adults and children in Peru, where incidence increased 8 percent per degree of temperature increase. Diarrheal foodborne diseases such as salmonellosis increase by 12 percent for each degree increase in weekly or monthly temperatures above 6°C.

The geographical range of malaria may contract in areas experiencing climate change–related drought, but elsewhere the range is expected to expand, lengthening the transmission season. Malaria has direct nutrition consequences, because infected individuals are at increased risk of anemia.

Finally, climate change is expected to increase the risk of emerging zoonotic diseases, that is, animal diseases that can be transmitted to humans. This is particularly relevant to continued increases in the global livestock population. These diseases also inhibit nutrient utilization via impaired absorption and cause increased nutrient requirements.

Nutrition Consequences

AR4 considers undernutrition linked to extreme climatic events one of the most important consequences of climate change. According to the United Nations Standing Committee on Nutrition, "Climate change affects food and nutrition security and further undermines current efforts to reduce hunger and protect and promote nutrition. Additionally, undernutrition in turn undermines the resilience to shocks and the coping mechanisms of vulnerable populations, lessening their capacities to resist and adapt to the consequences of climate change."

A 2009 study by G. C. Nelson and colleages estimates that 21 percent more children will be malnourished by 2050 than in a world without climate change.

Social Impacts of Climate Change

Although almost all human populations are or will be affected by climate change, certain demographic groups are particularly vulnerable. Most heavily rely on natural resources for food and livelihoods, for example, smallholder farmers, pastoralists, traditional societies, indigenous people, and coastal populations. These populations have extremely low incomes and adaptive capacity[3] and are often politically marginalized and are risk averse and have little or no access to social protection programs. They are also frequently concentrated in areas most exposed to climate change.

Poor rural women in developing countries who are wholly dependent on subsistence agriculture to feed their families are especially vulnerable. They frequently lack access to key productive assets and services such as land, water, rural infrastructure, credit, technology, and agricultural extension. They therefore have limited adaptive capacities and are highly dependent on climate-sensitive resources such as local water and food supplies. Climate change will likely exacerbate these gender inequalities.

Between 2010 and 2050, climate change–related factors could displace up to 200 million people. By increasing food and water scarcity, environmental degradation increases resource competition and consequent violent conflict; migration will tend to heighten this competition. In sub-Saharan Africa, where different ethnic groups often practice cropping and grazing, the advance of crops into pastureland (or vice versa) frequently results in conflict, as in the Senegal River Basin and northeastern Kenya. The Southern African Millennium Ecosystem Assessment suggests a bidirectional causal link between ecological stress and social conflict. Some analysts argue that violence in Darfur results in part from climate change and environmental degradation.

Drought and consequent livelihood losses also trigger population movements, particularly from rural to urban areas. Displacement to urban slums can mean increased competition for potable water and safe food and increased infectious disease incidence.

Adaptation and Mitigation Strategies

A revitalized twin-track approach can help address the challenges that climate change poses to food and nutrition security. This combines direct nutrition interventions with sustainable and climate-resilient and climate-smart agriculture and rural development programs, health and social protection schemes, disaster risk reduction, and community-based efforts to lessen vulnerability. Also, the global climate change agenda needs to explicitly address agriculture, food, and nutrition in order to develop integrated approaches for adaptation, mitigation, and sustainable development.

Adaptation

Adaptation strategies could be autonomous or planned. Many planned policy-based adaptations to climate change will interact with, depend on, or perhaps simply be a subset of broader policies on natural resource management, human and animal health, governance, and human rights.

Adaptation strategies for cropping systems include:

- Using heat- and drought-resistant inputs;
- Altering or reassessing fertilizer rates;
- Altering the timing of irrigation and other water management practices;
- Enhancing pest, disease, and weed management practices through integrated management approaches;
- Maintaining or improving quarantine capabilities and sentinel monitoring programmes; and
- Using seasonal climate forecasting to reduce risk.

Adaptations for livestock production include:

- Matching stocking rates with pasture production;
- Rotating pastures;
- Modifying grazing times;
- Altering forage and animal species/breeds;
- Altering mixed livestock/crop systems, including the use of adapted forage crops; and
- Ensuring adequate water supplies and using supplementary feeds and concentrates.

If widely adopted, these strategies have substantial potential to offset negative climate change impacts and take advantage of positive ones.

However, adaptation strategies have limitations. They are not gender-proofed: since these strategies build on existing systems, they may further exacerbate gender inequalities. Many women may be unable to access the benefits of adaptation. They have been largely absent in climate change and environmental management decision making. But adaptation and mitigation efforts need to take women's natural resource management knowledge into account. For example, the Clean Development Mechanism (CDM) of the Kyoto Protocol could target rural women for a range of low-emission technologies related to household energy, agricultural and food processing, forest management, and water pumping.

Adaptation strategies to protect food and nutrition security are also particularly complex and often involve trade-offs. For example, cassava production

Hot Spot

Lake Tanganyika

Two independent teams of scientists studying Central Africa's Lake Tanganyika, Africa's second-largest body of freshwater, have found a microcosm of crisis regarding global warming. The scientists have found that warming at the lake's surface has impaired the mixing of nutrients, reducing its population of fish. These reductions have affected the local economy. Fishing yields fell by a third or more during 30 years, with more declines anticipated. Heretofore, Lake Tanganyika's fish had supplied 25–40 percent of the protein consumed by neighboring peoples in parts of Burundi, Tanzania, Zambia, and the Democratic Republic of Congo.

Lake Tanganyika is a tropical body of water that experiences relatively high temperatures year-round, so the scientists were rather surprised to discover that further warming affected its nutrient balance so significantly. Like other deepwater lakes, however, Tanganyika's waters utilize temperature differences at various depths to mix water and nutrients. Such mixing is critical in tropical lakes with sharp temperature gradients that stratify layers of water with warm less-dense layers on top of nutrient-rich waters below.

"Climate warming is diminishing productivity in Lake Tanganyika," Catherine M. O'Reilly and colleagues wrote. "In parallel with regional warming patterns since the beginning of the 20th century," they continued, "a rise in surface-water temperature has increased the stability of the water column."

Bruce E. Johansen

is an attractive adaptation strategy, as the crop is hardy, easily grown, and drought resistant. But it is low in protein and can pose a threat to dietary quality if not consumed as part of a diverse diet. Planned agricultural adaptation to breed nutrient-rich food crops should thus be integrated with autonomous efforts to breed drought-resistant or water-tolerant varieties.

More evaluation is needed of the effectiveness and adoption of adaptation strategies, given the complex nature of agriculture decision making, the diversity of regional responses, and possible interactions among adaptation options and economic, institutional, human, and environmental health and cultural barriers to change.

While national adaptation policies and strategies are important, their local implementation will be the ultimate test of the effectiveness of adaptation. Crop and livestock productivity, market access, and the effects of climate change are all extremely location-specific. Community-based adaptation can help rural communities strengthen their capacity to cope with disasters, improve land-management skills, and diversify livelihoods.

The Costs of Adaptation

Even without climate change, increased public investments in agricultural science and technology are needed to meet the food demands of a world population expected to reach 9 billion by 2050. Many of these people will live in the developing world, have higher incomes, and desire a more diverse diet. It has been estimated that additional public investments of $7 billion annually in increased agricultural productivity are needed to raise calorie consumption enough to offset the negative impacts of climate change on children's health and well-being. These investments would focus on agricultural research and development, rural infrastructure, and irrigation.

Mitigation

Agriculture, forestry, and other land uses account for 30–35 percent of total greenhouse gas (GHG) emissions. However, agriculture and forestry also have huge mitigation potential.

Through 2030, the IPCC sees agriculture's mitigation potential mainly lying in improved crop and grazing land management to increase soil carbon sequestration. Additional options include degraded land restoration, improved rice cultivation, and better nitrogen fertilizer management to reduce nitrous oxide emissions.[4] Key livestock mitigation strategies include reducing emissions via improved diets and improved manure and biogas management.

Climate-smart agriculture supports sustainable agricultural activities that can potentially benefit development, food security, adaptation, and mitigation. Agricultural research can help create new technologies to facilitate agriculture-based mitigation. Research is under way to breed drought-tolerant staple crop varieties and varieties that use inputs more efficiently. New yield-enhancing technologies with fewer emissions are likely to be available before 2030.

Forest-related mitigation activities using available technologies can considerably reduce emissions and create synergies with adaptation and sustainable development. Curbing deforestation is highly cost-effective for reducing GHG emissions and can have a quick impact. Burning forests for agricultural production currently accounts for 20 percent of carbon emissions. Use of financial and market incentives to reduce burning and other deforestation is one strategy to improve forest management. New technologies are being developed to promote farming without deforestation. In Brazil, soil science breakthroughs have allowed farmers to grow crops in the low-fertility Cerrado without clearing rain forests.

Agroforestry—integration of trees and crops—is another mitigation strategy. It captures carbon, fixes nitrogen in the soil, and provides green manure, mulch, fodder, fruit, timber, fuel, medicines, and resins. In Southern Africa, tens of thousands of smallholders are planting self-fertilizing trees to increase soil fertility. Forest-based technologies likely to come on line before 2030 include improved tree species that increase biomass productivity and carbon sequestration.

Linking Adaptation, Mitigation, and Development

Sustainable development can reduce climate change vulnerability by enhancing adaptive capacity and resilience. National development plans should promote adaptation and mitigation strategies, for example, in land-use planning, infrastructure design, and disaster risk reduction schemes. Aid donors should assist developing countries in assessing capacity-building needs for devising strategies to address food security and nutrition challenges from climate change.

Marc J. Cohen, M. Cristina Tirado, Brian Thompson, Janice Meerman, and Noora-Lisa Aberman

Notes

1. Note that the IPCC assessments provide only weak information at the regional level and none on a national basis.
2. Range is due to multiple scenarios modeled by the IPCC's special report on emissions.
3. Adaptive capacity is the ability to change behavior in response to a changing climate. This can include devising and implementing solutions to protect livelihoods from negative climate change impacts or allow individuals or households to benefit from positive climate change impacts.
4. Ordinarily, nitrogen fertilizer tends to break down into nitrous oxide, a greenhouse gas that also contributes to ozone depletion and nitrate, which aids crop growth but also contaminates streams and groundwater, thereby threatening health and nutrition.

Further Reading

Biggs, R., E. Bohensky, P. V. Desanker, C. Fabricius, T. Lynam, A. Misselhorn, C. Musvoto, M. Mutale, B. Reyers, R. J. Scholes, S. Shikongo, and

A. S. van Jaarsveld. *Nature Supporting People: The Southern Africa Millennium Ecosystem Assessment.* Pretoria: Council for Scientific and Industrial Research, 2004.

Booth, S., and D. Zeller. "Mercury, Food Webs and Marine Mammals: Implications of Diet and Climate Change for Human Health." *Environmental Health Perspectives* 113(5) (2005): 521–526.

Campbell-Lendrum, D., A. Pruss-Ustun, and C. Corvalan. "How Much Disease Could Climate Change Cause?" In *Climate Change and Human Health: Risks and Responses,* edited by A. McMichael, D. Campbell-Lendrum, C. Corvalan, K. Ebi, A. Githeko, J. Scheraga, and A. Woodward, 133–159. Geneva: World Health Organization/World Meteorological Organization/UN Environment Programme, 2003.

Checkley, W., L. D. Epstein, R. H. Gilman, D. Figueroa, R. I. Cama, J. A. Patz, and R. E. Black. "Effects of El Niño and Ambient Temperature on Hospital Admissions for Diarrhoeal Diseases in Peruvian Children." *Lancet* 355 (2000): 442–450.

Clements, R. *The Economic Cost of Climate Change in Africa.* Warwickshire: Practical Action Consulting, 2009.

"Climate Change and Nutrition Security: Message to the UNFCCC Negotiators." United Nations System Standing Committee on Nutrition, 2010, http://www.unscn.org/files/Statements/Bdef_NutCC_2311_final.pdf.

Cline, W. R. *Global Warming and Agriculture: Impact Estimates by Country.* Washington, DC: Center for Global Development and Peterson Institute for International Economics, 2007.

Confalonieri, U., B. Menne, R. Akhtar, K. L. Ebi, M. Hauengue, R. S. Kovats, B. Revich, and A. Woodward. "Human Health." In *Climate Change 2007: Impacts, Adaptation and Vulnerability; Contribution of Working Group II to the Fourth Assessment Report of the Intergovernmental Panel on Climate Change,* edited by M. L. Parry, O. F. Canziani, J. P. Palutikof, P. J. van der Linden, and C. E. Hanson, 391–431. Cambridge: Cambridge University Press, 2007.

Consultative Group on International Agricultural Research. *Global Climate Change: Can Agriculture Cope?* Washington, DC: Consultative Group on International Agricultural Research, 2008. http://www.cgiar.org/impact/global/climate.html.

Easterling, W. E., P. K. Aggarwal, P. Batima, K. M. Brander, L. Erda, S. M. Howden, A. Kirilenko, J. Morton, J-F. Soussana, J. Schmidhuber, and F. N. Tubiello. "Food, Fibre and Forest Products." In *Climate Change*

2007: Impacts, Adaptation, and Vulnerability; Contribution of Working Group II to the Fourth Assessment Report of the Intergovernmental Panel on Climate Change, edited by M. L. Parry, O. F. Canziani, J. P. Palutikof, P. J. van der Linden, and C. E. Hanson, 273–314. Cambridge: Cambridge University Press, 2007.

FAO. *Agriculture, Forestry and Other Land Use Mitigation Project Database: An Assessment of the Current Status of Land-based Sectors in the Carbon Markets.* Rome: FAO, 2010.

FAO. "Expert Meeting on Climate-Related Transboundary Pests and Diseases Including Relevant Aquatic Species, Food and Agriculture Organization of the United Nations, 25–27 February 2008, Options for Decision Makers." http://www.fao.org/fileadmin/user_upload/foodclimate/presentations/diseases/OptionsEM3.pdf.

FAO. *FAO Profile for Climate Change.* Rome: FAO, 2009.

FAO. *Future Climate Change and Regional Fisheries: A Collaborative Analysis.* FAO Fisheries Technical Paper No. 452. Rome: FAO, 2003.

FAO. *Report of the Twenty-Fourth FAO Regional Conference for Africa.* Bamako: FAO, 2006.

FAO. *The State of Food and Agriculture, 2008.* Rome: FAO, 2008.

FAO. *The State of World Fisheries and Aquaculture, 2010.* Rome: FAO, 2010.

Fischer, G., M. Shah, F. N. Tubiello, and H. Van Velthuizen. "Integrated Assessment of Global Crop Production." *Philosophical Transactions of the Royal Society B* 360 (2005): 2067–2083.

Hitz, S., and J. Smith. "Estimating Global Impacts from Climate Change." *Global Environmental Change* 14(3) (2004): 201–218.

Katona, P., and J. Katona-Apte. "The Interaction between Nutrition and Infection." *Clinical Infectious Diseases* 46(10) (2008): 1582–1588.

Kovats, R. S., S. Edwards, S. Hajat, B. Armstrong, K. L. Ebi, and B. Menne. "The Effect of Temperature on Food Poisoning: Time Series Analysis in 10 European Countries." *Epidemiology and Infection* 132(3) (2004): 443–453.

Laczko, F., and C. Aghazarm, eds. *Migration, Environment, and Climate Change: Assessing the Evidence.* Geneva: International Organization for Migration, 2009.

LEAD [Livestock, Environment and Development LEAD Initiative]. *Livestock's Long Shadow: Environmental Issues and Options.* Rome: FAO, 2007. http://www.fao.org/docrep/010/a0701e/a0701e00.htm.

McMichael, A., D. Campbell-Lendrum, S. Kovats, S. Edwards, P. Wilkinson, T. Wilson, R. Nicholls, S. Hales, F. Tanser, D. Le Sueur, M. Schlesinger, and N. Andronova. "Global Climate Change." In *Comparative Quantification of Health Risks: Global and Regional Burden of Disease due to Selected Major Risk Factors,* Vol. 2, edited by M. Ezzati, A. Lopez, A. Rodgers, and C. Murray, 1543–1649. Geneva: World Health Organization, 2004.

Messer, E. "Climate Change and Violent Conflict: A Critical Literature Review." Oxfam America Research Backgrounder Series. Boston: Oxfam America, 2010. http://www.oxfamamerica.org/publications/climate-change-and-violent-conflict.

Metz, B., O. R. Davidson, P. R. Bosch, R. Dave, and L. A. Meyer, eds. *Climate Change 2007: Mitigation; Contribution of Working Group III to the Fourth Assessment Report of the Intergovernmental Panel on Climate Change.* Cambridge: Cambridge University Press, 2007.

Nelson, G. C., M. W. Rosegrant, J. Koo, R. Robertson, T. Sulser, T. Zhu, C. Ringler, S. Msangi, A. Palazzo, M. Batka, M. Magalhaes, R. Valmonte-Santos, M. Ewing, and D. Lee. *Climate Change Impact on Agriculture and Costs of Adaptation.* Washington, DC: International Food Policy Research Institute, 2009.

Nicholls, R. J., P. P. Wong, V. R. Burkett, J. O. Codignotto, J. E. Hay, R. F. McLean, S. Ragoonaden, and C. D. Woodroffe. "Coastal Systems and Low-lying Areas." In *Climate Change 2007: Impacts, Adaptation and Vulnerability; Contribution of Working Group II to the Fourth Assessment Report of the Intergovernmental Panel on Climate Change,* edited by M. L. Parry, O. F. Canziani, J. P. Palutikof, P. J. van der Linden, and C. E. Hanson, 315–356. Cambridge: Cambridge University Press, 2007.

Nori, M., J. Switzer, and A. Crawford. "Herding on the Brink: Towards a Global Survey of Pastoral Communities and Conflict." International Institute for Sustainable Development, 2005, http://www.iisd.org/pdf/2005/security_herding_on_brink.pdf.

OECD [Organisation for Economic Co-operation and Development]. *Bridge over Troubled Waters: Linking Climate Change and Development.* Paris: OECD, 2005.

Pachauri, R. K., and A. Reisinger, eds. *Climate Change 2007: Synthesis Report; Contribution of Working Groups I, II and III to the Fourth Assessment Report of the Intergovernmental Panel on Climate Change.* Geneva, Switzerland, 2007.

Parikh, J. K., and F. Denton. “Gender and Climate Change.” Climate Research Institute, 2002, http://www.cru.uea.ac.uk/tiempo/floor0/recent/issue47/t47a7.htm.

Parry, M., C. Rosenzweig, and M. Livermore. “Climate Change, Global Food Supply and Risk of Hunger.” *Philosophical Transactions of the Royal Society B* 360 (2005): 2125–2138.

Stern, N. *Stern Review on the Economics of Climate Change.* Cambridge: Cambridge University Press, 2006. http://www.hm-treasury.gov.uk/independent_reviews/stern_review_economics_climate_change/stern_review_report.cfm.

Umlauf, G., G. Bidoglio, E. H. Christoph, J. Kampheus, F. Krüger, D. Landmann, A. J. Schulz, R. Schwartz, K. Severin, B. Stachel, and D. Stech. “The Situation of PCDD/Fs and Dioxin-like PCBs after the Flooding of River Elbe and Mulde in 2002.” *Acta Hydrochimica et Hydrobiologica* 33(5) (2005): 543–554.

UNEP [United Nations Environment Programme]. *Sudan Post-Conflict Environmental Assessment.* Nairobi: United Nations Environment Programme, 2007.

Von Braun, J. *The World Food Situation: New Driving Forces and Required Actions.* Washington, DC: International Food Policy Research Institute, 2007. http://www.ifpri.org/pubs/fpr/pr18.pdf.

World Health Organization. *Ecosystems and Human Well-being: Health Synthesis; A Report of the Millennium Ecosystem Assessment.* Geneva: World Health Organization, 2005.

World Bank, FAO, and IFAD. *Gender in Agriculture Sourcebook.* Washington, DC: World Bank, 2008.

Yohe, G. W., R. D. Lasco, Q. K. Ahmad, N. W. Arnell, S. J. Cohen, C. Hope, A. C. Janetos, and R. T. Perez. “Perspectives on Climate Change and Sustainability.” In *Climate Change 2007: Impacts, Adaptation and Vulnerability; Contribution of Working Group II to the Fourth Assessment Report of the Intergovernmental Panel on Climate Change,* edited by M. L. Parry, O. F. Canziani, J. P. Palutikof, P. J. van der Linden, and C. E. Hanson, 811–841. Cambridge: Cambridge University Press, 2007.

Zimmerman, M., A. DePaola, J. C. Bowers, J. A. Krantz, J. L. Nordstrom, C. N. Johnson, and D. J. Grimes. “Variability of Total and Pathogenic Vibrio Parahaemolyticus Densities in Northern Gulf of Mexico Water and Oysters.” *Applied and Environmental Microbiology* 73(23) (2007): 7589–7596.

Forest Conservation, U.S.

Conservation is one of the easiest environmental concepts to grasp. It is based, of course, on the premise that it is not in humans' best interest to extinguish or even to deplete the resources on which they rely. This reality, though, went against some of the basic principles on which the United States was established and therefore was not automatically accepted. Before American policy could contend with complex issues such as climate change, the government needed to devise laws and regulations for ensuring conservation.

Colonial settlement of North America reinforced the concept that the continent teemed with such bountiful supplies of natural resources that it could never be depleted. With a fairly small population on the continent, little occurred that would dislodge this perception. After 1850, though, industrialization expanded the ability of Americans to harvest resources such as forests. In addition, economic success helped to create a wealthy class of Americans who began to think in alternative ways about America's natural resources. Therefore, an active debate began in the late 1800s over what responsibilities the federal government should take over managing the consumption of natural resources, such as North America's remarkable supply of trees.

Chopping Trees on the Plains and Arbor Day

On the Great Plains, where there were very few trees, the idea of development and conservation often faced off. An important fuel and building material, lumber played a crucial role in American development after 1850. With each press westward of European settlement, the supply of timber grew. The need for lumber, of course, grew as well. The boom in building, particularly in the American West, increased lumber development through 1900. For instance, Oregon doubled its population between 1890 and 1910. In the Pacific Northwest, timber production in the 1880s increased to 2.6 billion board feet per year. By 1910, Oregon and Washington ranked in the top three lumber producing states in the United States.

Forests, of course, could not replenish themselves as quickly as the trees were felled. In fact, clear-cutting often left forests with no ability to regenerate. This observation as well as timber shortages throughout Europe alerted the public to the need to conserve forests as early as the 1870s. In an effort to please all concerned, Congress passed the Timber Culture Act in 1873, which promoted tree planting as well as western settlement. This effort was partly spurred by the scientific theory of Joseph Henry of the Smithsonian Institution, who argued that planting trees would stimulate rainfall in the

Gifford Pinchot, photographed at his desk ca. 1945, was the first trained forester in the U.S. He oversaw policies emphasizing forest conservation and "sustained yield" management practices. (Library of Congress)

arid West. Clearly, settlers thought that they could fell the forests and still have sufficient a supply of timber to support the nation's future.

With Henry's theory in mind, Nebraskans in 1872 even went as far as implementing the first celebration of Arbor Day, a holiday intended for tree planting. Many other western states followed. By 1907, however, Arbor Day was observed by all of the United States. Although the belts of trees did not bring additional rain to the West, they did help with settlement by breaking wind and providing shade.

Conserving American Forests

The concept of conservation in the United States began in reference to forest reserves, which trace their origin to an 1876 act of Congress that created the Forest Service. Based on the lessons of European nations, this act authorized the Department of Agriculture to hire a forestry agent who would investigate the present and future supply of timber. In addition, the agent studied the methods employed by other nations attempting to manage or conserve timber supplies. Although federal forest lands were not set aside until 15 years later, the Division of Forestry was set up in 1881. The first head of the agency was a Prussian-educated forester named Bernhard E. Hough. Compiling many reports about American forest use past and present, Hough was very critical of American patterns of use. He urged the nation to adopt regulations to "secure an economical use."

Although the division began as primarily a disseminator of information, by the 1890s it provided forestry assistance to states and to private forest landowners. In 1891, Congress granted the president the authority to establish forest reserves from the existing public domain lands under the jurisdiction of the Interior Department. The Organic Administration Act of 1897 stipulated that forest reserves were intended "to improve and protect the forest within the reservation, or for the purpose of securing favorable water flows, and to furnish a continuous supply of timber for the use and necessities of the citizens of the United States." By the early 1900s, this office administered approximately 56 million acres. However, the agency was about to take a leading role in the broader American conservation movement.

Entrusting Forests to the Federal Forest Service

This interest in forest conservation arrived at the federal government during the 1870s. The formation of the Division of Forestry within the Department of Agriculture during this decade proved to be a fairly insignificant political step. A corrupt federal bureaucracy overwhelmed any bona fide efforts at conservation. Instead, the division was used to more readily sell off timbered tracts of land. Historians have noted that the Timber Culture Act allowed for 20 more years of unregulated clear-cutting. It was this intensification, though, that stirred public cries for reform.

When Bernard Fernow became the chief of forestry in 1886, he brought with him a commitment to the scientific management of American forests. Although the division continued its primary effort at disseminating public information, it soon added scientific experimentation to the division's responsibilities. President Benjamin Harrison established the first timber land reserve on March 30, 1891, and placed it under the control of the General Land Office rather than the Division of Forestry. The president now had the authority to set aside public land in the West. By 1897, 40 million acres in the northern Rocky Mountains had been set aside. In this year, Congress moved the national forests into the Department of the Interior's General Land Office.

Hot Spot

Western North America

Mortality rates have risen substantially even in previously healthy conifer forests of the western United States averaging more than 200 years of age, doubling during two to three decades as new trees often fail to replace those that die. Drought provoked by rising temperatures is a major reason for rising death rates of pine, fir, hemlock, and other trees, according to a study released early in 2009. The increasing mortality of trees also reduces the ability of the forests to absorb carbon dioxide.

"Summers are getting longer," said Nathan L. Stephenson of the U.S. Geological Survey (USGS), a coauthor of the study with Phillip van Mantgem, also of the USGS. "Trees are under more drought stress." The recent warming in the West "has contributed to widespread hydrologic changes, such as a declining fraction of precipitation falling as snow, declining water snow pack content, earlier spring snowmelt and runoff, and a consequent lengthening of the summer drought," the scientists wrote. "It's very likely that mortality rates will continue to rise," said Stephenson, adding that the death of older trees is rapidly exceeding the growth of new ones, analogous to a human community where the deaths of old people surpass the number of babies being born. "If you saw that going on in your hometown, you'd be concerned."

Bruce E. Johansen

Many of Fernow's scientific ideas were incorporated into the national forests by the first American trained in forestry, Gifford Pinchot. Pinchot replaced Fernow as director in 1898 and set out to force private timber firms to undertake cooperative forestry management on the reserves and reforestation programs on private lands.

Gifford Pinchot, America's First Forester

Forest conservation was the business of Gifford Pinchot, America's first native professionally trained forester. He had returned from forestry school in Germany to establish one of the first U.S. model forests, which lay on the Biltmore estate of Cornelius Vanderbilt in North Carolina. Pinchot took over administration of the national forests in 1898 and became the first chief of the renamed Forest Service in 1905, when the agency took over the forest reserves from the Department of the Interior's General Land Office. The agency was renamed the U.S. Forest Service, and ultimately forest reserves were renamed national forests. Pinchot voiced a refrain of the conservation movement when he instructed Secretary of Agriculture James Wilson that the goal of forest administration needed to be based on "the greatest good for the greatest number in the long run."

Pinchot's real fame came when his close friend Theodore Roosevelt became president in 1901. Together, they oriented the Forest Service toward the wise use of timber resources. This guideline was the forerunner of the multiple-use and sustained-yield principles that have guided forest management in recent years. These principles stress the need to balance the uses that are made of the major resources and benefits of the forests—timber, water supplies, recreation, livestock forage, wildlife and fish, and minerals—in the best public interest.

With Roosevelt's support, the 1905 act transferred the Forestry Division from the Department of the Interior to the Department of Agriculture and combined it with the larger Bureau of Forestry. Roosevelt and Pinchot more than doubled the forest reserve acreage in the two years following the merger, to a total of 151 million acres by 1907. This total included 16 million acres that were squeezed through after Congress placed limits on the president's authority to proclaim additional forest reserves. Without congressional approval, Roosevelt designated these areas in defiance of Congress. When he left office in 1909, there were 195 million acres of National Forest.

Pinchot organized the Forest Service quickly. In 1907 the reserves were renamed National Forests. During the following year, six district or regional offices were organized in the West for administering fieldwork. These National Forests, however, considered their priority to be ensuring that the nation would maintain an abundant timber supply in perpetuity. This conservation mandate concerned chopping wood and making roads and knew nothing of wilderness or preservation. National forests were similar to vaults holding a natural resource of vital importance to the nation and had little similarity to national parks.

The following is an excerpt from Pinchot's *A Primer of Forestry:*

THE SERVICE OF THE FOREST.

Next to the earth itself the forest is the most useful servant of man. Not only does it sustain and regulate the streams, moderate the winds, and beautify the land, but it also supplies wood, the most widely used of all materials. Its uses are numberless, and the demands which are made upon it by mankind are numberless also. It is essential to the well-being of mankind that these demands should be met. They must be met steadily, fully, and at the right time if the forest is to give its best service. The object of practical forestry is precisely to make the forest render its best service to man in such a way as to increase rather than to diminish its usefulness in the future. Forest management and conservative lumbering are other names for practical forestry. Under whatever name it may be known, practical forestry means both the use and preservation of the forest. . . .

FEDERAL FOREST RESERVES.

When the President was given the power to make forest reserves, the public domain still contained much of the best timber in the West, but it was passing rapidly into private hands. Acting upon the wise principle that forests whose preservation is necessary for the general welfare should remain in Government control, President Harrison created the first forest reserves. President Cleveland followed his example. But there was yet no systematic plan for the making or management of the reserves, which at that time were altogether without protection by the Government. Toward the end of President Cleveland's second Administration, therefore, the National Academy of Sciences was asked to appoint a commission to examine the national forest lands and report a plan for their control. The academy did so, and upon the recommendation of the National Forest Commission so appointed, President Cleveland doubled the reserved area by setting aside 13 additional forest reserves on Washington's Birthday, 1897.

The Cleveland forest reserves awakened at once great opposition in Congress and throughout the West, and led to a general discussion of the forest policy. But after several years of controversy widespread approval took the place of opposition, and at present the value of the forest reserves is rarely disputed, except by private interests impatient of restraint.

The recommendations of the National Forest Commission for the management of the forest reserves were not acted upon by Congress, but the law of June 4, 1897, gave the Secretary of the Interior authority to protect the reserves and make them useful. The passage of this law was the first step toward a national forest service. . . .

The forest reserves lie chiefly in high mountain regions. They are 62 in number, and cover an area (January 1, 1905) of 63,308,319 acres.

> They are useful first of all to protect the drainage basins of streams used for irrigation, and especially the watersheds of the great irrigation works which the Government is constructing under the reclamation law, which was passed in 1902. This is their most important use. Secondly, they supply grass and other forage for many thousands of grazing animals during the summer, when the lower ranges on the plains and deserts are barren and dry. Lastly, they furnish a permanent supply of wood for the use of settlers, miners, lumbermen, and other citizens. This is at present the least important use of the reserves, but it will be of greater consequence hereafter. . . .

Educating Foresters in the 20th Century

By the model set by the new Forest Service, Americans learned to view forests very differently than they once had. Pinchot made sure that applicants to work for the Forest Service needed to demonstrate a mix of formal education and practical aptitude. These new types of foresters, he hoped, would approach old problems differently. Primarily, science would now be included in efforts to solve some of the service's most pressing problems, including fires, overgrazing by cattle and sheep, soil disturbance and stream pollution caused by mining, and insect and disease impacts.

In 1908, the Forest Service established its first experimental station near Flagstaff, Arizona. Other stations were later added throughout the West and eventually in other regions. The Forest Service also established its laboratory to experiment with new products that could be manufactured from forests while also establishing the framework that would allow states to use the proceeds from any products made from their national forests to be used locally for new schools and other services. With science at its core, the Forest Service initiated some of the nation's first efforts to set aside wilderness and primitive areas. The most famous of these efforts was the Gila Wilderness in New Mexico, which was the first primitive area in the service set aside in 1930.

Forest research got a big boost in 1928 through the McSweeney-McNary Act, which authorized a broad permanent program of research and the first comprehensive nationwide survey of forest resources on all public and private lands. Tree planting in national forests was expanded under the Knutson-Vandenberg Act of 1930. The Forest Service also operated more than 1,300 Civilian Conservation Corps (CCC) camps in national forests during the 1930s New Deal. More than 2 million unemployed young men in the CCC program performed a vast amount of forest protection, watershed restoration, erosion control, and other improvement work, including the planting of 2.25 billion tree seedlings. Although the administration of forests would shift with the political wind, a core ethic had crept into the Forest Service that emphasized conservation of lumber supplies and, at times, administration of national forests as complex ecological sites. Conserving the nation's forests still continued to have controversial aspects.

Fire on Public Lands: To Burn or Not?

Forests do not exist like money in a vault, they continue to be part of natural systems. As such, forest conservation policy has been tested consistently, particularly by part of every forest's natural cycle: fire. According to historian Stephen Pyne,

> We hold a species monopoly over fire. With fire we claim a unique ecological niche: this is what we do that no other creature does. Our possession is so fundamental to our understanding of the world that we cannot imagine a world without fire in our hands. Or to restate that point in more evolutionary terms, we cannot imagine another creature possessing it.
>
> Yet while humans come genetically equipped to manipulate fire, we do not come programmed in its use. We apply and withhold it according to social institutions, cultural norms, perceptions of how we see ourselves in nature. Different people have created distinctive fire regimes, as they have distinctive literatures and architectures. In this way fire became both natural and cultural. If fire measures our ecological agency, so how we choose to apply and withhold it testifies to our understanding of that agency and the values, choices, and means by which we act. Fire enters humanity's moral universe, and thus into the scholarly realm of the humanities.

What is a forest fire? A necessary part of the natural processes of a forest? A scourge endangering anyone near a forest? A waste of important resources? Part of the American fire regime has been to interpret fire's meaning, particularly for places such as federal lands, including national parks. Changes in the definition of fire for such land elicits dramatic debate from many Americans, particularly residents living near such locales.

In the late 19th century, fires burned throughout the United States but particularly in the West. Lightning—and therefore nature—ignited some of these fires, but more and more were caused by human means, including the use of steam-powered trains throughout the countryside that left burning ash in their wake and settlers burning field fallow for pasturage. Although seasonal, fire occurrence was tolerated. In fact, in many areas it was thought that lands were uninhabitable without it. By the late 1800s, progressive conservationists argued for government intervention to put a stop to the burning.

During the 20th century, park administrators frequently debated the question of whether to burn or not to burn—in short, whether to allow fires in such locales to burn themselves, thereby seeing fire as an organic portion of nature, or to extinguish the fire and thereby perceive it as an artificial interloper in otherwise natural scenes. Pyne writes that

> The problem with fires became one of maldistribution—too much of the wrong kind of fire, not enough of the right kind. Most fires burned as wildfires, set by lightning, accident, or arson. Many lands suffered from a fire famine, the shock of having a process to which they had long adjusted abruptly removed. On many sites, natural fuels had ratcheted up to levels against which fire suppression stood helpless. Government fire agencies sought to reinstate fire, often at considerable cost and risk, even in the face of public skepticism.

Yellowstone has functioned as a case study for the changing approaches to fire. The nature that had been preserved in the national park was at least partly a product of fires, large-scale conflagrations sweeping across the park's vast volcanic plateaus, hot wind-driven fires torching up the trunks to the crowns of the pine and fir trees at intervals of several hundred years. Park officials did not choose to view fire in this way. However, by the 1940s, ecologists recognized that fire was a primary agent of change in many ecosystems, including the arid mountainous western United States.

Controlled burns were used in Yellowstone, and by the 1970s the park had implemented a natural fire management plan that allowed the process of lightning-caused fire to be allowed to continue if they started. In the first 16 years of Yellowstone's natural fire policy (1972–1987), 235 fires were allowed to burn 33,759 acres. Only 15 of those fires were larger than 100 acres, and all of the fires were extinguished naturally.

The incredible destruction of fires in 1988, though, changed everything. By July 15, only 8,500 acres had burned in the entire greater Yellowstone area. The fires did not become noticeable to visitors until the end of July, when they grew to engulf nearly 99,000 acres. In August, fires reached across more than 150,000 acres.

Great public debate followed. A symbol of preservation, Yellowstone was now a symbol of the ecology of fire. National Park Service studies showed that 248 fires started in greater Yellowstone in 1988; 50 of those were in Yellowstone National Park. Only 31 of them were allowed to burn. Fighting the fires in the region was estimated to have cost $120 million and to have involved a total of 25,000 firefighters. Ecosystem-wide, about 1.2 million acres were scorched; 793,000 (about 36 percent) of the park's 2,221,800 acres were burned. Subsequent studies, though, have also demonstrated nature's remarkable ability to recover.

Therefore, national parks and forests across the United States suspended and updated their fire management plans. The 1992 Yellowstone revised plan continued a wildland fire management plan but with stricter guidelines under which naturally occurring fires may be allowed to burn. Through continued public education, scientific research, and professional fire management, the U.S. government hopes to preserve the process of natural fire in Yellowstone while minimizing adverse effects on park visitors and neighbors, recognizing the inevitability of this force to continue shaping the landscape as it has for centuries.

Throughout the United States, after a terrible fire season in 1994, a revised federal policy emerged in December 1995 that was organized by controlled burning, which could help thin forests of deadwood and thereby eliminate excessive fuel for naturally caused fires. Under this new approach to fire, the National Park Service attempted burns and lost control of two in 2000. As a result, the secretary of the interior placed the National Park Service program under a moratorium.

Currently, federal administrators extinguish fires that threaten facilities and surrounding homes and businesses. Busy fire seasons in the far West in 2006–2007, however, have made each side of the fire argument flare up. In addition, scientists expect one of the outcomes of climate change to be an increase in drought in arid regions and, commensurately, a rise in forest fires.

Brian C. Black

Further Reading

Hays, Samuel P. *Conservation and the Gospel of Efficiency.* Pittsburgh: University of Pittsburgh Press, 1999.

Hirt, Paul W. *A Conspiracy of Optimism: Management of the National Forests since World War Two.* Lincoln: University of Nebraska Press, 1994.

Miller, Char. *Gifford Pinchot and the Making of Modern Environmentalism.* New York: Shearwater Books, 2004.

Nash, Roderick. *Wilderness and the American Mind.* New Haven, CT: Yale University Press, 1982.

National Interagency Fire Center. www.nifc.gov.

Fox, Stephen. *The American Conservation Movement.* Madison: University of Wisconsin Press, 1981.

Pinchot, Gifford. *Breaking New Ground.* New York: Island Press, 1998.

Pyne, Stephen J. *Fire in America: A Cultural History of Wildland and Rural Fire.* Princeton, NJ: Princeton University Press, 1982.

Steen, Harold K. *The U.S. Forest Service.* Seattle: University of Washington Press, 1976.

Williams, Michael. *Americans and Their Forests.* New York: Cambridge University Press, 1992.

Fossil Fuels: Current and Projected

Fossil fuels such as petroleum, natural gas, and coal take millions of years to form naturally. However, they will be completely used up in just a few centuries. In fact, most of those few centuries have already occurred. At this point in time, the resources of all three fossil fuels will be depleted within the next century.

Defining the Reserve

In order to understand the nature of fossil-fuel supplies, we must first establish the difference between resources and proven reserves.

Resources are the total amount of a fossil fuel that is believed to exist. Most resources have not even been discovered yet. Even when discovered, resources are often impossible to extract using today's technology or because of current economic conditions. Thus, the term "resources" is used when it is understood that future discoveries will be made, new technologies will be developed for future exploration, and prices will continue to rise to make all this futuristic thinking profitable.

"Proven reserves" is a much more down-to-earth term. Proven reserves have already been discovered and can be developed with known technology under current economic conditions. The quantity of proven reserves of oil and natural gas is much less than that of the estimated oil and natural gas resources.

The two terms do not apply to coal. Coal is much easier to explore, discover, and exploit. The amount of coal in the United States and in the world is well known and is called the Demonstrated Reserve Base. Of this amount, 50–55 percent can be extracted, which represents the Estimated Recoverable Reserves.

At the current rate of use and without imports, the U.S. supply of coal will satisfy the country's needs for 240 years, and the proven reserves of oil and natural gas will last 3.2 years and 8.9 years, respectively. From these scientific facts, it is easy to see why we import most of our oil today. It is clear why energy companies are continuously exploring new sources of oil and gas. With the assumed continuation of these new discoveries and with new drilling technologies, at the current rate of consumption total fossil fuel resources will allow oil to last 30 years and natural gas 65 years, although prices will be considerably higher.

The reality of supply makes it easy to see why coal must play a major role in the energy future of the United States. But coal won't really last 240 years. Coal has already started to replace the other fossil fuels for some uses, and the total use of fossil fuels is expected to grow for at least several decades, according to the U.S. Department of Energy. If all fossil fuels are considered exchangeable so that coal is used to replace oil and natural gas as they are depleted, the total U.S. resources of all fossil fuels will be depleted within 65 years if the current growth in energy consumption continues.

There is a total of 41,000 quads of fossil fuel resources in the world. At the current rate of consumption, these will be gone within 100 years. But when the growth in energy consumption, especially by the developing world, is considered, all of these fossil fuel resources will be depleted in just 60 years. It is worth noting that even with the increased consumption by the developing world that has been assumed, the per-capita energy consumption of the developing world would still only be 30 percent of the per capita energy consumption of the technologically developed world. If the developed world were to use as much energy per person as the rest of the world, all fossil fuels could be gone in as little as 30 years.

The reader might assume that since most of these end-of-energy predictions are looking 60 years or more into the future, he or she will not be affected. But the supply of fossil fuel energy will not simply keep flowing and then one day be turned off. The supply of fossil fuel energy will continue to increase for a number of years, peak, and then rapidly diminish. It is safe to assume that there will be at least some amount of each fossil fuel remaining 100, 200, or even 500 years from now. But there will be very few who will be able to afford this energy. Thus, from a practical viewpoint, most people will have to stop using these fossil fuels many years before they have actually run out.

This doesn't mean that society will be thrown into the Stone Age, although it is probably a good thing for people at least to be concerned about that. This may indeed be what happens if society doesn't properly prepare for a transition by changing wasteful consumption habits and creating new energy strategies that increase energy efficiency among consumers and producers of energy. With the proper use of renewable energy and with the very efficient use, through technology, of fossil fuels, society can continue to prosper for the foreseeable future.

Sources and Sectors of Current Energy Use

There are several main sources of energy. These include petroleum, coal, natural gas, nuclear fission, and various renewable energies (biomass, hydroelectric, wind, solar, and geothermal). The energy-using sectors fall into four broad categories: transportation, industrial, residential/commercial, and electric power. Electric power is available to each of the other three, although it is currently not used much for transportation.

Some raw energy sources can only be used for very limited purposes. Nuclear fission is the most obvious case, as it can only be used commercially for generating electricity. Petroleum is the most expensive form of raw energy and is mostly used for transportation due to the convenient portability of its liquid-fuel derivatives, gasoline and diesel. Despite common belief, petroleum is not used for generating electricity except in a very few special circumstances. Coal is used mostly for generating electricity but is also used in industrial settings for the smelting of metals and other industrial uses. Most renewable energies, except biomass, are also used only for electricity, although solar-thermal energy can also be used for space and water heating. In practice, the most flexible energy sources are natural gas and biomass. These are commonly used in nearly every consumer sector for electricity, heating, transportation, and industrial uses. In theory petroleum is also very flexible, yet due to its high cost and dependence on foreign sources, it is generally reserved only for transportation.

In an equilibrium situation, most sources of energy are priced approximately the same per energy unit. This is because it is usually possible for some, but not all, consumers to switch from one energy source to another source should the prices of two sources become unbalanced. In large commercial settings, these energy sources can be switched around rather quickly. In the past there were quite a few industries that could switch between natural gas and petroleum, almost with the flick of a switch. Thus, the wholesale price of these two fuels has been approximately the same. However, now that petroleum has increased in price by so much, there are very few industries that use petroleum when natural gas, coal, or electricity would suffice.

What is missing in this discussion, however, is how the raw energy is delivered to the consumer for use. There is nearly always some intermediate step in which the raw energy is modified into some other form to be used by the consumer. Without the intermediate step, the raw energy is utterly useless. Just imagine driving up to the gas station and getting 10 gallons of crude oil. Your car would be ruined. Petroleum is used mostly for transportation, but it must first be refined into the appropriate intermediate products: gasoline, diesel, jet fuel, and so forth. But these products don't have to be made from petroleum; they can also be made from coal, natural gas, or biomass. Thus, if the price of petroleum gets to be too high, producers of gasoline can just switch to a different fuel source to make their gasoline and can save money in the process. This is precisely what has happened recently. Gasoline made from biomass is now competitive with gasoline made from petroleum.

Imagine trying to turn on your lights when all you had was a pile of coal in your cellar. Most residential energy is consumed from electricity, which in turn comes mostly from coal. But the typical homeowner doesn't really care where the energy comes from as long as the lights turn on when the switch is flipped, the temperature of the house is comfortable, and everything works 99 percent of the time without the need for costly repairs. Thus, while most electricity is generated from coal, the coal itself is useless. Indeed, the homeowner would never know the difference if that electricity was generated from some other source of energy, such as natural gas, nuclear fission, wind, solar, or hydro. The same goes for a home's heating and cooling system.

Interchangeable Sources of Energy

Transportation fuels and electricity are the two main intermediate energy forms. Each can be produced from a variety of raw sources and then used in any consumer sector.

Transportation Fuels

Transportation fuels such as gasoline and diesel can often come from multiple sources of raw energy, such as petroleum, biomass, coal, or natural gas. A petroleum refinery can make gasoline using less petroleum and more biomass (ethanol) if the price of the biomass is less than petroleum. Of course, this will then increase the demand for and thus the price of the biomass feedstock, which sends ripples through the rest of the economy. In the current case of ethanol, the biomass product is mostly corn. When the price of petroleum spiked to $100 per barrel and even higher, the demand for corn-based ethanol also spiked, since it was a cheaper alternative to petroleum. This caused the corn demand and price to spike as well. The results were seen in higher prices not only for delicious sweet corn but also for dairy and meat products (cows are fed corn) and all other grain-based foods, as other grain prices increased too.

Gasoline can also be made from coal, using a process known as coal-to-liquids (CTL) or liquefaction. CTL is an old technology developed by Germany during World War II to provide fuels when access to petroleum became difficult. Today this technology is profitable, as it is cheaper to make gasoline from coal than from petroleum when oil is over $100 per barrel. This technology is not being pursued in the United States but has been in South Africa, where the equivalent of 60 million barrels of oil per year come from local coal. China is also pursuing CTL technology. Natural gas can be converted into a liquid fuel, but this is less advantageous than CTL. Transportation can also be fueled with electric batteries, hydrogen fuel cells, and even compressed air, but each of these does require some form of raw energy.

Electric-Power Generation

Electric-power generation is the other major intermediate energy form. Electricity is also special because it is the most useful form of energy and can be used to do just about anything with up to 100 percent efficiency. Electricity can even be used to leverage the generation of heat so that one unit of electricity can generate three or four units of heat. Other sources of energy, particularly raw sources of energy, are usually much less efficient when actually applied in some situations. However, producing electricity is not without its drawbacks. Generating electricity from any raw energy source is usually very inefficient. Electricity can be generated by wind and solar at only 15–20 percent efficiency, by coal and nuclear fission at about 35 percent efficiency, and by natural gas at up to 65 percent efficiency. But efficiency is not the only criterion. The cost of the raw energy is important. Wind and sunlight are free, whereas natural gas is very expensive. Complicating this even more is the cost to build and maintain the power plant.

Since electricity can be produced from so many different raw energy sources, electric utilities are very prone to shopping around for the lowest price for wholesale electricity. If the price of natural gas rises, then utilities will purchase less electricity from that source and more from sources such as coal, wind, and nuclear. During the 1990s and early 2000s, natural gas was particularly cheap. Prior to the turmoil of the 2008–2009 global recession, natural gas rates climbed, so that the focus was now on electricity from sources such as wind and coal. Another example is that the expected movement toward some sort of carbon-trading system is pushing the economics of electricity production toward wind, solar, and nuclear, since it is assumed that coal will become more expensive to use. But regardless of what raw source of energy is used to generate electricity, a kilowatt-hour at the electrical receptacle is still a kilowatt-hour, no matter what the source.

The Competitive Energy Marketplace

Although it is technically possible for any one raw source of energy to be substituted for any other source, this is not typically done for some consumption sectors. Two examples of this are transportation and commercial/residential heating. In the case of transportation there is some flexibility, as discussed above, but to be completely flexible, we would need to be able to drive our cars, trucks, trains, ships, airplanes, and so forth on coal, nuclear, wind, solar, or any raw energy source. And while this may be technically possible, it is very impractical. Nobody wants to drive a car that resembles a 1900 coal-burning steam locomotive, nor would we want a nuclear reactor in the trunk of our car.

Electricity is the key to making the many raw energy sources exchangeable. Electricity is the most flexible energy source and can be used very efficiently for just about any use. Electricity is already used as an intermediate energy for many devices, but transportation and commercial/residential heating are two exceptions in which electricity is not generally involved. Any raw energy source can be used to generate electricity. In most cases, generating electricity as an intermediate step before the energy is finally consumed is more efficient than using some raw energy source directly for consumption.

Very few homes or businesses in heating-dominated climates are heated with electricity. However, with today's prices for heating oil, natural gas, and propane, using technologies powered by electricity or solar is the cheapest and most efficient way to heat homes, businesses, and water. By using electricity or solar for this type of heating, the raw sources of natural gas and petroleum may also be changed to the raw sources of coal, nuclear, and various renewable sources. But it takes time to make these changes. We can't just wait until petroleum is gone and expect to flip a switch to make the necessary change. The technologies that must be employed are air-source heat pumps, ground-source heat pumps, and solar collection. Although these technologies will use less energy and have an overall lower cost over the long-term, they are also costly to install, and thus many people are reluctant to choose them. In fact, many consumers don't even know that these are the cheapest ways of heating a structure or that they can be used in practically any climate. Newer technologies also have fewer qualified technicians who can design, build, or install them. These factors combine to add a delay in the switch from one raw energy source to another, even when the newer energy source is cheaper and more efficient.

The delay is also true for the transportation sector. It takes time for newer technologies to break into the market. When gasoline prices hovered around three to four dollars per gallon in 2008, it would have been much cheaper to run vehicles on natural gas, electricity from a variety of raw sources, or even on liquefied coal fuels. It would take time, however, for such alternative-fueled vehicles to be designed, manufactured, and placed into service across the United States. For natural gas–powered vehicles, a large network of refueling stations would need to be built, in addition to the mass production of the vehicles themselves. For electric vehicles to be a reality, the production of batteries needed for electric vehicles needs to be ramped up, but this type of sudden large-scale production is difficult for any emerging product. Finally, the use of liquefied coal requires large production facilities to be built to convert solid coal into liquid fuels that can be refined into gasoline. All of these technologies are being pursued in other nations to a much larger degree than in the United States.

As certain fossil fuels are used up either locally or globally, the consumers of that raw energy will need to switch to some other raw energy source. This requires full competition between the various raw energy sources for all consumer sectors. In order for this to happen, it is necessary that all such sources feed into the same energy pool from which all the various energy consumers will draw. This is essentially the purpose of the national electric gird or, alternatively, a hydrogen-based economy.

In a hydrogen-based energy economy, some form of raw energy would need to be used to produce this hydrogen, which could then be piped around the country in a network of hydrogen pipelines. The main advantage of using hydrogen is that energy in the form of hydrogen can be stored during times of low demand and high production and then used during times of high demand and low production. An example of the benefit of storage is that solar energy could then be stored for use at night. The shortcomings of a hydrogen-based energy economy is that hydrogen is currently much too expensive to produce, store, and convert back into useful energy.

If the national electric grid were significantly updated to have a much larger capacity, this would also provide the infrastructure for all the raw energy sources to compete with each other. The infrastructure for an electricity-based energy economy is much closer to the currently existing infrastructure than is the infrastructure for a hydrogen-based energy economy. Thus, the cost of an expanded electric grid is modest. Such large electricity-based networks would also increase the impact of renewable energies such as wind and solar because it is likely that the wind would be blowing somewhere in the country at any given time, producing energy that would be available to someone anywhere in the country. It also allows solar energy to be collected in regions with plentiful sunshine and then consumed in regions with less sunshine. The shortcomings of an expanded electric grid is that it does not offer any inherent energy storage.

Shifting infrastructure for use by other resources is extremely costly and difficult. One difficulty is that of unintended consequences. Consider CTL technology, which allows coal to compete with petroleum in the transportation sector. When the price for petroleum exceeds $100 per barrel, making synthetic gasoline from coal becomes cheaper than making gasoline from petroleum. But to use CTL technology, a huge investment must first be made to build one or many CTL plants at a cost of billions of dollars. In the energy industry this type of investment is not unheard of, and many energy companies can readily fund this type of investment when they choose to do so. However, it will take years for the investment to pay off with the profits of the synthetic gasoline. When a CTL plant is built, the overall supply of gasoline will increase, and thus by simple supply and demand economics the price of petroleum will decrease. This in turn decreases the price of traditional gasoline. At the same time, the CTL plant will also increase the demand for coal, causing the price of coal and therefore the price of synthetic gasoline to rise. These two factors together make the CTL technology less competitive with petroleum. In fact, if the price swings are large enough, they can cause the CTL plant to lose money and go bankrupt, even though it was viable before it started production.

Such a case of new competing technologies going bankrupt is not just a theoretical scenario. Many alternative energy companies went bankrupt because of this situation, and in fact there are concerns that many of the ethanol plants that sprang up in the early 21st century will go bankrupt due to the increased price of biomass feedstock and the falling price of petroleum in late 2008.

Raw Energy on the International Market

All fossil fuels can be exported and imported on the international market. But petroleum is by far the fuel that is traded the most. This is because the qualities that make petroleum an ideal transportation fuel on land also make it an ideal fuel to transport via ship. Being a liquid, petroleum can easily be loaded and unloaded on a ship with pumps, and it is a very energy-dense fuel. Although coal can also be loaded and unloaded, it is not quite as easy to do, and coal is not nearly as energy-dense as petroleum. It would take a much larger ship to transport the same amount of energy in the form of coal as in the form of petroleum.

Natural gas can also be imported via ship, but it is expensive. The natural gas must be cooled to cryogenic temperatures and carefully loaded and unloaded. There are relatively few locations that have the ability to handle imports or exports of natural gas. But due to the high demand for this relatively clean fuel, the international market for natural gas is increasing.

The United States currently imports about 30 percent of its total raw energy. Nearly all of that is crude oil, and a small but growing amount is natural gas. The United States cannot produce enough petroleum for its own needs, and it never will be able to supply its own needs even if there were a massive expansion of domestic drilling. The reason for these imports is that oil is the raw energy source that is the hardest to replace with some other energy source, due to transportation's need for gasoline. If transportation could be fueled by some other energy source, such as electricity, then the vast quantity of this imported oil would no longer be needed.

There is also the volatility of energy prices that we have come to know all too well. Energy prices routinely go up and down. This is caused not only by supply and demand but also by external factors such as economic strength and weather. The internal effects of supply and demand tend to be long-term effects and are very substantial when global demand nears peak global supply. From the demand side, as prices go up, consumers will change their energy consumption habits. This may be by purchasing a more fuel-efficient vehicle, driving less, carpooling, adding insulation to their homes, buying more efficient appliances, and so forth. Not everyone will do this, but enough people will so that the nation's demand for energy will decrease, and therefore energy prices will decrease as well or at least not increase as much. There is always a delay between the prices going up and the demand going down, because it takes time for enough individuals to change their consumption. There was a fine example of this during the spring and summer of 2008: as gasoline prices escalated, the sales of small cars went up, the sales of trucks went down, and the actual consumption of gasoline decreased. In the autumn of 2008 the external effect of world economic strength took over, and prices plummeted due to lower energy demand caused by the global recession.

From the supply side, when prices are high, new technologies for producing renewable energy or for tapping harder-to-reach fossil fuels are suddenly profitable and worthwhile. But it takes some time to get these new sources of energy into the market. Once in the market, the supply of energy will increase and the price will fall, all other things being equal (such as zero inflation and zero-demand growth). A very serious problem arises as a result of this. If the price falls too much, then some of those new technologies will no longer be profitable, and the companies involved could face ruin if too much money had been invested. Because of this, many new technologies are not pursued until they are very profitable, not just barely profitable.

The external effects on energy prices include weather, the strength of the U.S. and global economies, and the strength of the U.S. dollar and investment markets. Weather may be unusually warm or cold, which has a temporary impact on the demand for natural gas needed to produce electricity for air-conditioning or heating and on the demand for petroleum for home heating oil. Severe weather, particularly in the Gulf of Mexico, can have an impact on the supply of natural gas and petroleum from that region as well as on refining petroleum into gasoline and home heating oil. Weather effects will usually not impact prices for longer than a few months.

The major external influence on the price of energy is the strength of the regional and global economy. The first reason for this is that when the economy of the United States or the world is strong and growing rapidly, lots of energy will be needed in order to produce the products that make up the expanding gross domestic product of the nation or the world. Thus, the demand for energy will be high whenever the economy is prospering, and the demand will likewise be low whenever the economy is weak or even in recession. This happened in the autumn of 2008 to a very large degree. The largest and fastest drop in oil prices resulted from the sudden slowdown of the global economy.

Closely related to this is the fact that worldwide, petroleum is priced in U.S. dollars and is traded on the open market. Because oil is priced in U.S. dollars, whenever the U.S. economy is weak compared to the rest of the world, the value of the U.S. dollar will fall in currency exchanges. This makes the price of petroleum and other globally traded energy sources rise due just to the value of the U.S. dollar. In essence, it becomes easier for other nations to buy petroleum and harder for the United States to buy petroleum whenever the U.S. dollar is low.

One last externality on the price of energy, particularly petroleum, is that it is considered a commodity. Investors are constantly looking for the right combination of the safest place to invest their money and get the greatest yield on their investment. Whenever the investment markets (stocks, bonds, real estate) look weak, investors may move funds out of these markets and put them into the commodity markets (gold, metals, food, petroleum, and so forth). Just like anything else, when the investment demand for commodities is high, their prices will also be high, regardless of whether there is consumer demand for the same commodity.

Factoring in Pollution and Climate Change

Energy production requires the use of natural resources, and this inherently leads to pollution. The amount of pollution is not the same for all energy sources or even for the same energy source in different situations. For example, it is possible to burn coal with very low emissions of pollutants or, as is usually the case, with relatively high emissions. Even renewable energy sources such as wind and solar require the production and installation of wind turbines or solar panels. The energy generated may be pollution-free, but the production and installation of the equipment requires an initial use of fossil fuels and thus causes some one-time emission of pollutants.

Water pollution can be a concern, especially with coal mining, dumping of coal ash residue, and, to a lesser extent, oil and gas drilling. Water pollution is usually a localized effect except for acid rain. Air pollution is a broad concern relating to the production of energy. Common air pollutants are sulfur dioxides (SO_x), Nitrous Oxides (NO_x), fine particulate matter (PM_{10}), volatile organic compounds (VOC), heavy metals, and carbon dioxide (CO_2). Until recently, carbon dioxide was not legally considered a pollutant susceptible to Environmental Protection Agency (EPA) regulations; however, that changed in 2007, and it now has the same legal status as other pollutants, although the regulations concerning allowable emissions have not yet been determined.

Each type of pollution tends to have its own environmental and health impact. Heavy metals are emitted mostly from the burning of coal. Such heavy metals include mercury, lead, arsenic, and even radioactive uranium and thorium. In fact, there is a greater emission into the environment of radioactive uranium from the normal operation of coal-fired power plants than from nuclear power plants. Mercury is of particular concern, since it is easily absorbed into the human body. After the mercury is emitted into the air, it then precipitates down and contaminates rivers, streams, and lakes. Fish then naturally concentrate this mercury so that they become too toxic to eat.

Photochemical smog and ground-level ozone are mostly problems in cities, where they are made worse by warm weather. Smog is a combination of smoke and fog in which the solid particulates in smoke are suspended in the fog. Ground-level ozone is the result of a series of chemical reactions that occur in the atmosphere, starting with the emission of VOCs and NO_x, particularly from vehicles. This ground-level ozone is bad and is very different from the good stratospheric ozone that is needed to protect Earth from harmful ultraviolet radiation. Ground-level ozone is a significant health problem and is responsible for thousands of deaths in the United States each year from asthma, bronchitis, and COPD.

Sulfur dioxides react with air and water vapor to form sulfuric acid, which then is washed out of the atmosphere by precipitation. This is the cause of acid rain. Acid rain and particulate matter are responsible for most of the health problems related to electric-power generation. It is estimated that 20,000 to 30,000 Americans die each year from electric-power plant pollution, mostly due to coal-burning power plants. Such deaths may be heart attacks, lung cancer, asthma, and so forth, and of course it is impossible to know exactly who these 20,000 to 30,000 people are from among the millions of Americans who die each year.

Carbon dioxide is only of concern as a greenhouse gas causing global climate change. Other major greenhouse gases are methane, nitrous oxides, and chlorofluorocarbons (CFCs). Although CFCs are a far more powerful greenhouse gas on a per-molecule basis than CO_2, CO_2 is still more important because there is far more CO_2 produced by humans than CFCs. All fossil fuels release CO_2 when they are burned, and unless this CO_2 is captured and sequestered, it will increase the greenhouse effect and cause global climate change. Carbon dioxide does not pose the same immediate threat as other pollutants (e.g., killing 25,000 people per year), but it does pose a more severe long-term threat with global climate change. Global climate change is usually expressed as the temperature change that Earth will see based on global yearly average temperatures. According to the latest United Nations IPCC prediction, Earth will be about 2–4.5°C (4–8°F) warmer within the next 100 years. The range depends mainly on how much and how fast the world is able to curb the emission of carbon dioxide and on the uncertainty of climate modeling.

This average temperature is not the main problem with climate change. More significant effects of global climate change include changing weather patterns, such as extreme summer and winter temperatures, the timing of growing seasons, rising sea levels, episodes of severe weather, and, more importantly, annual precipitation patterns. These climatic changes are expected to lead to worldwide mass extinctions.

It is known with great certainty that precipitation patterns around the world will change, but it is not certain exactly how they will change. Some places that currently receive enough rain for crop production will become so arid that crops can no longer be sustained. Other places will receive more rain, enabling more crops to be grown, but will also be at risk of flooding. In both cases, there will be economic turmoil because centuries' worth of farming infrastructure will need to be relocated and in many cases moved to other nations or continents.

Mass extinctions will occur all over the planet as weather and temperature patterns change too quickly for species to adapt. Under normal circumstances, animal species can move relatively quickly to keep up with changing and moving climate patterns. But in many cases, forest fragmentation will prevent this. Some species, such as malaria-carrying mosquitoes, will be pushed into currently malaria-free populated regions. This will increase the occurrence of many diseases. Equally important, the plant species with which the animals have evolved to coexist are much slower to move. It is easy to conceive how an animal can migrate many miles in one day or even cross an entire continent in just one year. However, a tree can't just pull up its roots and start walking. A tree migrates by having its seeds spread relatively short distances in random directions. Those seeds then have to wait a decade or more to grow into mature trees to repeat the process. A tree cannot migrate the necessary hundreds or even thousands of miles per century. Thus, without the required plant species, an animal that has moved with the changing weather patterns will not be able to thrive. It is estimated that up to 75 percent of all species on Earth will go extinct with global climate change. This mass extinction will seriously decrease the biodiversity of the planet, which is what provides the backbone of life's resiliency on this rock we call Earth.

Thus, while global climate change may not cause thousands of human deaths per year now, it is quite possible that it will become a leading cause of death over the next century in the form of starvation, malnourishment, displacement resulting from coastal flooding, and diseases of all types. The annual death rate due to global climate change could easily dwarf that which is due to all of the classic forms of air pollution.

Conclusion: Carbon Sequestration and Future Use

Factoring in the considerations of climate change with the inability to immediately retrofit the entire energy sector has led developers to consider some radical ideas.

Most pollution can be reduced using technology, but doing so is costly and can even reduce the net amount of energy produced. In the case of coal-burning electric-power plants, the emissions that need to be controlled are particulates (PM), SO_x, NO_x, and CO_2. It is a fairly simple process to eliminate larger particulate matter (over 100 microns in size) using a cyclone filter.

But the smaller and more dangerous particulate matter under 10 microns in size requires the use of electrostatic precipitators. In order to remove the SO_x, wet and dry flue scrubbers can be used. To remove NO_x, selective catalytic reactors are required. All this pollution-control technology can be expensive, especially to install in existing power plants. To bring the pollution controls of an old power plant up to best modern technological standards costs about $1 billion per unit.

This dollar figure seems like an awfully large burden to impose on a power plant until this number is put into perspective. For a typical power plant unit, this amounts to an increase of about 1 cent per kilowatt-hour of electricity generated, compared to the average retail price of 12 cents per kilowatt-hour of electricity. Thus, consumers would see only an extra 10 percent charge for electricity. To put this in different terms, the $1 billion cost of pollution controls will be passed on to consumers. But the consumers will also benefit from the cleaner environment and will be the beneficiaries of about $3 billion in health care savings. Thus, the societal investment in pollution controls is well worth the expense.

It is also possible to reduce the amount of pollution at a coal-fired power plant by burning the coal in a very different way. Old furnaces can be converted into advanced pulverized furnaces, fluidized-bed furnaces, or even integrated-gasification combined-cycle (IGCC) turbines. Updating the way the coal is burned has the advantage not only of helping to reduce pollution before it is even created but also of increasing the overall efficiency of the power plant. The most advanced of these IGCC turbines is also being considered as a candidate plant design in which to test carbon sequestration technology.

It is also possible to reduce the amount of carbon dioxide that is emitted into the air using carbon capture and sequestration. The technologies for doing so are much less evolved than for other pollutants. It is fundamentally impossible to extract energy from fossil fuels without creating carbon dioxide. Thus, the only solution that reduces CO_2 emissions is capturing the CO_2 and then doing something to keep it out of Earth's atmosphere.

There are currently no full-scale working examples of carbon capture and sequestration in the U.S. energy industry. However, there is much research in this area, and there are a few full-scale test projects that are set to begin construction soon. In addition, there are plenty of long-standing examples of CO_2 being pumped into the ground to enhance the production of oil and gas from old wells. But these projects have always been about how to pump more oil and gas out of the ground and not about putting as much CO_2 into the ground as possible and keeping it there.

There is a variety of carbon sequestration ideas that are being developed. Most methods involve geologic storage of pressurized CO_2. Carbon dioxide can be stored in depleted oil and gas wells, salt domes, unminable coal beds, deep aquifers dissolved in the ocean, and even pools of liquid CO_2 created along deep seabeds. Early Department of Energy estimates are that carbon sequestration will add an extra two to five cents per kilowatt-hour to the cost of electricity. This would amount to an increase of about 15–30 percent on the retail rate of electricity. It is worth noting that early estimates of the cost of pollution-control equipment to reduce SO_x emissions turned out to be very inflated.

The cost of reducing pollution is related to the amount of pollution reduction that can be attained. Simple equipment can be small and cheap. But the best pollution-control equipment is usually very large, cumbersome, and expensive. This makes it ill-suited to the transportation sector but well suited for the electric-power industry and other large industrial consumers. Power plants and other industrial plants do not have to move about on wheels, so adding several hundred tons of equipment doesn't matter. Furthermore, power and industrial plants generally operate for several decades. Thus, expensive equipment that can be financed over a long period of time makes sense for this sector. This does not make sense in the transportation sector, where the purchase price of a vehicle is only financed over five years or less.

As with other pollution controls, carbon sequestration is best suited for the electric power industry and other large industrial sectors. Carbon sequestration technology will probably be impossible to deploy in the transportation sector as long as fossil fuels are used directly in the vehicle. However, if a shift is made to electric-powered or hydrogen-powered vehicles, then carbon can be sequestered during the production of the electricity or hydrogen. Then the vehicles can be driven with no carbon emissions.

None of these sequestration methods has been tested on a large scale, so it is impossible to compare the cost, effectiveness, or environmental impact of each sequestration method.

Richard Flarend and Brian C. Black

Further Reading

Adams, David Arthur. *Renewable Resource Policy: The Legal-Institutional Foundation.* Washington, DC: Island Press, 1996.

Andrews, Richard N. L. *Managing the Environment, Managing Ourselves.* New Haven, CT: Yale University Press, 1999.

Aurand, Harold W. *Coalcracker Culture: Work and Values in Pennsylvania Anthracite, 1835–1935.* Harrisburg, PA: Susquehanna University Press, 2003.

Berry, Wendell. *A Continuous Harmony.* New York: Shoemaker Hoard, 1972.

Black, Brian. *Petrolia: The Landscape of America's First Oil Boom.* Baltimore: Johns Hopkins University Press, 2000.

Bradsher, Keith. *High and Mighty: SUVs; The World's Most Dangerous Vehicles and How They Got That Way.* New York: PublicAffairs, 2002.

Brennan, Timothy J., et al. *A Shock to the System: Restructuring America's Electricity Industry.* Washington, DC: Resources for the Future, 1996.

Brower, Michael. *Cool Energy: Renewable Solutions to Environmental Problems.* Rev. ed. Cambridge, MA: MIT Press, 1992.

Buckley, Geoffrey L. *Extracting Appalachia: Images of the Consolidation Coal Company, 1910–1945.* Akron: Ohio University Press, 2004.

Crosby, Alfred. *Children of the Sun.* New York: Norton, 2006.

Cunfer, Geoffrey. *On the Great Plains: Ag and the Environment.* College Station: Texas A&M University Press, 2005.

Darst, Robert G. *Smokestack Diplomacy: Cooperation and Conflict in East-West Environmental Politics.* Cambridge, MA: MIT Press, 2001.

Daumas, Maurice, ed. *A History of Technology and Invention,* Vol. 3, *The Expansion of Mechanization, 1450–1725.* New York: Crown, 1969.

Francaviglia, Richard. *Hard Places.* Iowa City: University of Iowa Press, 1997.

Freese, Barbara. *Coal: A Human History.* New York: Perseus, 2003.

Gelbspan, Ross. *The Heat Is On: The Climate Crisis.* Reading, MA: Perseus, 1995.

Gordon, Richard, and Peter VanDorn. *Two Cheers for the 1872 Mining Law.* Washington, DC: Cato Institute, April 1998.

Gordon, Robert B., and Patrick M. Malone. *The Texture of Industry.* New York: Oxford, 1994.

Gorman, Hugh. *Redefining Efficiency: Pollution Concerns.* Akron, OH: University of Akron Press, 2001.

Horowitz, Daniel. *Jimmy Carter and the Energy Crisis of the 1970s.* New York: St. Martin's, 2005.

Hughes, Thomas. *American Genesis.* New York: Penguin, 1989.

Hughes, Thomas. *Networks of Power: Electrification in Western Society, 1880–1930.* Baltimore: Johns Hopkins University Press, 1983.

McNeil, John R. *Something New under the Sun: An Environmental History of the Twentieth-Century World.* New York: Norton, 2001.

Miller, B. *Coal Energy Systems.* Burlington, MA: Elsevier Academic, 2005.

Montrie, Chad. *To Save the Land and People: A History of Opposition to Surface Coal Mining in Appalachia.* Chapel Hill: University of North Carolina Press, 2003.

Moorhouse, John C., ed. *Electric Power: Deregulation and the Public Interest.* San Francisco: Pacific Research Institute for Public Policy, 1986.

Nye, David. *Consuming Power.* Boston: MIT Press, 1984.

Nye, David. *Electrifying America.* Boston: MIT Press, 1999.

Nye, David. *Technological Sublime.* Boston: MIT Press, 1996.

Platt, Harold. *Electric City.* Chicago: University of Chicago Press, 1991.

Poole, Robert W., Jr., ed. *Unnatural Monopolies: The Case for Deregulating Public Utilities.* Lexington, MA: Lexington Books, 1985.

Smil, Vaclav. *Energy in China's Modernization: Advances and Limitations.* Armonk, NY: M. E. Sharpe, 1988.

Smil, Vaclav. *Energy in World History.* Boulder, CO: Westview, 1994.

Smith, Duane. *Mining America: The Industry and the Environment, 1800–1980.* Lawrence: University Press of Kansas, 1987.

Stearns, Peter N. *The Industrial Revolution in World History.* Boulder, CO: Westview, 1998.

Wheelwright, Jeff. *Degrees of Disaster: Prince William Sound; How Nature Reels and Rebounds.* New Haven, CT: Yale University Press, 1996.

White, Richard. *Organic Machine.* New York: Hill and Wang, 1996.

Yergin, Daniel. *The Prize: The Epic Quest for Oil, Money & Power.* New York: Free Press, 1993.

Fossil Fuel Use and Industrialization

Between 1500 and 1750, there were great technological developments but no genuine revolution. In an era in which scientific and technological innovations were frowned upon and when energies and monetary support were focused on exploring the globe, it is relatively remarkable that any developments occurred at all. After 1750, a new concentrated interest on energy resources and expansion changed everything for the human species. Such a dramatic shift in patterns of living is referred to as a revolution.

The Industrial Revolution refers to the era from 1750 to the present in which a spirit of innovation connected with national economic interests to allow a few select nations to leap ahead of all others in the world. The primary tool that these industrializing nations shared was mastery of Earth's supplies of latent energy. Most often, the power in these fuels was realized through burning, which eventually proved to have a variety of by-products, including alteration of Earth's climate.

Mining Our Hydrocarbon Past

In terms of energy production, the Industrial Revolution marked the moment when humans turned to the flexibility and concentrated energy within minerals such as coal. Created from the remains of plants that lived and died about 100 million to 400 million years ago when parts of Earth were covered with huge swampy forests, coal had been mined by humans since the era of ancient Rome. Formed over millennia, though, coal cannot replenish itself. Therefore, coal is classified as a nonrenewable energy source. The energy we get from coal and petroleum today derives from the energy that plants absorbed from the sun millions of years ago. Plants, just like all living things, store energy from the sun. In plants, this process is known as photosynthesis. After the plants die, this energy is released as the plants decay.

Under conditions favorable to coal formation, however, the decay process is interrupted. The plants' energy is not lost, and the material retains its stored solar energy. Most often, geologists believe that this occurs as the dead plant matter fell into the swampy water, and over the years a thick layer of dead plants lay decaying at the bottom of the swamps. Over time, the surface and climate of Earth changed, and more water and dirt washed in, halting the decay process. The weight of the top layers of water and dirt packed down the lower layers of plant matter. Under heat and pressure, this plant matter underwent chemical and physical changes, pushing out oxygen and leaving rich hydrocarbon deposits. What once had been plants gradually turned into coal. This coal then is compacted into pockets within Earth.

Seams of coal—ranging in thickness from a fraction of an inch to hundreds of feet—represent plant growth from thousands of years prior. One important coal seam, which is known as the seven-foot-thick Pittsburgh Seam, may represent 2,000 years of rapid plant growth. One acre of this seam is estimated to contain about 14,000 tons of coal, which is enough to supply the electric power needs of 4,500 American homes for one year.

Although petroleum would become a vital cog in portions of the industrial era, coal was the prime mover that achieved most of the work. Of course, coal deposits are scattered throughout the globe; however, northeastern Pennsylvania holds a 500-square-mile region that is unique from any other. When coal was formed over 1 million years ago, northeastern Pennsylvania accelerated the process with a violent upheaval known as the Appalachian Revolution. Geologists speculated that the mountains folded over and exerted extra pressure on the subterranean resources. In northeastern Pennsylvania, this process created a supply of coal that was purer, harder, and of higher carbon content than any other variety. Named first with the adjective "hard," this coal eventually became known as anthracite. Geologists estimate that 95 percent of the supply of this hard coal in the Western Hemisphere comes from this portion of northeastern Pennsylvania.

This supply defined life in the state during the late 1800s. Thousands of families of all different ethnic backgrounds moved to mining towns to support themselves by laboring after coal. In other areas, mills and factories were built that relied on the coal as a power source. In between, the railroad employed thousands of workers to carry coal and raw materials to the mills and finished products away from them.

Coal would alter every American's life through the work it made possible. Although coal was found in a few Mid-Atlantic states, Pennsylvania possessed the most significant supplies and therefore became ground zero for the ways that coal culture would influence the nature of work and workers' lives in the United States. The rough-hewn coal communities that sprouted up during the anthracite era reflected the severe organization that defined labor in the coalfields. An elite class of coal owners and operators often lived in magnificent Victorian mansions, while their immigrant laborers lived in overcrowded company-owned patch towns. The class disparity was perpetuated by a steady change in ethnic laboring groups. Waves of European families arrived to live and work in the company towns found throughout the Appalachian Mountains. The original miners from Germany and Wales were soon followed by the Irish and later the Italians, Poles, and Lithuanians.

Despite difficult living conditions and ethnic discrimination from more established groups, these diverse ethnic groups ultimately created vibrant enclaves. In each patch town, each ethnic group built churches, formed clubs, and helped others from their nation of origin to get a start in the fields.

Energy for Life

Of course, humans have not always dug open pits in an effort to get coal, nor will we acquire energy this way forever. We can say, however, that humans will always have a relationship with energy sources. Because energy is not a tangible object that can be picked up and held, reliance on energy may seem like a somewhat vague idea. Rather than being defined by what it is, energy is defined in terms of what it can do. Energy is defined scientifically as the capacity to do work and transfer heat. Energy can be the work of boiling water or sawing a log, but energy is also the flow of heat from a hot object to a colder one.

Human methods for harvesting energy have taken many forms, beginning with the earliest use of fire and the agricultural revolution. Although one might not readily categorize it with energy use, agriculture relies on photosynthesis—the conversion of sun's energy by plants—into food that humans can eat.

It was through the harvest of energy-producing beings—whether plants or animals—that humans first became intimately involved in Earth's energy cycle. As hunter-gatherers and even as farmers after the agricultural revolution, humans had a limited impact on the planet's biological systems and could very likely have continued to live in this fashion in perpetuity. However, the 1700s brought significant changes in the size and prevailing lifeways of the human population. Innovations in energy use provided the foundation for this new way of life. Dependence on such energy sources as coal and other fossil fuels grew so significantly in the wake of the Industrial Revolution that by the end of the 20th century, extremely costly fuel production methods, such as open-pit mining and mountaintop removal, became economically justifiable.

The phenomenal economic and population growth of the 20th century was made possible largely by harvesting energy from onetime nonrenewable sources. Historian Alfred Crosby described our approach to energy management from a macroscopic level in his 2006 book *Children of the Sun* in this fashion:

> In the past two centuries we have . . . been burning immense, almost immeasurable, quantities of fossilized biomass from ages long before our species appeared. Today, as ever, we couldn't be more creatures of the sun if we went about with solar panels on our backs. . . . In the last half century our demand for energy has accelerated to the verge of exceeding what is produced and can be produced by conventional ways of harvesting sun source energy.

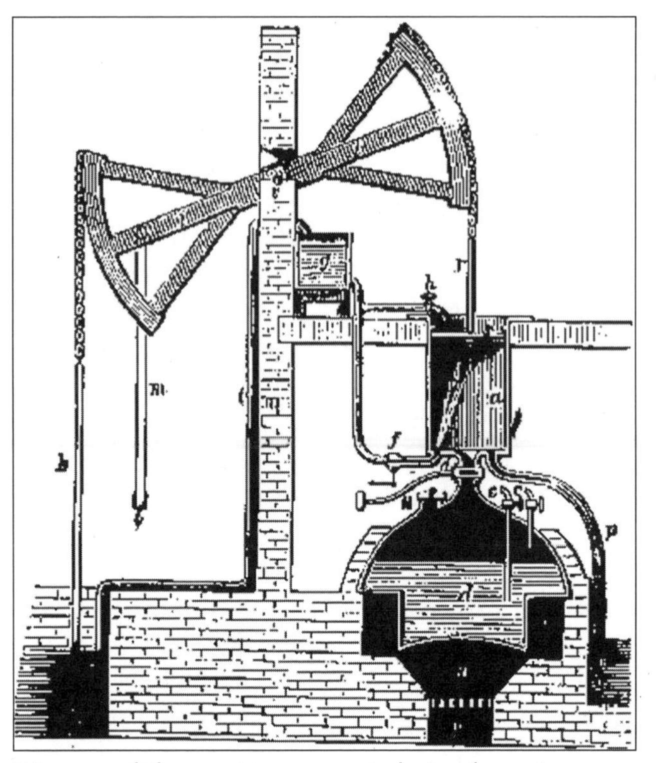

Diagram of Thomas Newcomen's design for a steam-engine-powered pump ca. 1705. Newcomen's engine was the first effective pump for draining mines of water. The steam engine, though, became the flexible prime mover behind numerous industrial activities.

The use of this fossilized biomass, though, can be considered neither sustainable nor renewable. As the science behind energy management has crept out of the laboratory and into other realms of thought, the need to integrate new forms of renewable energy into our future has become increasingly apparent. By drawing on the ethics of ecology and physics to understand our energy-intensive existence based on the industrial era, historians can help promote the cultural changes necessary to transition us to more sustainable modes of energy use. Human choices about the exploitation or reliance on certain resources represent an ethic, an ethic that in turn is evidenced in their culture.

Prime Movers Power Steam Engines

What historians of technology refer to as the "great transition" is not necessarily the emergence of the Industrial Revolution in the mid-1700s. In order to reach that revolution, a great transition was necessary in intellectual thought and in the availability of energy resources. Biomass fuels such as wood and charcoal had been in use for centuries, but they did not necessarily support an entirely new infrastructural system of machines. Coal, on the other hand, emerged as a prime mover during the 1600s and did exactly that.

After England experienced serious shortages of wood in the 1500s, domestic coal extraction became the obvious alternative. Most of the existing coalfields in England were opened between 1540 and 1640. By 1650, the annual coal output exceeded 2 million tons. It would rise to 10 million tons by the end of the 1700s.

Mining technology, of course, needed to be quickly developed. In addition, coal possessed power possibilities far in excess of previous resources. Thus, new industrial capabilities became possible. Primary among these was the steam engine.

The basic idea of the steam engine grew from the intellectual exploration of the great intellects of the scientific revolution. The idea of the piston, which was the basis of the engine, only came about after the realization of the existence of Earth's atmosphere. Although other societies had thought about the concept of an atmosphere and pressure holding things to Earth, it was Europeans who began to contemplate the possibilities of replicating this effect in miniature.

In the mid-1600s English engineers began contemplating a machine that utilized condensation in order to create a repeating vacuum to yield a source of power. The first model of such a device is attributed to Denis Papin, who in 1691 created a prototype piston that was moved within a cylinder using steam. This device remained unreliable for use, though, because the temperature could not be controlled. In 1712 Thomas Newcomen used atmospheric pressure in a machine that he alternatively heated and cooled in order to create the condensation pressure necessary to generate force. Additionally, Newcomen's engine was fairly simple to replicate by English craftsmen. Employed to pump out wells and for other suction purposes, the Newcomen engine spread to Belgium, France, Germany, Spain, Hungary, and Sweden by 1730.

Although it lacked efficiency and could not generate large-scale power, the Newcomen engine was the first economically viable machine to transfer thermal energy into kinetic energy. Through its wider application, the late 18th and early 19th centuries was a transitional period, with animate muscular energy being almost entirely replaced by inanimate hydraulic-based energy. Steam engines converted coal's energy into mechanical motion but still remained very limited in their application. James Watt created an engine that did not require cooling in 1769, which then allowed for the spread of the use of steam. Ultimately, then, during this same era water-powered milling was replaced by inanimate fossil fuel–based energy in the form of steam power.

Coal Spreads a New Industrial Era

America of the early 1800s still relied on energy technologies that would be considered sustainable and alternative to fossil fuels. The transition, though, had begun as industrialists expanded the use of charcoal, which created an infrastructure that could be expanded to include additional energy sources. Some of these resources, however, were complicated to harvest and manage. Their acquisition demanded entirely separate technological innovations as well as shifts in the accepted patterns of human life.

In the early 1800s, timber or charcoal (made from wood) filled most Americans' heating and energy production needs. This changed rather suddenly during the War of 1812, which pitted the United States against Great Britain in a conflict over trade. The war started in 1812 and ended in stalemate in 1815. The root of the conflict was the rights of American sailors who were being impressed to serve in the Royal Navy. The major military initiative of Britain during the war, though, was more related to trade: the British blockade of ports such as Philadelphia nearly crumbled the economy of the young republic.

The blockades of the War of 1812, though, became instrumental in moving the United States more swiftly toward its industrial future. Depleting fuelwood supplies combined with the British blockade to create domestic interest in using anthracite or hard coal, particularly around Philadelphia. Historian Martin Melosi writes that "When war broke out . . . [Philadelphia] faced a critical fuel shortage. Residents in the anthracite region of northeastern Pennsylvania had used local hard coal before the war, but Philadelphia depended on bituminous coal from Virginia and Great Britain." Coal prices soared by more than 200 percent by April 1813. Philadelphia's artisans and craftsmen responded by establishing the Mutual Assistance Coal Company to seek other sources. Anthracite soon arrived from the Wilkes-Barre, Pennsylvania, area. After the war, industrial use of hard coal continued to increase slowly until 1830. Between 1830 and 1850, the use of anthracite coal increased by 1,000 percent.

As the Industrial Revolution swept from Europe and into other parts of the world in the early to mid-1800s, the nations most susceptible to its influence were rich in raw materials and committed to the individual freedom of economic development. In these considerations, the United States led the world.

Thanks to the American interest in free enterprise and the astounding supplies of raw materials, including coal and later petroleum, the United States became the industrial leader of the world by the early 1900s—after only four or five decades fully committed to industrialization. Economic prosperity, massive fortunes for a few, and employment for nearly everyone who wanted to work were a few of the outcomes of American industry. Another outcome from the intense use of the natural environment exerted by industrialization was degradation.

In the industrial era that stretched from 1850 to 1960, many industrialists were willing to create long-term environmental problems and messes in the interest of short-term gain. Some of these gains came in the form of unparalleled personal fortunes. Other benefits included long-standing economic development for communities and regions around the United States. However, this economic strategy took shape on the back of the harvest, manipulation, and exploitation of natural resources. This ethic of extraction was felt to some degree in any industrial community, but possibly it was most pronounced in mining areas, particularly those areas mining for energy resources such as coal and petroleum.

As American society committed to a primary course of development that was powered by fossil fuels, much of the evidence of extraction and production was viewed as a symbol of progress. Few checks and balances existed to demand care and conservation. In the 19th century, the environmental consequences of mining for these hydrocarbon resources buried deep in Earth was of little concern. Most often, industries were viewed almost solely for the economic development that they made possible.

New Tools of the Industrial Era: Railroads

In addition to stimulating the development of mining in locales such as Pennsylvania, industrial development contributed to and even fed the development of related ancillary undertakings. More and more industries became essential to everyday American lives. Throughout American history, transportation was one of the most important applications of energy use. In the case of coal, the use of the railroad made coal supplies accessible while also involving coal's energy in innumerable other activities during the 1800s.

The planning and construction of railroads in the United States progressed rapidly during the 19th century. Some historians say that it occurred too rapidly. With little direction and supervision from the state governments that were granting charters for construction, railroad companies constructed lines where they were able to take possession of land or on ground that required

Young miner with headlamp at the Turkey Knob Mine in West Virginia, ca. 1908. Legally, children had to be 12 years old to work, but many underage children worked in the coal mines. (Library of Congress)

the least amount of alteration. The first step to any such development was to complete a survey of possible passages.

Before 1840, most surveys were made for short passenger lines that proved to be financially unprofitable. Under stiff competition from canal companies, many lines were begun only to be abandoned when they were partially completed. The first real success came when the Boston and Lowell Railroad diverted traffic from the Middlesex Canal in the 1830s. After the first few successful companies demonstrated the economic feasibility of transporting commodities via rail, others followed throughout the northeastern United States.

The process of constructing railroads began with reconstruing humans' view of the landscape. Issues such as grade, elevation, and passages between mountains became part of a new way of mapping the United States. Typically, early railroad surveys and their subsequent construction were financed by private investors. When shorter lines proved successful, investors began talking about grander schemes. These expansive applications of the railroad provided the infrastructure for remarkable commercial growth in the United States, expanding the impact of the Industrial Revolution.

By the 1850s, though, the most glaring example of this change was coal-powered railroads. The expanding network of rails allowed the nation to expand commercially. Most important, coal-powered railroads knitted together the sprawling United States into a cohesive social and commercial network. Although this could be seen in microscopic examples, including cities such as Pittsburgh and Chicago to which railroads brought together the raw materials for industrial processes such as steel making, on the macroscopic scale railroads allowed American settlement to extend into the western territories.

It was a cruel irony that the industrial era that evolved in the late 1800s relied intrinsically on transportation. Long, slender mountains stretched diagonally across Appalachian regions such as Pennsylvania, creating an extremely inhospitable terrain for transporting raw materials. Opening up the isolated and mountainous region required the efforts of a generation of capitalists and politicians, who used their resources and influence to create a transportation network that made the coal revolution possible. Canals were the first step in unlocking the great potential of the coalfields. Soon, though, industrialists focused on a more flexible transportation system that could be placed almost anywhere. Railroads quickly became the infrastructure of the industrial era. Knitting together the raw materials for making iron, steel, and other commodities, railroads were both the process and product of industrialization.

The iron rails produced in anthracite-fueled furnaces extended transportation routes throughout the nation. This revolution in transportation led to corresponding revolutions in the fueling of industries and the heating of urban residences, which in turn required more and more miners and laborers.

Although each of these social and cultural impacts of the railroad altered American life, it was, after all, primarily an economic enterprise. Primitive as it was, the antebellum railroad entirely remade American commerce. Americans needed to entirely remake ideas of prices and costs. Previously, prices had factored in the length of time involved with transporting goods via turnpikes, the steamboat, and the canal. From the start, railroad rates were significantly cheaper than wagon rates. The increasing systemization of the railroad process made low costs even more possible.

The possibility of railroads connecting the Atlantic and Pacific coasts was soon discussed in Congress, and this initiated federal efforts to map and survey the western United States. A series of surveys showed that a railroad could follow any one of a number of different routes. The least expensive, though, appeared to be the 32nd Parallel route. The Southern Pacific Railroad was subsequently built along this parallel. Of course, this decision was highly political: southern routes were objectionable to northern politicians, and the northern routes were objectionable to the southern politicians.

Although the issue remained politically charged, the Railroad Act of 1862 put the support of the federal government behind the transcontinental railroad. This act helped to create the Union Pacific Railroad, which subsequently joined with the Central Pacific Railroad at Promontory, Utah, on May 10, 1869, and signaled the linking of the continent.

Railroading became a dominant force in American life in the late 19th century, and one of its strongest images was its ability to remake the landscape of the entire country. Following 1880, the railroad industry reshaped the American-built environment and reoriented American thinking away from a horse-drawn past and toward a future with the iron horse.

New Tools of the Industrial Era

Andrew Carnegie and Steel Manufacture

Railroads and the reliance on fossil fuels enabled the implementation of complex industrial undertakings at a scope and scale never seen before. Although iron manufacture increased in scale with the more intense model of industrialization after 1850, steel is possibly the best example of this new era's capabilities. Using railroads as its linking device, Andrew Carnegie perfected the process of steel manufacturing and created one of the greatest fortunes in history.

Into 1 pound of steel, observed Carnegie, went 2 pounds of iron ore brought 1,000 miles from Minnesota, 1.3 pounds of coal shipped 50 miles to Pittsburgh, and one-third of a pound of limestone brought 150 miles from Pittsburgh. Rivers and railroads brought the material to the Carnegie Steel Works along Pittsburgh's Monogahela River, where Bessemer blast furnaces fused the materials into steel. One of the greatest reasons for the rapid rise of American industry was its flexibility compared to that of other nations. Railroading could be integrated immediately into various industries in the United States, which, for instance, allowed American industry to immediately embrace the new Bessemer steel-making technology. Other nations, such as Britain, needed to shift from previous methods.

One innovation contributed to another in the late industrial era. Inexpensive energy made it feasible to gather the disparate materials that were necessary to make steel. Steel was stronger and more malleable than iron, which made possible new forms of building. Carbon levels make the bulk of the distinction between the two metals. Experiments with removing the oxygen content of pig iron required more heat than ordinary furnaces could muster. The Bessemer invention created a Bessemer blow, which included a violent explosion to separate off additional carbon and produce the 0.4 percent oxygen level that was desirable for steel.

New tasks, such as running the Bessemer furnace, created specialized but also very dangerous jobs. Working in the steel mill created a new hierarchy for factory towns. In the case of steelmaking, hot or dangerous jobs such as working around the Bessemer furnace eventually fell to African American workers.

Electricity and the Evolution of the Energy Industry

Industrial applications of energy shaped the industrial era; however, by later in the 1800s, coal, in the form of electricity, was also remaking the everyday lives of many Americans. On the whole, new energy made from fossil fuels altered almost every American's life by 1900. In 1860 there were fewer than 1.5 million factory workers in the country; by 1920 there were 8.5 million. In 1860 there were about 31,000 miles of railroad in the United States; by 1915 there were nearly 250,000 miles. The energy moving through such infrastructure would not remain limited to the workplace.

In the 19th century, energy defined industry and work in America but did not necessarily impact everyday cultural life. This would change dramatically by the end of the 1800s with the development of technology to create, distribute, and put to use electricity. Although electricity is the basis for a major U.S. energy industry, it is not an energy source. It is mostly generated from fossil fuel (coal, oil, natural gas), hydroelectric (waterpower), and nuclear power. The electric utilities industry includes a large and complex distribution system and as such is divided into transmission and distribution.

Following experiments in Europe, the electrical future of the United States fell to the mind of Thomas Edison, one of the nation's great inventors. In 1878 Joseph Swan, a British scientist, invented the incandescent filament lamp, and within 12 months Edison made a similar discovery in America. Edison used his DC generator to provide electricity to light his laboratory and later, in September 1882, to illuminate the first New York street to be lit by electric lamps. From this point, George Westinghouse patented a motor for generating alternating current. Society became convinced that its future lay with AC generation. This, of course, required a level of infrastructural development that would enable the utility industry to have a dominant role over American life.

Once again, this need for infrastructural development also created a great business opportunity. George Insull went straight to the source of electric technology and ascertained the business connections that would be necessary for its development. In 1870, Insull became a secretary for George A. Gourand, one of Edison's agents in England. Then, Insull came to the United States in 1881 at age 22 to be Edison's personal secretary.

By 1889 Insull became vice president of the Edison General Electric Company in Schenectady, New York. When financier J. P. Morgan took over Edison's power companies in 1892, Insull was sent west to Chicago to become president of the struggling Chicago Edison Company. Under Insull's direction, Chicago Edison bought out all its competitors for a modest amount after the Panic of 1893. He then constructed a large central power plant along the Chicago River at Harrison Street. The modest steam-powered electricity-generating operation would serve as Insull's springboard to a vast industrial power base.

By the early 1900s, Insull's Commonwealth Edison Company made and distributed all of Chicago's power. Insull connected electricity with the concept of energy and also diversified into supplying gas. Then he pioneered the construction of systems of dispersing these energy sources into the countryside. The energy grid was born. It would prove to be the infrastructure behind each American life in the 20th century. Through the application of this new technology, humans now could defy the limits of the sun and season. The greatest application of this—and a symbol of humans' increased reliance on fossil-fueled power—is the lightbulb.

For decades, inventors and businessmen had been trying to invent a source of light that would be powered by electricity. Primarily, their experiments emphasized positioning a filament in a vacuum. The electric current, then, was sent through in hopes of making the filament glow. The filaments consistently failed, disintegrating as soon as the current reached them.

In 1878, Edison decided to concentrate his inventive resources on perfecting the lightbulb. Instead of making his filament from carbon, he switched to platinum, which was a more resilient material. In 1879 he obtained an improved vacuum pump called the Sprengel vacuum, and it proved to be the catalyst for a breakthrough. Using the new pump, Edison switched back to the less-expensive carbon filaments. Using a carbonized piece of sewing thread as a filament in late October, Edison's lamp lit and continued to burn for 13.5 hours. Edison later changed to a horseshoe-shaped filament, which burned for over 100 hours. Edison had invented a practical lightbulb, but, more important, he cleared the path for the establishment of the electrical power system that would revolutionize human existence.

It was this power system that became Edison's real achievement and created the market that would beget a huge new industry destined to affect the lives of every American. The nature of everyday life became defined by activities made possible by electric lighting as well as the nearly endless amount of other electrically powered items. The lightbulb was a critical innovation in the electrification of America; however, it also helped to create the market that stimulated efforts to perfect the industry of power generation.

At the root of power generation, of course, was the dynamo. The dynamo was the device that turned mechanical energy of any type into electrical power. When Edison started working on the lightbulb, the most effective dynamo produced electricity at approximately 40 percent of the possible efficiency. He developed a dynamo that raised this efficiency to 82 percent.

Together, these technological developments made it possible for Edison to start providing electricity commercially to New York City. By September 1882 he had opened a central station on Pearl Street in Manhattan and was eventually supplying electricity to a one-mile-square section of New York. These areas became futuristic symbols for the growing nation.

A New Scale and Scope for Energy: Black Gold

Coal provided the basic infrastructure for the Industrial Revolution. Through its impact on the factory system, American life changed radically. But a similar dependence derived from the primary energy resource that followed in the wake of coal: petroleum. The intricacy of petroleum to American life in the 1990s would have shocked 19th-century users of "Pennsylvania rock oil." Most farmers who knew about the oil in the early 1800s knew seeping crude as a nuisance to agriculture and water supplies. These observers were not the first people to consider the usefulness of petroleum, which had been a part of human society for thousands of years. Its value grew only when European Americans offered the resource their commodity-making skills.

As the oil's reputation grew, settlers to the region gathered oil from springs on their property by constructing dams of loose stones to confine the floating oil for collection. In the mid-1840s, one entrepreneur noticed the similarity between the oil prescribed to his ill wife and the annoying substance that was invading the salt wells on his family's property outside Pittsburgh, Pennsylvania. He began bottling the waste substance in 1849 and marketed it as a mysterious cure-all available throughout the northeastern United States. Although he still acquired the oil only by skimming, Samuel Kier's supply quickly exceeded demand because there was a constant flow of the oil from the salt wells. With the excess, he began the first experiments with using the substance as an illuminant, or substance that gives off light. The culture of expansion and development was beginning to focus on petroleum.

From this point forward, petroleum's emergence became the product of entrepreneurs—except for one important character: Edwin L. Drake of the New Haven Railroad. In 1857, the company sent Drake to Pennsylvania to attempt to drill the first well intended for oil. The novelty of the project soon had worn off for Drake and his assistant Billy Smith. The townspeople irreverently heckled the endeavor of a "lunatic." During the late summer of 1859, Drake ran out of funds and wired to New Haven, Connecticut, for more money. He was told that he would be given money only for a trip home—that the Seneca Oil Company, as the group was now called, was done supporting him in this folly. Drake took out a personal line of credit to continue, and a few days later, on August 29, 1859, he and his assistant discovered oozing oil.

Throughout its history, petroleum has exhibited wide fluctuations in price and output. The boom-and-bust cycle was even underwritten by the courts in the case of *Brown v. Vandergrift* (1875), which established the laissez-faire development policy that became known as the rule of capture. The oil could be owned by whoever first pulled it from the ground—captured it. The rush to newly opened areas became a race to be the first to sink the wells that would bring the most oil up from its geological pockets.

After the American Civil War, the industry consistently moved toward the streamlined state that would allow it to grow into the world's major source of energy and lubrication during the 20th century.

During the 19th century, petroleum's most significant impact may have been on business culture. The culture of the industry that took shape would change land use and ideas of energy management throughout the world. John D. Rockefeller and Standard Oil first demonstrated the possible domination available to those who controlled the flow of crude oil. Rockefeller's system of refineries grew so great at the close of the 19th century that he could demand lower rates and eventually even kickbacks from rail companies. One by one he put his competitors out of business, and his own corporation grew into what observers in the late 1800s called a trust (what today is called a monopoly). Standard Oil's reach extended throughout the world and became a symbol of the Gilded Age, when businesses were allowed to grow too large and benefit only a few wealthy people. Reformers vowed that things would change.

The laissez-faire era of government regulation of businesses, particularly energy companies such as Standard Oil, came to an end when progressive reformers took a different view of the government's role in American life.

President Theodore Roosevelt, who took office in 1901, led the progressive interest to involve the federal government in monitoring the business sector. In the late 1890s, muckraking journalists had written articles and books that exposed unfair and hazardous business practices. Ida Tarbell, an editor at *McClure's* who had grown up the daughter of a barrel maker in Titusville, took aim at Rockefeller. Her *History of the Standard Oil Company* produced a national furor over unfair trading practices. Roosevelt used her information to enforce antitrust laws that would result in Standard Oil's dissolution in 1911. Rockefeller's company had become so large that when broken into subsidiaries, the pieces would grow to be Mobil, Exxon, Chevron, Amoco, Conoco, and Atlantic, among others.

Even after Standard Oil's dissolution in 1911, the image of its dominance continued. Standard Oil had led the way into international oil exploration, suggesting that national borders need not limit the oil-controlling entity. Throughout the 20th century, large multinational corporations or singular wealthy businessmen attempted to develop supplies and bring them to market. Their efforts combined with consumer desire to make petroleum the defining energy resource of the 20th century. Similar to coal, though, the real revolution in consumption required basic changes in supply and the scale and scope by which petroleum could be used in American life.

Cheap Oil Sets the Tone for Our High-Energy Existence

The revolution in the supply of petroleum began with international expansion; however, it was a domestic source that truly defined petroleum's role in Americans' high-energy existence. Although new drilling technologies helped to increase supply, entire new regions were required to be developed. By 1900, companies such as Standard Oil sought to develop new fields all over the world. In terms of the domestic supply of crude, though, the most significant breakthrough came in Texas. With one 1901 strike, the limited supply of crude oil became a thing of America's past. It is no coincidence, then, that the century that followed was powered by petroleum.

This important moment came in eastern Texas, where without warning the level plains near Beaumont abruptly give way to a lone rounded hill before returning to flatness. Geologists call these abrupt rises in the land domes because hollow caverns lie beneath. Over time, layers of rock rise to a common apex and create a spacious reservoir underneath. Often, salt forms in these empty geological bubbles, creating a salt dome. Over millions of years, water or other material might fill the reservoir. At least, that was Patillo Higgins's idea in eastern Texas during the 1890s.

Higgins and very few others imagined such caverns as natural treasure houses. Higgins's intrigue grew with one dome-shaped hill in southeastern Texas. Known as Spindletop, this salt dome—with Higgins's help—would change human existence.

Texas had not yet been identified as an oil producer. Well-known oil country lay in the eastern United States, particularly western Pennsylvania. Titusville, Pennsylvania, introduced Americans to massive amounts of crude oil for the first time in 1859. By the 1890s, petroleum-derived kerosene had become the world's most popular fuel for lighting. Edison's experiments with electric lighting placed petroleum's future in doubt; however, petroleum still stimulated boom wherever it was found. But in Texas? Every geologist who inspected the "Big Hill" at Spindletop told Higgins that he was a fool.

With growing frustration, Higgins placed a magazine advertisement requesting someone to drill on the Big Hill. The only response came from Captain Anthony F. Lucas, who had prospected domes in Texas for salt and sulfur. On January 10, 1901, Lucas's drilling crew, known as roughnecks for the hard physical labor of drilling pipe deep into Earth, found mud bubbling in their drill hole. The momentary explosion turned to a prolonged roar, and suddenly oil spurted out of the hole. The Lucas geyser, found at a depth of 1,139 feet, blew a stream of oil over 100 feet high until it was capped nine days later. During this period, the well flowed an estimated 100,000 barrels a day—well beyond any flows previously witnessed. Lucas finally gained control of the geyser on January 19. By this point, a huge pool of oil surrounded it. Throngs of oilmen, speculators, and onlookers came and transformed the city of Beaumont into Texas's first oil boomtown.

The flow from this well, named Lucas 1, was unlike anything witnessed before in the petroleum industry: 75,000 barrels per day. As news of the gusher reached around the world, the Texas oil boom was on. Land sold for wildly erratic prices. After a few months, over 200 wells had been sunk on the Big Hill. By the end of 1901, an estimated $235 million had been invested in oil in Texas. This was the new frontier of oil; however, the industry's scale had changed completely at Spindletop. Unimaginable amounts of petroleum—and the raw energy that it contained—were now available at a low enough price to become part of every American's life.

It was the businessmen who then took over after Higgins and other petroleum wildcatters. Rockefeller's Standard Oil and other oil executives had managed to export petroleum technology and exploited supplies worldwide. The modern-day oil company became a version of the joint-stock companies that had been created by European royalty to explore the world during the period of mercantilism of the 1600s. Now, though, behemoth oil companies were transnational corporations, largely unregulated and seeking one thing: crude oil. Wherever black gold was found, oil tycoons set the wheels of development in motion. Boomtowns modeled after those in the Pennsylvania oil fields could suddenly pop up in Azerbaijan, Borneo, or Sumatra.

As gushers in eastern Texas created uncontrollable lakes of crude, no one considered the idea of shortage or conservation. Even the idea of importing oil was a foreign concept. California and Texas flooded the market with more than enough crude oil, and then from nearly nowhere, Oklahoma emerged in 1905 to become the nation's greatest oil producer. Now, however, what was to be done with this abundant, inexpensive source of energy?

Conclusion: The High-Energy Life

The high-energy lifestyle that humans adopted during the 20th century has enabled staggering accomplishments and advancements. The costs of extensive use of fossil fuels has also come with its share of serious impacts, some of which we are just learning about in the 21st century. This essay began with humans living during the agriculture revolution, living within the natural cycles of energy that begin with the sun. The Industrial Revolution made such natural forms of energy alternatives to the primary use of energy made from burning fossil fuels. Particularly in the United States, we made cheap energy part of our life and threw caution to the wind. The 20th century became a binge of cheap energy and all that it made possible.

At the dawn of the 21st century, Americans have come to recognize that the great energy resources of the industrial era were exhaustible: that the supplies of coal, petroleum, and natural gas were finite. Throughout the 20th century, though, amid the frenzy of energy decadence and its associated economic and social development, a growing chorus alerted consumers and politicians to the temporality of reliance on hydrocarbon-derived energy.

As one might imagine, the call for the use of alternative modes of power often went against the grain of basic ideas of American progress and success. By the end of World War II, the emergence of the United States as the global economic, military, and cultural leader was largely predicated on an existence of cheap energy. The infrastructure that emerged to support this society made certain that power from hydrocarbons involved some of the greatest economic, political, and social players of the century. To fight against such embedded interests might have seemed folly; however, the persistence and innovation of such minority voices has largely set the stage for a new era of energy use in the 21st century. Alternative energy ideas used as early as the 1100s would rise again as the most sensible and sustainable ways of creating power.

Brian C. Black

Further Reading

Black, Brian. *Petrolia: The Landscape of America's First Oil Boom.* Baltimore: Johns Hopkins University Press, 2000.

Brinkley, Douglas. *Wheels for the World: Henry Ford, His Company and a Century of Progress.* New York: Viking, 2003.

Crosby, Alfred. *Children of the Sun.* New York: Norton, 2006.

Gordon, Richard, and Peter VanDorn. *Two Cheers for the 1872 Mining Law.* Washington, DC: Cato Institute, 1998.

Gordon, Robert B., and Patrick M. Malone. *The Texture of Industry.* New York: Oxford University Press, 1994.

Gorman, Hugh. *Redefining Efficiency: Pollution Concerns.* Akron, OH: University of Akron Press, 2001.

Greene, Ann. *Horses at Work.* Cambridge, MA: Harvard University Press, 2009.

Hughes, Thomas. *American Genesis.* New York: Penguin, 1989.

Hughes, Thomas. *Networks of Power: Electrification in Western Society, 1880–1930.* Baltimore: Johns Hopkins University Press, 1983.

Hunter, Louis C., and Lynwood Bryant. *A History of Industrial Power in the United States, 1780–1930,* Vol. 3, *The Transmission of Power.* Cambridge, MA: MIT Press, 1991.

Ise, John. *The United States Oil Policy.* New Haven, CT: Yale University Press, 1926.

Melosi, Martin. *Coping with Abundance.* New York: Knopf, 1985.

Mokyr, Joel. *Twenty-Five Centuries of Technological Change.* New York: Harwood Academic, 1990.

Mokyr, Joel, ed. *The Economics of the Industrial Revolution.* Totowa, NJ: Rowman and Allanheld, 1985.

Montrie, Chad. *To Save the Land and People: A History of Opposition to Surface Coal Mining in Appalachia.* Chapel Hill: University of North Carolina Press, 2003.

Moorhouse, John C., ed. *Electric Power: Deregulation and the Public Interest.* San Francisco: Pacific Research Institute for Public Policy, 1986.

Motavalli, Jim. *Forward Drive: The Race to Build "Clean" Cars for the Future.* San Francisco: Sierra Club Books, 2001.

Mumford, Lewis. *Technics and Civilization.* New York: Harcourt, 1963.

Norton, Peter D. *Fighting Traffic.* Boston: MIT Press, 2008.

Nye, David. *Consuming Power.* Boston: MIT Press, 1984.

Nye, David. *Electrifying America.* Boston: MIT Press, 1999.

Nye, David. *Technological Sublime.* Boston: MIT Press, 1996.

Oliens, Roger M., and Dianna Davids. *Oil and Ideology: The American Oil Industry, 1859–1945.* Chapel Hill: University of North Carolina Press, 1999.

Opie, John. *Nature's Nation.* New York: Harcourt Brace, 1998.

Poole, Robert W., Jr., ed. *Unnatural Monopolies: The Case for Deregulating Public Utilities.* Lexington, MA: Lexington Books, 1985.

Rifkin, Jeremy. *The Hydrogen Economy.* New York: Penguin, 2003.

Sheppard, Muriel. *Cloud by Day: The Story of Coal and Coke and People.* Pittsburgh: University of Pittsburgh Press, 2001.

Smil, Vaclav. *Energy in China's Modernization: Advances and Limitations.* Armonk, NY: M. E. Sharpe, 1998.

Smil, Vaclav. *Energy in World History.* Boulder, CO: Westview, 1994.

Smith, Duane. *Mining America: The Industry and the Environment, 1800–1980.* Lawrence: University Press of Kansas, 1987.

Stearns, Peter N. *The Industrial Revolution in World History.* Boulder, CO: Westview, 1998.

Yergin, Daniel. *The Prize: The Epic Quest for Oil, Money & Power.* New York: Free Press, 1993.

Fracking

The development of Marcellus shale into natural gas after 2007 relied on a number of new and emerging technologies. In each case, these innovations helped energy companies to transform a well-known geological formation—shale layers deep below Earth's surface—into natural gas that might be harvested and used as a relatively clean alternative to many other fossil fuels. Although this process might appear sensible and straightforward, one of these technologies—hydraulic fracturing—makes harvesting gas from shale deeply controversial.

The debate over hydraulic fracturing (or fracking as it is called) is focused on states currently working to harvest gas from the Marcellus shale formation, which is located primarily in the Mid-Atlantic region of the United States. Although there are a variety of problematic critiques related to fracking, its threat to underground water supplies may be the most acute. Unfortunately, threats to water supplies are an intrinsic portion of the fracking process.

The remarkable ability to frack deep below Earth's surface begins with horizontal drilling. In the late 20th century, most oil and gas wells were drilled as vertical holes down through rock formations such as the Marcellus. Although a vertical gas or oil well could drain an area of about 10 to 40 acres, dense rock such as shale would not release its fluid or gas from a vertical hole drilled through it. Even though the Marcellus shale can be more than 900 feet thick, it does not achieve sufficient profile to release a significant amount of gas. Horizontal drilling creates a hole several thousand feet horizontally across the shale. After horizontal drilling creates such a hole, a hydraulic fracturing job, or frack job, is done on the newly drilled well. These techonologies continue to evolve, and their product is therefore referred to as unconventional natural gas.

After casing or heavy pipe has been run into the newly drilled hole to line it, extremely high-pressure pumps are hooked up, and fluids (mostly water and sand) are forced down into the shale to open cracks and fissures so that more natural gas can flow out of the formation. The fractures allow the gas to seep into the well bore for collection. When the gas is produced in the shale and becomes pressurized, the pressure splits the rock. The Marcellus shale is chock-full of natural fractures, making it easier for gas companies to extract gas. "Fractures are good for business," says one industry scientist. "If you have enough natural fractures, it's much cheaper to drill a production well than if you do not have them."

Hydraulic fracturing is the primary completion process used in the Marcellus shale. It has been the subject of controversy lately, with environmental groups protesting the drilling of new wells in the environmentally sensitive Chesapeake Bay watershed, among other areas. Wells in the Marcellus formation are as deep as 9,000 feet, and the productive zone is far below any water aquifers. A properly cased gas well, lined with thick steel pipe and cemented into place, will ensure a safe frack job. State and federal agencies monitor the process of casing and testing new wells and hydraulic fracturing. The source of contention among environmental groups is the number of chemicals pumped into the formation along with frack water and sand. Up to 50 percent of this fluid may remain in the ground; however, it may take millions of years for any of it to migrate upward to water-bearing zones. If

Protestors, such as these in Ponca City, Oklahoma in 2004, sought to draw attention to the lack of regulation over pollution created by fracking. This group of area landowners, Ponca Tribe members and officials with the paper, Allied-Industrial, Chemical and Energy Workers Union, said they were protesting what they perceived as a lack of response from state and federal environmental regulators. (Sue Ogrocki/AP/Wide World Photos)

proper safety measures are undertaken, hydraulic fracturing can be a safe means of producing Marcellus shale gas wells.

The impact of hydrofracking on surface water and groundwater is cause for great concern. About 5 million gallons of water per well are used to make fracking fluid, a mixture of water, sand, and hundreds of other possible chemicals, including such hazards as benzine, formaldehyde, arsenic, and diesel fuel. Some frack water stays in the ground, and some is returned to the surface.

This flowback accumulates higher concentrations of salts and minerals and sometimes radioactive contamination. Wastewater treatment plants cannot remove many of these pollutants. Some flowback is reused for fracking, but more is stored in huge artificial ponds. Spills occur during water transport and storage. Improperly cased wells leak frack water and methane into the water table, polluting public and private wells. Pennsylvanian waterways drain into the Chesapeake Bay, so gas drilling increases the bay's pollution problems. Storm water erosion carries increased silt loads from dirt roads and drill sites. Minerals from frack water discharged into streams and rivers increase total dissolved solids in the bay.

Some regulations are in place to reduce the problems caused by drilling for shale gas. New laws are being proposed and vigorously debated.

Currently in Pennsylvania, setbacks are required between gas wells and water sources. Drillers must acquire permits, submit plans, post bonds, and notify landowners whose water supply is within 1,000 feet of a gas well. Gas wells must be cased to prevent contamination of the groundwater table. Pennsylvania's Oil and Natural Gas Act prevents local municipalities from regulating drilling activities but allows local zoning of well placement.

In 2010, a bill proposed the repeal of exemptions from clean water and environmental impact regulations allowed to gas drillers by the 2005 Federal Energy Policy Act and required disclosure of frack water chemicals, but the bill died in committee. In 2011, SB 1100 proposed somewhat stiffer requirements to safeguard water sources and surface owners' rights and requires impact fees to mitigate landowner and community problems. The bill imposes no gas revenue tax, contrary to the wishes of most Pennsylvanians.

The history of coal mining shows the long-term costs of shortsighted extraction policies. Pennsylvania alone has 9,000 abandoned mines and acid mine drainage that pollutes 2,500 miles of streams. The projected cleanup cost is $15 billion. Awareness of these consequences can help Pennsylvanians choose better ways to benefit from our newest energy boom.

Brian C. Black

Further Reading

Andrews, Anthony. *Unconventional Gas Shales.* Washington, DC: BiblioGov, 2010.

"History of the Shale." WHYY, September 29, 2010, http://whyy.org/cms/news/health-science/2010/09/29/history-of-the-shale/46987.

McGraw, Seamus. *The End of Country.* New York: Random House, 2011.

Penn State Marcellus Center for Outreach and Research. http://marcellus.psu.edu/.

FutureGen

FutureGen is an energy technology demonstration project in the United States that was initially conceptualized by various U.S. energy experts to be the first near-zero–emissions commercial-scale coal-fired power plant. This project was initially conceptualized to simultaneously demonstrate carbon capture and storage (CCS) technology, hydrogen production technology, and the advanced coal power plant technology known as the integrated gasification combined cycle (IGCC). Due to political changes and concerns about the cost of the project, the plans have been reduced, and a less ambitious version of the original FutureGen vision is currently moving forward.

The FutureGen project was announced publicly in February 2003 and was presented as the flagship program for the George W. Bush administration's strategy on clean-coal technology development and climate change mitigation. Despite almost a decade of planning and political strategizing, FutureGen has not yet been built, and the project's design and structure have undergone two major restructuring initiatives.

FutureGen is a public-private partnership between the U.S. Department of Energy (DOE) and the FutureGen Industrial Alliance, Inc., a nonprofit consortium of some of the world's largest coal and energy companies. In addition to cost and risk sharing (initially the DOE was to contribute approximately 74 percent of the $1.5 billion project and the alliance the remainder), the alliance has had responsibility for the design, construction, and operation involved in integrating advanced technologies into the plant. The DOE has been responsible for a major portion of the financing, for independent oversight, and for coordinating participation from international governments.

An extensive competitive site-selection process occurred throughout 2006 and 2007, during which the 12 potential sites initially identified were narrowed down to 4 semifinalists: 2 in Texas and 2 in Illinois. Recognizing the potential economic and political benefits to their states of hosting FutureGen, both Texas and Illinois invested in competing for the project by providing technical justification to explain why their state should be selected and also used public engagement initiatives to raise awareness and acceptance about the project's potential benefits. In January 2008, the alliance announced that Mattoon, Illinois, was to be the site for the FutureGen plant.

A few weeks later, responding to the increasing projected project costs, the Bush administration's secretary of energy, Samuel Bodman, announced a restructuring of the FutureGen project and the withdrawal of the federal commitment to contribute funds to the project as it was initially conceptualized. This 2008 restructuring of the project from a research demonstration program to a near-term commercial demonstration program was justified as a more cost-effective way to advance CCS by redirecting funds to demonstrate components of CCS on multiple existing plants rather than investing in the construction of a new single large plant. The restructuring also altered the DOE cost-sharing from the original 74 percent to a maximum of 50 percent per demonstration project. Then in the summer of 2009, the Barack Obama

Hot Spot

Samsø Island

The 4,000 residents of largely agricultural Samsø Island in Denmark accepted a challenge from the Danish government during the 1990s to convert to a carbon-neutral lifestyle, and by 2008 they had largely accomplished the task, with no notable sacrifice of comfort. Farmers grow rapeseed and use the oil to power their machinery; homegrown straw is used to power centralized home-heating plants; solar panels heat water and store it for use on cloudy days, of which the 40-square-mile island has many; and wind power provides electricity from turbines in which most families on the island have a share of ownership. The island is now building more wind turbines to export power.

During the late 1990s, most of the 4,000 Samsingers' homes were heated with oil imported on tankers, and their electricity was generated from coal and imported from the Danish mainland on cables. The average resident of the island produced 11 tons of carbon dioxide a year.

In 1997 the Danish Ministry of Environment and Energy sponsored a renewable energy contest that the city government entered after an engineer who did not live on the island filed a proposal with the mayor's consent. Quite to everyone's surprise, Samsø was picked as Denmark's renewable energy island. The designation carried no prize money, so nothing happened for a few months. After that, residents established energy cooperatives and began to tutor themselves in such things as insulation and wind power.

Bruce E. Johansen

administration's secretary of energy, Steven Chu, announced a revival of the initial FutureGen project and pledged government support. In mid-2009, the alliance purchased a site in Mattoon, Illinois, and groundbreaking for the new power plant was expected in 2010. However, in August 2010 the DOE announced FutureGen 2.0—another major restructuring of the project. The redefined limited scope of FutureGen 2.0 includes demonstrating oxyfuel combustion in a nearby existing power plant rather than construction of a new state-of-the-art power plant. Responding to this announced change and to the associated reduced significance and reduced international and national prestige of the project, the community of Mattoon withdrew its involvement in FutureGen. Mattoon expressed disappointment and an unwillingness to participate if the community would only be providing the geological storage location for the captured CO_2. In the autumn of 2010 in conjunction with the DOE, the FutureGen Alliance began a new site selection process and in February 2011 the FutureGen Industrial Alliance announced its selection of Morgan County, Illinois, as the preferred location for the FutureGen 2.0 CO_2 storage site, visitor center, and research and training facilities.

Although FutureGen has not yet been built, the project has become widely known and has received significant media attention, thereby facilitating learning by multiple actors. One particular aspect of the FutureGen case, the competitive nature of the site selection process, has been critical in enhancing social learning about the project, about CCS, and about the need for climate change mitigation technologies. Owing to this site-selection process, multiple entities (the DOE, the FutureGen Industrial Alliance, and the individual states) were simultaneously investing in and promoting public awareness and support for the project and CCS technology. The FutureGen Industrial Alliance invested in a stakeholder engagement team that conducted more than 200 interviews and group meetings with local communities to listen to concerns, address questions, and explain the project, and both Illinois and Texas invested resources and instructed relevant agencies to promote and prepare their state for the potential project. This multipronged effort during the competitive site-selection process broadened the scope of social learning and awareness about both CCS and the global challenge of climate change. One CCS educational representative from Illinois explained that the FutureGen project is so widely recognized in the state by the general public that it has become a frequent reference within public discourse about CCS.

An explicit mechanism designed to enhance learning in both the initial and current versions of the FutureGen project is a proposed on-site public-access visitor center that will welcome local as well as national and international visitors. This visitor center has been planned to involve a public demonstration of all activities conducted at the site. In the original project design, with the power plant and CO_2 storage in the same location, this proposal would have allowed visitors to witness a range of plant activities, from the coal being unloaded to the pipelines delivering the captured CO_2 to the storage injection site. With FutureGen 2.0, plans for the visitor center are associated with the CO_2 storage location rather than with the power plant.

From the onset, the FutureGen project was designed to have an international influence. The FutureGen Industrial Alliance consists of multinational companies, and the project is designed to facilitate learning throughout the global CCS community. Recognizing that China is another global leader in coal use and has the world's third-largest coal reserve base, FutureGen partnered strategically from the outset with the largest coal-fueled utility in China, the China Huaneng Group, particularly in reference to the IGCC design component of the project. Although FutureGen was initially designed to be the first full-scale integrated IGCC-CCS demonstration project in the world, China has developed its own very similar project. With the delays and the restructuring of FutureGen, the Chinese GreenGen project is now further along in demonstrating CCS.

The initial FutureGen project was designed to serve as a backbone that would connect other smaller-scale CCS R&D efforts and thereby enable the integration and testing of multiple different technological components. The goal was to encourage researchers and research groups worldwide to come to FutureGen to utilize different slipstreams for their R&D efforts in order to take advantage of the potential for testing new capture technologies. With the more limited scope of FutureGen 2.0, this ambition has been scaled back. Another critical aspect of the project is that all information and technical advancements developed within FutureGen are to be publicly available so that intellectual property issues do not produce limitations on learning.

Throughout the almost decade-long FutureGen planning process, different kinds of learning have occurred. Key players in the project have been learning about the unstable and increasing projected project costs, the political instability of government support for the project, and the ways in which various publics have responded to the project. In addition, the high-profile nature of the proposed project has resulted in social learning and increased awareness about climate change and the climate-mitigating potential of emerging energy technologies, particularly CCS.

The notion of the United States as the home of the first large-scale fully integrated coal-fired power plant with CCS emerged from an aspect of U.S. national identity whereby the country is seen as a technologically advanced world leader. From the onset, FutureGen was framed as a critical project that would demonstrate to the world how a state-of-the-art near-zero–emissions coal-fired power plant could be designed and built. This framing of international technological leadership is also associated with both political and economic benefits. The strength of such international framing as well as the economic benefits associated with FutureGen can be seen in the logo of the FutureGen for Illinois organization: "The world needs FutureGen, FutureGen needs Illinois."

Given the presentation of FutureGen in public and political discourse, the proposed project has itself become a framing device; that is, the project signifies U.S. leadership in technological advances on climate change mitigation.

The framing also promotes the notion that a zero-emissions coal-fired power plant is possible and thereby supports the concept of clean coal.

Jennie C. Stephens

Further Reading

Bielicki, J., and J. C. Stephens. *Public Perception of Carbon Capture and Storage Technology Workshop Report.* Energy Technology Innovation Policy Group Workshop Series. Cambridge, MA: Harvard Kennedy School, 2008.

de Coninck, H., J. Stephens, and B. Metz. "Global Learning on Carbon Capture and Storage: A Call for Strong International Cooperation on CCS Demonstration." *Energy Policy* 37(6) (2009): 2161–2165.

Department of Energy. *FutureGen, Integrated Hydrogen, Electric Power Production and Carbon Sequestration Research Initiative.* Washington, DC: Department of Energy, Office of Fossil Energy, 2004.

FutureGen Alliance. http://www.futuregenalliance.org/.

FutureGen Alliance. "Guidance to Prospective Offerors (10-5-2010)." 2010, http://www.futuregenalliance.org/media/FGA_Guidance_100510_Final.pdf.

FutureGen for Illinois. http://www.futuregenforillinois.com/.

Global CCS Institute. *Strategic Analysis of the Global Status of Carbon Capture and Storage: Report 1, Status of Carbon Capture and Storage Projects Globally.* 2009. http://cdn.globalccsinstitute.com/sites/default/files/publications/5751/report-1-status-carbon-capture-and-storage-projects-globally.pdf.

IPCC. *IPCC Special Report on Carbon Dioxide Capture and Storage: Intergovernmental Panel on Climate Change, Working Group III.* Cambridge: Cambridge University Press, 2005.

Stephens, J. C. "Technology Leader, Policy Laggard: Carbon Capture and Storage (CCS) Development for Climate Mitigation in the U.S. Political Context." In *Caching the Carbon: The Politics and Policy of Carbon Capture and Storage,* edited by J. Meadowcroft and O. Langhelle, 22–49. Cheltenham, UK: Edward Elgar, 2009.

G

Gaia Hypothesis

The Gaia hypothesis refers to a range of scientific models that begin with the premise that organisms interact with and modify nonliving elements of Earth's ecosystem in ways that foster and maintain conditions favorable for the continuation of life. The Gaia hypothesis was developed by James Lovelock, an atmospheric scientist and chemist working in collaboration with Lynn Margulis, a microbiologist. The concept appeared in *New Scientist* in 1975 and was more fully elaborated in the 1979 book *Gaia: A New Look at Life on Earth.* The term "Gaia" is a reference to the Greek goddess of Earth and a metaphor for Lovelock's assertion that the planetary system functions in many ways like a self-regulating organism.

The Gaia hypothesis highlights the apparent harmony and interdependence between organisms and their environment. It also emphasizes the stability of the Earth system as a whole, which has supported life in one form or another for billions of years and sustained more advanced multicellular organisms for hundreds of millions of years. Many elements of the Gaia hypothesis are based on well-established scientific consensus, though other aspects remain controversial and contested. It is important to recognize the positive impact of the Gaia hypothesis in furthering a holistic systems-based approach to environmental science. However, we cannot overlook the areas where Gaia models and empirical evidence part ways. These points of divergence have profound implications for our understanding of climate systems and what we might expect from climate change.

Following Lovelock's initial publications, the Gaia hypothesis was the subject of much criticism, confusion, and controversy both within the scientific community and more broadly as the idea mixed and mingled with New Age mysticism. Some of the more fundamental problems were addressed at the First Gaia Conference in 1988, organized by climatologist Stephen Schneider and the American Geophysical Union. There, Lovelock presented a refined version of the Gaia hypothesis. Using a computer simulation called Daisyworld, he attempted to demonstrate that the self-regulating function of Gaia could arise through biological mechanisms, without conscious intent. This was a direct response to criticism that the idea was teleological, assuming purpose of design in natural systems, and it did indeed improve the reception of the Gaia hypothesis among scientists. Physicist and philosopher James Kirchner also helped clarify the debate by pointing out that the Gaia hypothesis was being used to describe many different propositions with different degrees of plausibility. Kirchner went on to identify five forms of Gaia, which he located along a spectrum of "weak Gaia" to "strong Gaia." He argued that the weaker forms are well supported by evidence and prior scholarship, while the stronger forms are irrefutable and therefore not very useful as scientific hypotheses.

Weaker forms of the Gaia hypothesis assert that organisms collectively have a significant impact on Earth's environment (Influential Gaia) and that therefore life and the environment evolve interdependently (Coevolutionary Gaia). These forms are well integrated into concepts such as ecosystems, the biosphere, and biogeochemical cycles, which all encompass interactions between organisms and abiotic elements of the atmosphere, hydrosphere, cryosphere, and lithosphere. Stronger forms of the Gaia hypothesis claim that the planetary ecosystem can be modeled as a living organism or superorganism (Geophysiological Gaia) or that the effect of life acting on the environment is to optimize the conditions that sustain and perpetuate life (Optimization Gaia). The strongest forms have largely been abandoned by scientific inquiry, although they retain metaphorical and spiritual significance to many proponents of Gaia.

Between these extremes is a form of the Gaia hypothesis that Kirchner has labeled Homeostatic Gaia, which proposes that interactions between organisms and abiotic systems are regulated by negative feedback mechanisms that have the net effect of stabilizing the global environment. It is Homeostatic Gaia that has remained a focus of research and debate. To a large extent, this more sophisticated Gaia model reflects a broad shift in climate science and biogeochemical research toward a systems-based approach to the global environment. But it goes well beyond the conventional wisdom that climate and life have evolved together and that the Earth system exhibits macrolevel behaviors as a result of ongoing interactions between biotic and abiotic factors of the ecosphere. Homeostatic Gaia is not a world of unpredictable, complex feedback mechanisms. It is a world where a particular kind of feedback (negative) is predominant, such that the environment is stabilized within a particular range deemed beneficial to life.

Homeostatic Gaia is not well supported by research concerning greenhouse gas emissions and climate change. Feedback mechanisms between organisms and the environment can be either stabilizing (negative) or destabilizing (positive), and the consequences of these interactions can be positive or negative for any given population or species. There is no reason to assume that negative Gaian feedback is typical; in fact, there is a growing body of evidence that suggests that the combined effect of biologically mediated feedback loops will amplify (rather than mitigate against) climate change and accelerate the accumulation of greenhouse gases in the atmosphere.

Despite the lack of evidence for a Gaian homeostasis, it is clear that Earth exhibits planetary-scale feedback mechanisms. Ice core records demonstrate that atmospheric fluctuations of chemistry and temperature have stayed within a set range throughout the last four glacial cycles. But it is also clear that this range cannot be considered a hard and fast limit; anthropogenic emissions have caused concentrations of greenhouse gases that far exceed the upper bound of the last 400,000 years. Whether this is because Gaia's self-regulating function has been compromised by human activity or because Earth is a much more complicated system than a thermostat is beside the point. In either case, it is possible and even likely that positive feedback loops will intensify the warming of the planet and the effects of climate change beyond current projections. A belief that biogeochemical cycles and interactions are by nature beneficial to life will not help us grapple with the full complexity of the Earth system or understand its response to these unprecedented changes.

Adrienne LaPierre

Further Reading

Kirchner, J. W. "The Gaia Hypothesis: Fact, Theory, and Wishful Thinking." *Climate Change* 52 (2002): 391–408.

Kleidon, A. "Beyond Gaia: Thermodynamics of Life and Earth System Functioning." *Climate Change* 66(3) (2004): 271–319.

Lovelock, J. E. *Gaia: A New Look at Life on Earth.* 1979; reprint, Oxford: Oxford University Press, 2000.

Lovelock, J. E., and L. Margulis. "Atmospheric Homeostasis by and for the Biosphere: The Gaia Hypothesis." *Tellus* 26(1) (1974): 2–10.

Gender and Climate Science

Climate change is a gendered issue. Research illustrates that women, who make up the majority of the world's poor, are disproportionately affected by climate change. They often rely more than men on natural resources for survival and are bound by gendered roles and responsibilities that involve tasks such as collecting water and fuel. These tasks are made more difficult by the impacts of climate change, especially in the least developed countries. On top of these vulnerabilities, most women and poor men around the world have been left out of climate change debates and policy discussions. These discussions strongly rely on science and technology—fields that are inherently masculinist in their approach and are dominated by educated men, generally from the most powerful countries. The fact that the gendered fields of science and technology play such an important role in defining the problem of and solutions to climate change has important impacts on the lives of women and men around the world.

Femininst Critiques of Science and Technology

Feminist theorists and gender and development practitioners critique the fields of science and technology as being gendered biases on two fronts: the underrepresentation of women in science and technology and the masculinist ideologies underpinning science and technology.

Women are significantly underrepresented in science and technology fields around the world. Often this dearth of women is blamed on the lower literacy or education levels of girls, particularly in developing regions. Additionally, though, the male dominance in these fields is fostered through gendered stereotypes that portray males as being better at rational skills such as problem solving and that associate females with nurturing and domestic responsibilities. The close relationship between science and technological advancement allows developed countries to dominate the field and makes it more difficult for communities and people with more diverse forms of traditional knowledge to join these conversations. Feminist theorists argue that the lack of diversity within climate science leads to biases and oversights that omit certain lines of questioning and data interpretation.

Feminist critiques are concerned not only with who is engaging in science and technology but also what kinds of approaches these fields use to define and address problems. They argue that both science and technology are dominated by masculinist ideologies that prioritize rationality over emotion, abstract thought over bodily experience, and large-scale market-oriented solutions over other options. Feminists are generally concerned with the powerful notions of objectivity and truth that science claims to bring to the table, which effectively frames climate change discussions in a way that makes more diverse types of knowledge and experience seem illegitimate and irrelevant. For example, researchers in the field argue that framing climate change through science as an urgent global crisis sets up quick-fix

In many African nations, refugees from extreme weather and political instability have become more frequent. This photo shows women returning to a camp with firewood for internally displaced persons in Darfur, Sudan, on August 17, 2006. (Derk Segaar/IRIN)

macrotechnological and economic solutions as the only options for addressing this crisis and does not allow room or time to consider how these solutions will affect environments and less powerful groups of individuals around the world in the long term.

Gender Implications of Climate Science and the Use of Technology

Science and technology play an important role in causing, defining, and addressing climate change in ways that are both overtly and more subtly gendered.

First, in terms of contributing to greenhouse gas emissions, a limited amount of gender-disaggregated data indicates that on an individual level, men are responsible for more emissions than women. This is largely due to men's comparative wealth, role in industry, and access to technology. For example, men own the majority of vehicles worldwide, and women are more likely to use public transportation where available. Understanding these differential relationships to technology based on social categories (such as gender and poverty) will be important when attempting to implement emissions reduction strategies.

Second, initial research shows that women are generally more concerned with climate change as an environmental problem than men and are wary of the large-scale technological fixes promoted in international meetings such as carbon sequestration, carbon offset and trading, and genetically modified organisms. Women are more likely than men to see individual behavioral change, sustainable consumption, and reliance on the precautionary principle as legitimate reactions to climate change. Despite these concerns, the male-dominance in and masculinist approaches to the field described above have led to climate policies that generally focus on traditionally male livelihoods and economic and technological solutions.

Third, gender and development practitioners argue that the implementation of these technological solutions must be examined in terms of gendered access to them and the impacts these solutions have on the everyday lives of people in different communities. They recommend that technological transfer to developing countries be accompanied by skills training, access to credit, and community participation so that women and men can take advantage of these technologies and ideally influence the development of the technologies in the first place. However, even when these recommendations are considered, the impacts that technological transfer will have on gendered power relations are complex and difficult to predict. For example, the introduction of solar panels and battery packs in a village in Bangladesh in order to reduce reliance on bio and fossil fuels resulted in women being expected to undertake more income-generating work using the additional evening light. This extended their workday well beyond men's, further entrenching gender inequalities. Even less technologically dependent reactions to climate change face these problems. For example, some researchers worry that a focus on sustainable household consumption will increase women's work burdens, as they are most often responsible for domestic consumption and household management around the world.

In sum, gender theorists and practitioners argue that technological responses to climate change have gendered implications that must be considered. Ideally, climate change mitigation technologies or adaptation strategies would work hand in hand with the empowerment of women and other disadvantaged groups so that one does not occur at the expense of the other. In addition, attention must be paid to the implications of defining the problem of and solutions to climate change through male-dominated and masculinist fields. Moving forward, climate science must be opened up to include more diverse participants and types of knowledge in order to produce more just and equitable policies and solutions.

Roberta Hawkins

Further Reading

Harding, S. *Whose Science, Whose Knowledge? Thinking from Women's Lives.* Ithaca, NY: Cornell University, 1991.

Kitetu, C. W., ed. *Gender, Science and Technology: Perspectives from Africa.* Dakar: CODESRIA, 2008.

MacGregor, S. "A Stranger Silence Still: The Need for Feminist Social Research on Climate Change." *Sociological Review* 57 (2009): 124–140.

Röhr, U., M. Hemmati, and Y. Lambrou. "Towards Gender Equality in Climate Change Policy: Challenges and Perspectives for the Future." In *Women, Gender and Disasters: Global Issues and Initiatives,* edited by E. Enarson and P. G. Dhar Chakrabarti, 289–303. Washington, DC: Sage, 2009.

Terry, G., eds. *Climate Change and Gender Justice.* London: Practical Action Publishing, Oxfam, 2009.

Geochemical Carbon Management

Many approaches to carbon management have been proposed to store the carbon released from the burning of fossil fuels in a reservoir other than the atmosphere. One important category of carbon management approaches is the methods that rely on geochemical reactions. In general, carbon management approaches can be categorized either by location (i.e., land surface, ocean, or geosphere) or by the form of storage (i.e., whether biological, geochemical, or physical processes are the primary mechanism for storage). Within this categorization, biological approaches rely on the photosynthetic process to capture and convert atmospheric CO_2 into organic carbon, geochemical approaches rely on a chemical reaction to transform the carbon in gas-phase CO_2 into dissolved or solid-phase carbon, and physical approaches rely on barriers that confine gas-phase CO_2 in a location other than the atmosphere. Geochemical approaches have received less attention than physical or biological approaches.

Geochemical approaches include the possibility of engineering geochemical processes to accelerate the formation of carbonate minerals and to increase the ocean's carbon storage capacity. Geochemical approaches could also include the chemical processes involved in utilizing CO_2 in the production of chemical products. Other industrial processes, including cement manufacture and steel production, provide potential niche opportunities for cost-effective chemical storage of CO_2, but these have a limited scale of potential CO_2 reductions. Chemical approaches deserve attention, given the scale of the CO_2 problem, the need to develop a portfolio of potential mitigation strategies, and the potential obstacles in feasibility and effectiveness associated with many of the physical and biological carbon storage approaches.

The carbon storage approach that has received the most attention is the physical approach of directly injecting CO_2 gas into geologic formations, including depleted oil and gas reservoirs, unminable coal seams, and deep saline aquifers. The mobility of the injected gas and the potential risks associated with leakage of the CO_2 back into the active biosphere are, however, serious unresolved concerns, as are uncertainties in storage capacity. While biologic storage is the approach that involves the least engineering, the short (decade-scale) storage time frame limits the potential of biological storage, as does recent research suggesting that the biosphere may soon become a net source rather than a net sink of atmospheric carbon due to changes in climate. Additionally, negative environmental impacts of large-scale biological storage associated with ecologically precarious monoculture plantations and the replacement of native forests with faster-growing species could be devastating.

Geochemical approaches involving the conversion of CO_2 gas to carbonate minerals or dissolved bicarbonate bypass many limitations associated with physical and biological approaches; these approaches have the potential to reduce risks of leakage, expand geographic limits, and minimize requirements for integration with existing infrastructure by allowing for the direct removal of CO_2 from the atmosphere. A need for more serious consideration of these chemical approaches has been recognized, but major challenges exist.

Multiple Geochemical Approaches to Carbon Management

Proposals to engineer chemical carbon storage can be considered methods of accelerating the components of the natural carbon cycle that act on the 10,000- and/or 1 million-year timescale. Many of these approaches involve attempts to accelerate the weathering of magnesium (Mg) or calcium (Ca) silicate minerals so that the cations can either be available to form carbonates or to increase the alkalinity of the oceans. In discussions of enhancing silicate mineral weathering to facilitate carbon storage, Mg silicate minerals are often targeted for two critical reasons: (1) Mg silicate minerals are found in larger and more concentrated deposits than are Ca silicate minerals, so they are more accessible for large-scale mining, and (2) Mg silicates are more reactive than Ca silicates.

Making Solid Carbonate Minerals

Carbon may be immobilized by reacting carbon dioxide (a weak acid) with geologically available bases containing metals such as magnesium or calcium to produce carbonate minerals, a process known as carbonation. In the presence of suitable bases, carbonates are the lowest-energy (or, more precisely, lowest-enthalpy) state for carbon, lower than carbon, which is fully oxidized as CO_2. The formation of carbonates from CO_2 is therefore exothermic. Carbonates are geochemically stable, so carbon within these minerals can be immobilized for millennia, eliminating the safety and environmental concerns associated with leakage of confined gas-phase CO_2.

Industrial carbonation, engineered facilities designed to make solid carbonates to dispose at or near Earth's surface, could mimic a natural process; the carbon released from the lithosphere by burning fossil fuels will naturally be returned to the lithosphere over millennia as carbonates are generated by weathering. Carbonation may therefore be described as the creation of an opposite carbon flux, the anthropogenic return of carbon to the lithosphere.

Ca and Mg are the most common elements to form stable carbonates, so attention has focused on the potential formation of Ca and Mg carbonates. Iron (Fe) is another common element that forms stable carbonates, and although the potential of forming Fe carbonates has been recognized, the engineered acceleration of iron carbonate formation has not yet been extensively explored.

In principle, there are three potential sources of cations required for carbonation: the oceans, ultramafic igneous rocks, and cation-rich subsurface brines. In practice, extracting cations from the oceans to make carbonates on land does not make sense because the net effect would be to increase atmospheric CO_2, as the removal of ocean alkalinity would decrease the ocean's store of dissolved carbon. In fact, the opposite process—dissolving calcium carbonate, releasing CO_2 to the atmosphere and dumping the calcium oxide (CaO) in the oceans to increase oceanic uptake—is a plausible method of removing atmospheric CO_2. Both deposits of ultramafic igneous rocks (Ca and Mg containing silicate minerals) and subsurface brines rich in Ca and Mg are, however, potentially viable sources of cations for carbonation.

Using Silicate Minerals as Cation Sources for Making Solid Carbonate Minerals

Ca and Mg silicate minerals are abundant; calcium makes up 3.9 percent of Earth's continental crust, while Mg makes up 2.2 percent. Deposits of Ca and Mg silicates exceed quantities of fossil fuels, so enhanced weathering of these minerals could theoretically compensate for all the CO_2 entering the atmosphere from the burning of fossil fuels. The kinetics of the chemical reactions critical to carbonation, particularly the weathering of silica minerals to provide the source of cations, are extremely slow. Research in this area therefore has focused on the acceleration of these reactions at elevated temperatures and pressures.

Research on accelerating the kinetics of aqueous silicate mineral weathering and subsequent carbonation has focused more intently on Mg silicates rather than Ca silicates because they are more easily accessible and found in more concentrated deposits. Olivine (Mg_2SiO_4) and serpentine ($Mg_3Si_2O_5[OH]_4$) are the most abundant Mg silicates, although serpentine is more abundant and accessible than olivine and olivine has a higher molar concentration of Mg than serpentine and reacts more rapidly, so olivine is the mineral studied most extensively. The chemical reactions for aqueous-phase conversion of olivine to the Mg carbonate, magnesite, are shown in figure 1.

In this and similar reactions, the kinetics are typically limited by dissolution of the silicate to produce aqueous metal ions (reaction 1) rather than by the subsequent reaction of the metal ions with bicarbonate to produce carbonates (reaction 2) (see figure 1). During simultaneous dissolution and carbonation, interactions among the species involved in both reactions can alter the overall rate of carbonation, that is, the formation of the carbonate phase can coat the silicate phase, hindering dissolution. To accelerate these naturally slow reactions to timescales that might be manipulated to reduce atmospheric CO_2 concentrations, various different mechanical and thermal pretreatments in addition to high temperature and pressure reaction conditions have been examined.

$$Mg_2SiO_{4(s)} + 4CO_2 + 2H_2O \rightarrow SiO_2 + 4HCO_3^- + 2Mg^{2+} \quad (1)$$

$$2Mg^{2+} + 4HCO_3^- \rightarrow 2MgCO_3 + 2H_2O + 2CO_2 \quad (2)$$

$$Mg_2SiO_{4(s)} + 2CO_2 \rightarrow 2MgCO_{3(s)} + SiO_2 \quad (3)$$

Figure 1 Chemical reactions for aqueous-phase conversion of olivine to the Mg carbonate (magnesite)

Using Brines as Cation Sources for Making Solid Carbonate Minerals

The idea of using brines, groundwaters rich in accumulated cations, as cheap and easily accessible sources of Ca and Mg as inputs to an industrial process to make carbonates has been given some consideration. Within the context of long-term biogeochemically cycling, dissolved Ca and Mg cations in brines are simply products of already weathered material, so isolating these cations to be reacted with CO_2 requires considerably less effort than extracting the cations from silicate minerals. This approach, because it bypasses the silicate weathering step, avoids the slow kinetics of silicate dissolution and the associated problem of carbonate formation coating the silicate surface, hindering silicate dissolution, and slowing down the carbonation process.

This idea was developed with the suggestion that calcium-rich saline evaporitic brines could be pumped from the subsurface to a chemical plant located near power plants or oil refineries, and calcium carbonate could be precipitated with the CO_2 from the power plant and the calcium chloride solution of the salty brine. This process could be viewed as the conversion of one acid, carbonic acid, into another acid, hydrochloric acid. With each mole of carbonate precipitated, two moles of hydrochloric acid would be formed; within this context, the challenge becomes how to effectively dispose of large amounts of hydrochloric acid (the quantities would be so large that usage in secondary industries would be negligible). Waste brine and spent acid could be injected back to the subsurface, while surplus carbonates could be disposed of in coal pits. The hydrochloric acid would eventually dissolve subsurface bases making salts, so the whole process may be viewed as a two-step method of solution mining of base minerals. Saline brines are often located in the same sedimentary basins as oil and gas reserves, so transportation costs associated with getting CO_2 and the cations in the same location could be minimal.

Barriers to Industrial Carbonation

The most important barrier to using industrial carbonation as a carbon management technique is the slow kinetics associated with extracting the required base cations, reactions that occur very slowly in nature. Elevated temperatures and pressures are the most obvious means to accelerate the reactions, but high-pressure reaction vessels are expensive, and a successful process will only be possible if very high reaction rates can be achieved so that the high capital cost of the pressure vessels can by offset by high material throughputs. Likewise, while heat can accelerate the reactions, the thermal energy inputs will need to be small (per mole of carbon) in comparison to the energy released by combustion of fossil fuels that generated the carbon. Although there is no fundamental physical barrier, results to date do not allow the design of industrial processes that would be cost competitive with other means of managing carbon. Preliminary estimates of the cost of an engineered olivine carbonation process, for example, have exceeded $250 per ton of carbon, roughly on order of magnitude larger than the costs of geologic disposal by direct injection of CO_2 into the subsurface. (This figure is the cost per ton avoided; it is larger than the cost per ton carbonated because it accounts for the carbon emissions arising from energy used in the process.)

Beyond kinetics and cost, other challenges include accessing the required bases in sufficient quantity and disposing of the carbonates both with acceptable cost and acceptable environmental impact. Carbonating all the CO_2 from a 1 gigawatt (GW) coal-fired power plant would require approximately 55 kiloton (kt) of mineral per day, roughly five times larger than the mass of coal consumed by the plant. For a given quantity of electricity, it would be necessary to extract, process, and dispose of a mass of silicates and carbonates that are several times larger than the mass of coal. Offsetting this unfavorable mass ratio is the fact that coal is often found in relatively narrow seams that require displacement of a considerably larger mass of overburden than the mass of coal, whereas magnesium silicates are, in some locations, found in very large deposits, which would require displacement of a relatively small amount of overburden per unit of silicate. It is plausible that in favorable circumstances the total mass of materials (including overburden) that would need to be moved for the carbonation process would be smaller than that required for the coal. The same might be true of the amount of surface area disturbed by mining. Countering this possibility is the difficulty of finding places where coal deposits and magnesium silicates are collocated. In North America, for example, ultramafic deposits are found along both coasts rather than in the interior, where most coal deposits and coal-fired power plants are located. Systematic analyses of the availability of collocated deposits of coal and ultramafic rocks and of the costs and environmental impacts of the mining activities required for large-scale industrial carbonation are needed to judge the extent to which these factors might constrain the use of industrial carbonation.

Increasing the Store of Dissolved Carbon in the Oceans by Increasing Alkalinity

More than 99 percent of dissolved carbon in the oceans is present in the form of carbonate or bicarbonate; these are the ocean's dominant anions, and their

$$2\times[CO_3^{2-}] + HCO_3^-] \cong [A]$$

Figure 2 Alkalinity is equivalent to the sum of charges of the ocean's dominant anions, carbonate (CO_3^{2-}) and bicarbonate (HCO_3^-)

total charge is effectively equal to the supply of cations, called the alkalinity, as shown in the equation in figure 2.

At constant alkalinity, shifts in the ratio between the doubly-charged carbonate ions and the singly charged bicarbonate ions can increase the total amount of dissolved carbon and change the pH. The current rise in atmospheric CO_2 has resulted in an increase in oceanic uptake of CO_2, and because of the relatively constant alkalinity during this period, surface bicarbonate ion concentrations have increased while carbonate ion concentrations have decreased, resulting in a reduction of surface-ocean pH. If the alkalinity of the oceans was increased, the ocean's store of dissolved carbon could be increased without shifting the carbonate/bicarbonate ratio and changing the ocean's pH. Increasing the ocean's alkalinity therefore has the potential to increase the ocean's store of dissolved carbon while also restoring ocean pH to preindustrial levels.

Two distinct strategies of increasing the store of dissolved carbon in the oceans by increasing the alkalinity of the oceans have evolved: (1) accelerating calcium carbonate mineral dissolution and adding the dissolution products to the surface oceans, a set of approaches that manipulates the 10,000-year timescale marine carbon chemistry process, and (2) adding alkalinity directly to the oceans without adding additional carbon either from noncarbonate sources or from carbonate dissolution with CO_2 capture. This second strategy is an attempt to accelerate both the million-year weathering timescale by enhancing the release of alkalinity from minerals and the 10,000-year carbonate chemistry processes by mitigating against the increasing acidity of the surface oceans and its impact on the distribution of inorganic dissolved carbon.

Each approach increases the ocean alkalinity. As atmospheric CO_2 has been increasing over the past century, ocean alkalinity has been relatively stable; at constant alkalinity, increases in partial pressure of carbon dioxide (pCO_2) result in an increase in acidity and associated shifts of the distribution of dissolved carbon from CO_3^{2-} to HCO_3^-. By adding alkalinity, the ocean's capacity to store dissolved inorganic carbon will increase, and the associated change in acidity will be reduced. These approaches therefore have the potential to simultaneously reduce atmospheric CO_2 and also reduce the ocean acidification that is occurring.

The first strategy, adding the dissolution products of either $CaCO_3$ or $MgCO_3$ to the oceans, would contribute two units of alkalinity for each unit of carbon, allowing a net increase in the ocean's uptake of atmospheric CO_2. Although this approach has more favorable kinetics than silicate mineral dissolution, which is associated with the second strategy, twice as many moles of carbonate material than silicate mineral material would be required to achieve the same amount of additional oceanic carbon storage.

Barriers to Increasing Ocean Alkalinity

A potential challenge associated with adding alkalinity is that sudden increases in the local concentration of base cations, carbonate, or hydrogen ion (pH) are likely to result in precipitation of carbonates or oxides (oxide precipitation can occur with a spike in pH); this precipitation would remove alkalinity from the water, defeating the original purpose. Effective dispersion design would be critical. The addition of rapidly dissolving sodium carbonates, for example, would cause a spike in the carbonate ion concentration, which in the already supersaturated surface waters could lead to significant carbonate formation. The addition of magnesium or calcium hydroxides could result in major pH spikes that could cause the formation of calcium or magnesium oxides. These could be avoided with well-designed injection and dispersion methods, but no systematic analysis has addressed these issues.

Potential environmental consequences to marine biota of increasing surface alkalinity are speculative, and research addressing this issue is needed. Impacts can be divided into local short-term impacts and widespread longer-term impacts. On the local scale, possible sensitivity of marine organisms to pH changes might require the engineering of systems that avoid sudden large perturbations in pH. The degree of this impact will depend on the form of alkalinity to be added; that is, a change in pH would be greater for magnesium hydroxide ($Mg[OH]_2$) than for magnesium carbonate ($MgCO_3$) and the marine organisms that live in the alkalinity distribution areas. On a larger scale, the addition of the base cations would not be expected to impact overall marine ecology because the relative perturbation to the natural high concentrations of Ca and Mg ions in seawater would be small. The associated increases in pH throughout the world's oceans could reestablish the preindustrial ocean pH, so this could be viewed as an additional benefit of the approach. Ocean pH has been decreasing, so a slight increase in pH associated with an increase in alkalinity could be viewed as restoring the oceans. It is not clear, however, that all of the environmental impacts of the current reductions in pH could be mitigated by a corresponding increase in pH.

Magnesium has been shown to inhibit the formation of $CaCO_3$ by organisms, so it is possible that an increase in oceanic magnesium would further reduce calcification rates. Another potential impact of adding alkalinity would be a dramatic increase in particulate solid material raining down through the ocean. Carbonate particles could affect seawater turbidity and organic carbon export, so it is possible that systems would need to be engineered to minimize the introduction of particulates into the marine environment.

In Situ Geochemical Approaches to Carbon Management

Facilitating these same geochemical reactions in situ—that is, in locations where the minerals exist naturally—encompasses another set of carbon management approaches. By directly injecting captured CO_2 into geologic formations with mineralogical conditions that would enhance the formation of carbonates and increase the solubility of the gaseous CO_2 in the groundwater, geochemical reactions could reduce concerns about leakage.

Both enhanced carbonate formation in calcium and magnesium silicate-rich aquifers and carbonate dissolution in submarine carbonate deposits are potentially promising approaches that harness the combined effectiveness of the physical trapping of the CO_2 gas and the geochemical reactions that transform the CO_2 gas into dissolved or solid carbon. These approaches involve identifying locations for direct injection that have conditions favoring the chemical reactions discussed above that will convert the carbon in CO_2 gas into the solid or dissolved form.

Expanding on the idea of using cation-rich evaporitic brines, some recent discussion has focused on injecting CO_2 underground into cation-rich brines, where some of the CO_2 will dissolve in the brine and some will form carbonates. Deep brine formations have received a lot of attention in discussions of physical underground storage options because they have the largest potential storage capacity of all the candidate underground storage sites; however, the minimal geologic characterization of aquifers compared to oil and gas formations increases concerns about leakage risks. Incorporating brine-water chemistry and the potential for carbonate formation into the underground site-selection process could allow for maximized chemical storage, reducing leakage risks associated with physical storage alone.

It has been long acknowledged that the formation of carbonates will occur to some degree during the direct injection of CO_2 gas into underground aquifers with Ca and Mg silicates mineralogy. Injecting CO_2 into deep aquifers is well recognized as a promising location for carbon storage because of the large potential capacity for storage and the co-occurrence of deep aquifers and fossil fuels within sedimentary basins, but the potential extent and scale of dissolution of underground rocks and subsequent formation of carbonates are not well understood.

One challenge associated with carbonate formation in aquifers is the reduction of effective porosity of the aquifer and the armoring of the remaining surface area to prevent future dissolution. This armoring may reduce further carbonation but could minimize the leakage potential of the gaseous CO_2 escaping from the underground reservoir. Techniques to measure and observe in situ carbonation have also been recently developed to facilitate understanding and quantification of the potential for this approach.

Each of these in situ reactions in an aquifer is dependent on the rate of dissolution of CO_2 in the pore water. How much of the CO_2 gas will be dissolved into the groundwater depends on the specific chemistry of the groundwater as well as the hydrological flow. Indeed, increasing the flow of brines by groundwater pumping could be a way to accelerate CO_2 dissolution.

Another in situ scheme that relies on geochemical reactions and physical trapping of CO_2 gas is the injection of CO_2 into carbonate sediments on the ocean floor at depths sufficient to ensure that the CO_2 is negatively buoyant. In geological reservoirs and in the ocean, the density of CO_2 increases with increasing depth/pressure. In geologic reservoirs, the increasing temperature means that the density of CO_2 is always less than that of water, whereas in the ocean, in which temperatures are cold and roughly independent of depth below the thermocline, there is a depth, typically about 3.5 kilometers, below which the density of CO_2 exceeds that of water. Sediment deposits on the abyssal plain are relatively thin and poorly consolidated; if they were located at shallow depths, they would be poor candidates for geologic storage, as CO_2 would probably leak to the sediments into the ocean. But at depths below which CO_2 is negatively buoyant, the risk of leakage is greatly reduced, as CO_2 would tend to migrate downward through the porous media.

Leakage from such sediments is still possible, however, because the sediments are unconsolidated and would probably be fractured by injection of CO_2, which could then migrate upward. In carbonate sediments, however, dissolution and subsequent addition of alkalinity to the pore fluid would tend to neutralize CO_2 that was released, providing a second trapping mechanism. Finally, release of CO_2 to the oceans will be impeded by a third trapping mechanism, the formation of CO_2 hydrates with pore waters. Among the major uncertainties of this approach to be investigated in currently ongoing research is the potential capacity associated with deep carbonate sediments and the actual integrity and security of CO_2 gas injected into these deposits.

Jennie C. Stephens

Further Reading

Caldeira, K., and G. Rau. "Accelerating Carbonate Dissolution to Sequester Carbon Dioxide in the Ocean: Geochemical Implications." *Geophysical Research Letters* 27(2) (2000): 225–228.

Kheshgi, H. S. "Sequestering Atmospheric Carbon Dioxide by Increasing Ocean Alkalinity." *Energy* 20(9) (1995): 915–922.

Lackner, K. S. "Carbonate Chemistry for Sequestering Fossil Carbon." *Annual Review of Energy and the Environment* 27 (2002): 193–232.

Stephens, J. C., and D. W. Keith. "Assessing Geochemical Carbon Management." *Climatic Change* 90(3) (2008): 217–242.

Geoengineering

The term "geoengineering" represents a set of potential approaches to climate change mitigation that involve manipulating Earth's systems in large-scale intentional ways. Geoengineering is generally defined as a large-scale intentional alteration of Earth's energy balance to counter anthropogenic influences.

Among the different climate mitigation approaches considered to be geoengineering are proposals to deflect incoming solar radiation by injecting sulfate aerosols into the atmosphere or installing mirrors in the atmosphere. Other definitions of geoengineering include various approaches to engineering ways to enhance the removal of carbon dioxide from the atmosphere.

Geoengineering has become a controversial topic primarily due to perceived risks and unintended consequences of large-scale manipulation of Earth's systems as well as concerns about the governance, implementation, and management of such global-scale actions. Many scientists and policy makers are adamantly opposed to considering geoengineering or supporting consideration and research related to geoengineering. Yet other scientists who are aware of the risks of climate change and the seemingly insurmountable social and political challenges to reduce greenhouse gas emissions at the scale that is required to mitigate climate change claim that it would be irresponsible for society not to explore the potential of geoengineering as a possible last-ditch solution to the rising threats of climate change.

Geoengineering has gained increasing attention in the past 5 to 10 years or so because of the lack of discernable progress in slowing the growth of CO_2 emissions. Some scientists and engineers who understand the dangers of climate change and recognize the political and social challenges of reducing greenhouse gas emissions have been calling for increased research on novel approaches to carbon management, including geoengineering. A lively dialogue and debate about the role and responsibility of scientists in considering and researching geoengineering approaches to confronting climate change has emerged since the 2006 publication of P. J. Crutzen's speculative paper on the possibility of injecting sulfur aerosols into the stratosphere to offset Earth's warming with cooling derived from engineered enhanced albedo from the sulfur particles. One of the main objections to this and other geoengineering ideas that seek to alter Earth's radiative balance is that they attempt to treat the symptoms rather than the cause, and thus they ignore and may even exacerbate other environmental problems caused by high atmospheric CO_2 concentrations. Given these important concerns and objections as well as recent calls for combined mitigation and geoengineering approaches, increased attention to and consideration of other approaches that can directly reduce atmospheric CO_2 concentrations are appropriate.

Some reviews of geoengineering approaches define two major categories: (1) sunlight-reducing approaches and (2) infrared-altering approaches. Sunlight-reducing approaches attempt to restore Earth's energy balance by reducing incoming solar radiation. Infrared-altering approaches attempt to restore Earth's energy balance by reducing the greenhouse gases that are accumulating in the atmosphere and by absorbing the outgoing infrared radiation.

Among the most explored and discussed geoengineering approach is the idea of injecting sulfate aerosols into the atmosphere to deflect some incoming solar radiation. Sulfate aerosols reflect incoming UV radiation, reducing the amount of direct sunlight at the surface. This increases the total planetary albedo and is known as the direct effect of aerosols. Aerosols also act as cloud condensation nuclei and therefore change the albedo, lifetime, and water content of clouds. This is the indirect effect of aerosols, which is less predictable and not highly understood. A geoengineered world could include a stratosphere that is injected with enough sulfate aerosols to increase radiative forcing, cool the planet, and counteract climate warming. The theory is based on volcanic eruptions that temporarily cool the planet when large amounts of sulfur particles are emitted, although the cooling is regional, and higher latitudes tend to experience warming. A geoengineering scenario is different because it requires continual injections of aerosols or precursors, which is unprecedented, and the long-term impacts are unknown.

The contribution of Working Group I of the Intergovernmental Panel on Climate Change (IPCC) fourth assessment report discusses in detail that the uncertainty in the radiative forcing potential for sulfate aerosols is higher than previously thought. In addition, when the IPCC mentions this approach, the authors state clearly that geoengineering options remain "largely speculative," with the risk of unknown side effects. This option is the last on the list of short- to medium-term mitigation approaches.

Among the many concerns associated with the idea of injecting sulfate aerosols into the atmosphere to reduce incoming solar radiation to combat climate change is the fact that such an approach does not lessen any of the other impacts of heightened CO_2 concentrations. For example, introducing large quantities of aerosols into the atmosphere will not moderate many regional climate changes and will do nothing to counter ocean acidification that is a direct result of elevated carbon dioxide concentrations in the atmosphere.

Another major concern that has been raised is related to the specific carbon mechanisms for sulfate injection. In order to inject the amount of aerosols necessary to counteract climate warming, one estimation suggests that an aircraft injection scenario would require 1 million flights per year, with thousands of aircraft operating continually. This mechanism would clearly have major environmental impacts. Also, the complexities associated with aerosol particle size, concentration, and coagulation processes are not well understood, which provides additional concerns associated with geoengineering proposals.

One of the most widely discussed geoengineering approaches that can be categorized as an infrared-altering approach is the idea of iron fertilization of the oceans to enhance the capacity of the oceans to absorb carbon dioxide from the atmosphere. Iron is a rate-limiting nutrient that currently restricts the growth of plankton. If there were more plankton in the surface oceans performing photosynthesis and taking up atmospheric carbon dioxide, the flux of carbon dioxide into the atmosphere from fossil fuel burning would be counterbalanced to a greater degree. Another approach to geoengineering removal of carbon dioxide from the atmosphere involves designing and installing artificial trees on land that could take carbon dioxide out of the atmosphere.

Despite major concerns and controversy, geoengineering has gained increased attention and support. Several scientific conferences have recently included sessions on geoengineering, and several scientific associations are commissioning reports on geoengineering. In addition to scientific, technical, and environmental uncertainty, there are major social and political questions associated with what entity or organization would be responsible for deciding upon, implementing, and monitoring any geoengineering strategy and how adjustments and modifications would be managed.

Jennie C. Stephens and Julia Lendhart

Further Reading

Crutzen, P. J. "Albedo Enhancement by Stratospheric Sulfur Injections: A Contribution to Resolve a Policy Dilemma?" *Climatic Change* 77(3–4) (2006): 211–219.

Keith, D. W. "Geoengineering the Climate: History and Prospect." *Annual Review of Energy and the Environment* 25 (2000): 245–284.

Mercer, A., D. W. Keith, and J. D. Sharp. "Public Understanding of Solar Radiation Management." *Environmental Research Letters* 6 (2011): 1–9 doi: 10.1088/1748-9326/6/4/044006.

Robock, Alan. "20 Reasons Why Geoengineering Might Be a Bad Idea." *Bulletin of the Atomic Sciences* 64(2) (2008): 14–18.

Wigley, T. M. L. "A Combined Mitigation/Geoengineering Approach to Climate Stabilization." *Science* 314(5798) (2006): 452–454.

Geoengineering and Public Policy

Climate change geoengineering, defined by the Royal Society as "deliberate large-scale manipulation of the planetary environment to counteract anthropogenic climate change," was until recently considered outside the mainstream of climate policy making. However, the feckless response of the world community to addressing climate change has dramatically changed the landscape in the past few years. While there is a solid consensus among scientists and policy makers that temperature increases of 2°C or greater above preindustrial levels will visit extremely serious impacts on human institutions and ecosystems, the current commitments made by the world community to reduce greenhouse gas emissions put us on track for temperature increases of 2.5–4.2°C by the end of this century, with further increases thereafter. As a consequence, there is increasingly serious consideration of the potential role of geoengineering as a means to avoid a climate emergency, such as rapid melting of the Greenland and West Antarctic ice sheets, or as a stopgap measure to buy time for effective emissions mitigation responses. However, little research has been conducted on potential geoengineering options, and no major research programs are currently in place.

Categories of Climate Geoengineering

Climate geoengineering options can be divided into two broad categories: solar radiation management (SRM) methods and carbon dioxide removal (CDR) methods.

Solar Radiation Methods

SRM methods focus on reducing the amount of solar radiation absorbed by Earth by an amount sufficient to offset the increased trapping of infrared radiation by rising levels of greenhouse gases. In more popular parlance, these schemes essentially put a dimmer switch on the sun. Researchers have concluded that the warming effects associated with a doubling of atmospheric carbon dioxide concentrations could be offset by reducing solar radiation inflows by 1.8 percent. The primary SRM geoengineering schemes that have been discussed to date are stratospheric sulfur aerosol injection, cloud albedo enhancement, and space-based systems.

Perhaps the most widely discussed climate geoengineering option is enhancement of planetary albedo (surface reflectivity of sun's radiation) using stratospheric aerosols. While most proposals have focused on the use of sulfur, other potential options include aluminum, hydrogen sulfide (H_2S), carbonyl sulfide, ammonium sulfide, soot, and engineered nanoscale particles. A recent study concluded that the amount of sulfur emissions required to compensate for projected warming by 2050 would be 5–16 teragrams (TgS) per year, increasing to 10–30 TgS per year by the end of the century. Potential delivery vehicles for stratospheric sulfur dioxide injection include aircraft, artillery shells, stratospheric balloons, and hoses suspended from towers. Supporters of stratospheric aerosol injection tout the fact that it could prove to be an extremely cheap option, perhaps costing only a few billion dollars annually.

However, sulfur aerosol injection schemes could pose negative ramifications. Reductions in evaporation associated with deployment could substantially weaken Asian and African monsoons, threatening the food and water supplies of billions of people. Moreover, there is substantial concern that injection of sulfur particles into the stratosphere could imperil recovery of the ozone layer by catalyzing chemical reactions that deplete ozone. This could delay recovery of the ozone layer in the Antarctic by between 30 and 70 years, substantially increasing the incidence of skin cancer and other maladies associated with UV-B radiation. Finally, unless greenhouse gas emissions were substantially reduced after deployment of a sulfur aerosol injection scheme or any other SRM approach, termination of such a scheme could result in warming of more than 20 times current rates. This would be a consequence of the buildup of carbon dioxide that had accrued in the atmosphere in the interim, with its suppressed warming effect, as well as the temporary suppression of climate-carbon feedbacks.

Cloud albedo enhancement geoengineering schemes contemplate dispersing seawater (NaCl) droplets approximately one micrometer in size in marine stratiform clouds. These droplets would be sufficiently large to act as

We need energy and environmental policy. That is critical. We need to create an environment for technology to flourish. In a sense, I think of our company as a distributor of new technologies in many ways—an implementer, a tester. We use our system as a beta site. We also recognize that we have to retire and replace, and that reality is right in front of us, because it's almost as daunting to do that in the next 40 years as it is for the Chinese to fill the nuclear demand. So, in that sense, we share that in common.

But what's missing—and I think this is why I'm delighted to be here tonight—is that we need to connect the dots. We need to connect the dots between energy, environmental issues, and the economy. We need to connect those dots. We can't afford to think linearly. We need to think holistically in terms of how these pieces fit together in the future. And we must address all these issues contemporaneously. And we have to take a portfolio approach. We can't take anything out of the equation.

Jim Rogers, Duke Energy, 2010

cloud condensation nuclei when they rise into the bases of stratiform clouds and shrink through evaporation to about half their original size. Increases in cloud condensation nuclei increase cloud droplet numbers and decrease cloud droplet size. This enhances overall droplet surface area and results in an increase in cloud albedo. Studies indicate that a 50–100 percent increase in droplet concentration of all marine stratiform clouds by mechanical generation of sea salt spray could offset warming associated with a doubling of atmospheric carbon dioxide. As is the case with sulfur dioxide injection schemes, the cost of this approach could be extremely low, perhaps no more than $2 billion.

As is the case with sulfur dioxide injection, cloud albedo enhancement could have substantial impacts on regional precipitation patterns. Recent studies indicate that some areas characterized by low precipitation, such as sub-Saharan Africa and eastern Australia, could actually benefit from increased rainfall associated with cloud albedo enhancement. However, certain areas of South America, including in the Amazon, could experience disastrous declines in precipitation, as much as 50 percent in some regions.

The final primary SRM option is the positioning of sun shields in space to reflect or deflect incoming solar radiation, potentially reducing radiation inflows by 1.8 percent. Several proposals involve placing reflectors in near-Earth orbits, including placement of 55,000 mirrors in random orbits or the creation of a ring of dust particles guided by satellites at altitudes of approximately 1,200–2,400 miles. An alternative approach could be to establish a cloud of spacecraft with reflectors in a stationary orbit near a gravitationally stable point between Earth and the sun. While deployment of this technology, if proven to be feasible, would take decades to deploy, atmospheric temperatures would respond within a few years after it was in place. Moreover, some proponents contend that the potential side effects would be less significant and more predictable than alternative geoengineering options.

However, deployment of space-based systems could prove extremely challenging. As indicated earlier, some configurations of sunshades could prove unstable and thus ultimately sail out of orbit. Low Earth-orbit systems could also face tracking problems, posing the threat that mirrors could collide. Space-based systems would also present imposing logistical challenges. Estimates of the number of flyers that would need to be produced range from 5 trillion to 16 trillion. This would require an unprecedented scale of production and could take a century to produce. The cost of deployment would also be extremely high, pegged at approximately $5 trillion by one major proponent, though this is arguably far less than the damages associated with climate change under a business-as-usual scenario. Finally, space-based systems could lead to precipitation declines in some regions and could leave space littered with trillions of flyers potentially spread over 60,000 miles or more, posing a threat to Earth-orbiting aircraft.

Carbon Dioxide Removal Methods

Carbon dioxide is the primary anthropogenic greenhouse gas in the atmosphere contributing to climate change. Carbon dioxide removal geoengineering options seek to remove carbon dioxide from the atmosphere and are intended to cool the planet by reducing the absorption of heat in the atmosphere. There are three subcategories of CDR schemes: those that seek to enhance uptake and storage by terrestrial biological systems, those that seek to enhance uptake and storage by oceanic biological systems, or those that use physical, chemical, or biochemical engineered systems.

Ocean iron fertilization (OIF) techniques seek to stimulate the production of phytoplankton through the addition of iron to ocean regions that are allegedly deficient in this micronutrient. The southern ocean is the predominant high-nutrient low-chlorophyll region in the world and thus the primary focus for proponents of OIF. Phytoplankton takes up carbon dioxide from seawater to carry out photosynthesis and to build up particulate organic carbon (POC). Ultimately part of the POC sinks to the deep ocean, where it can be stored for a century or more.

Proponents of OIF have contended that this approach could reduce atmospheric concentrations of carbon dioxide by between 50 and 107 parts per million in 100 years. However, a number of field trials in recent years have severely undercut these estimates, indicating that potential reductions might be more on the order of 15 parts per million. There are also serious concerns about potential negative impacts from OIF deployment, including decreasing primary production in large regions of temperate oceans as a consequence of the exportation of nutrients and the potential production of harmful algal blooms.

A second CDR approach is air capture. Air capture is an industrial process that captures carbon dioxide from ambient air, producing a pure stream of carbon dioxide that can be used or sequestered terrestrially or in oceans. Most potential technologies would use sorbent materials to capture carbon dioxide, such as solid amines, or highly or moderately alkaline solutions. One study concluded that we could capture 650 gigatonnes of carbon by 2100, using 35,000 air capture facilities with a footprint of less than 300 square kilometers.

Assuming there is adequate storage options for captured carbon dioxide, air capture could be a highly desirable choice because it does not appear to pose side effects of the magnitude of many other geoengineering options. However, this option could prove to be cost-prohibitive. Recent estimates peg the cost of air capture at $20 trillion per 50 parts per million of sequestered carbon dioxide. Some proponents contend, however, that technological costs could be substantially reduced over time and would compare favorably with mitigation options.

A third CDR option, mineral sequestration, would seek to accelerate the natural weathering process, producing a reaction between silicate rocks (perhaps olivine or serpentine) and carbon dioxide that forms solid carbonate and silicate materials. The reaction consumes one carbon dioxide molecule for each silicate molecule, with storage of carbon as a solid mineral. While some proponents contend that this approach could store all the carbon that is available in fossil fuels, they also acknowledge the imposing costs of such schemes, probably rendering this option unviable in all but the long term.

Governance Issues

Developing an effective international governance framework for geoengineering testing or deployment is critical. As indicated above, while many nations might benefit from such schemes, others might suffer disproportionately from potential negative side effects. Moreover, the relatively low cost of many of these options means that a single nation, small group of nations, or even a wealthy individual could launch a geoengineering program without consulting others.

Perhaps the most logical institution to govern climate geoengineering activities is the United Nations Framework Convention on Climate Change (UNFCCC), a treaty with 195 party states and tasked with reducing greenhouse gas emissions and helping the world to adapt to climate change. However, the focus of the UNFCCC is on reducing greenhouse gas emissions and enhancing mechanisms that accumulate and store carbon. Thus, while the UNFCCC would clearly have jurisdiction over carbon dioxide removal geoengineering schemes, it is not clear whether it could govern solar radiation management approaches, since they would neither seek to control greenhouse gas emissions nor sequester carbon. Additionally, given the failure of the parties to the UNFCCC to effectively control the greenhouse gas emissions of major emitting states, it is hard to be sanguine about its prospects to regulate the activities of countries that might believe that deployment of geoengineering technology might be critical to protecting their interests.

Another possible avenue for international regulation of geoengineering could be the Convention on the Prohibition of Military or Any Other Hostile Use of Environmental Modification Techniques (ENMOD). However, ENMOD has only been ratified by 73 countries. Moreover, the treaty only prohibits the "military or any other hostile use of environmental modification techniques" (Article 1.1) and expressly preserves the right to use such techniques for "peaceful purposes" (Article 3.1). Thus, any nation deploying geoengineering technologies might successfully argue that it had no hostile intent toward other countries in so doing, even should incidental harm occur.

There are other international regimes that might assert jurisdiction over one or more geoengineering options. For example, both the Convention on Biological Diversity and the International Maritime Organization have sought to circumscribe ocean iron fertilization activities, though the enforceability of these decisions is unclear. Ocean iron fertilization schemes as well as any other geoengineering options that might affect ocean ecosystems might also be subject to pertinent provisions of the United Nations Convention on the Law of the Sea that seek to protect ocean environments from pollution. The Montreal Protocol on Substances that Deplete the Ozone Layer might be pertinent to aerosol injection schemes if they would have negative impacts on stratospheric ozone levels. The Long-Range Transboundary Air Pollution Convention might constrain intentional releases of sulfur aerosols into the atmosphere. Many national laws could also be applied to geoengineering schemes, such as the Clean Air Act and the Clean Water Act in the United States.

William C. G. Burns

Further Reading

Blain, S., et al. "Effect of Natural Iron Fertilization on Carbon Sequestration in the Southern Ocean." *Nature* 446 (April 2007): 1070–1074.

Brovkin, V., et al. "Geoengineering Climate by Stratospheric Sulfur Injections: Earth System Vulnerability to Technological Failure." *Climatic Change* 92 (2009): 243–259.

Bunzl, M. "Researching Geoengineering: Should Not or Could Not?" *Environmental Research Letters* 4 (October–December 2009): n.p.

MacCracken, M. "Geoengineering: Worthy of Cautious Evaluation?" *Climatic Change* 77 (2006): 235–243.

Royal Society. *Geoengineering the Climate: Science, Government and Uncertainty.* London: Royal Society, 2009, http://royalsociety.org/Geoengineering-the-climate/.

Virgoe, J. "International Governance of a Possible Geoengineering Intervention to Combat Climate Change." *Climatic Change* 95 (2009): 103–119.

Geography and Climate Change

Geography is a scientific discipline studying human and physical patterns and processes on the surface of Earth. Two main branches of geography are human geography and physical geography. According to the Association of American Geographers, "human geography is concerned with the spatial aspects of human existence" (i.e., "how people and their activity are distributed in space, how they use and perceive space, and how they create and sustain the places that make up the earth's surface"), while "physical geographers study patterns of climates, land forms, vegetation, soils, and water." In 1963 William D. Pattison highlighted his four traditions of geography, attempting to define the discipline: (1) spatial tradition, (2) area studies tradition, (3) man-land tradition, and (4) earth science tradition. The spatial tradition is related to determination and display of important geographic attributes of reality through mapping, for example, distance, form, direction, and position. Two dominant themes of this tradition are geometry and movement. The area studies tradition focuses on the nature of places, regions, and locations as well as their character and their differentiation from other places.

Hot Spot

Darfur

Stephan Faris, writing in the April 2007 edition of the *Atlantic,* traced the conflict between ethnic groups in Darfur to decreasing rainfall and increasing aridity that has increased competition for land, a likely product of climate change. "By the time of the Darfur conflict four years ago, scientists had identified another cause [in addition to ethnic conflict]," wrote Faris.

Bruce E. Johansen

The man-land tradition is the study of nature's influence on people and of the effects that human beings have on the physical environment. The core of the man-land tradition is the interconnections between man and physical environment. The earth science tradition embraces the study of natural processes in or on Earth's atmosphere, lithosphere, hydrosphere, and biosphere and the associations and the interactions among these physical systems locally, regionally, and globally. Human geography generally consists of the first three traditions applied to human societies, and physical geography is the fourth tradition while mixed with the first and second traditions. Example subfields of geography, differing mainly by research subject matter, are cultural geography, economic geography, urban geography, historical geography, climatology, geomorphology, hydrology, and biogeography. Indeed, the geographic advantage lies in an enhanced understanding of relationships between people and the environment, the importance of spatial variation, processes operating at multiple and interlocking geographic scales, and the integration of spatial and temporal analysis. The issues and potential impacts of climate change show the interconnection of people, places, regions, and environment across multiple geographic and political scales. The discipline of geography is central to current climate change research. Geographers from a wide variety of subfields are very active in the study of global warming and climate change, and in fact they were among the first scientists to alarm us that human-induced changes to the environment were starting to threaten the balance of life itself. Many geographers have contributed to the summary reports of the Intergovernmental Panel on Climate Change (IPCC), and those topic areas are climate system, impacts and response options, and economic and social dimensions of climate change.

Geographers use both qualitative and quantitative methods and techniques in their research. Qualitative methods refer to such methods of inquiry as in-depth interviews, questionnaire surveys, participant observation, focus groups, and life history collection. Quantitative methods are the use of scientific concepts, computational mapping, mathematical modeling, and statistical analysis techniques to understand geographical phenomena. Since the 1990s, geographic information systems (GIS) and remote sensing have been widely applied by geographers as powerful devices to collect, store, describe, analyze, and visualize complex, spatially distributed data. Spatial statistics and geostatistics provide geographers with important frameworks for developing advanced techniques of spatial analysis, modeling, and visualization. Some geographers use questionnaire surveys to assess and understand personal concerns, risk perceptions, and knowledge regarding climate change from various groups of people. The technical abilities of geographers in spatial and statistical analyses are vital in the climate change issue. GIS, remote sensing, and spatial statistics are increasingly becoming valuable tools for geographers and scientists from other disciplines to study regional and global climate change issues, for example, mapping, analyzing, and visualizing spatial patterns of climate change risk perceptions and impacts.

The distinctive intuition and skills of geographers, especially human geographers, can be used to make human sense of climate change. Human geographers have the ability to reconnect culture with climate, understand and disseminate the idea of climate across various scales, and delineate the knowledge claims of climate science and policy. The concept of climate, defined in purely physical terms and constructed from meteorological observations, can only be understood when its physical dimensions are interpreted by their cultural meanings and contexts. The meaning of climate varies in different geographic places, to different people, and at different times. Human geographers can also ensure that local meanings of climate retain their integrity across scales and are indexed to the physical dimensions of weather. In addition, human geographers can critically examine how climate change knowledge, for example, those from various assessments of the IPCC, is spatially produced and consumed within different institutional and cultural settings. This provides a spatially contingent view of climate change knowledge.

One widely cited assertion is that human geographers are uniquely positioned to examine the social, cultural, ethical, and political impacts of climate change. Human geographers have recently started to study how individuals and communities understand climate and climate change at the local landscape level, as a lived personal experience combined with geographies, lay

I grew up in St. Michael, where there are 400 people in the village; and I grew up going to fish camp and doing a lot of cultural activities. I also was brought up by a bunch of my elders, which they taught me to respect others the way I want to be treated and also how to live off subsistence; and, hopefully, someday I get to teach that to my children and my grandchildren.

The global warming effects that I have experienced personally is coastal erosion where my families' houses are falling into the bay and also the graveyards that we have, like the Russian orthodox graveyards, are also falling into the bay; and it is where all of our whole family goes and plays. And also it is really dangerous.

But we have also been having decreasing in subsistence food, like our moose, our fishing or just seals and whales, all the native foods that we eat off the land. The berry picking spots that we go to every single year are not there anymore. The hunters are more endangered in the wintertime because they go out on the ice, and a lot of them have fallen through.

[. . .]

Just through my lifetime, I have seen so many changes in our community that it hurts not to be able to have our—it is really scary to lose our tradition, our culture. We have been living here for thousands of years, and it is not just that we are losing our food, it is losing our homes and—because we are spiritually connected and emotionally and physically connected to our homes, and there are so many communities that are in trouble.

It is an emergency. We need to take action now because—I don't know if you have heard about the Shishmaref. Their whole community has to move, and it has taken so much money just to relocate 500 people.

And we need to take action now. This is going to impact my future, all of our futures because we have to leave our homes, our traditions, where our ancestors taught us how to take care of ourselves from traditional culture lifestyles.

Charlee Lockwood, congressional testimony, 2007

knowledge, and participative practices. Climate change is not only a global problem but is also a local one. Correspondingly, climate change knowledge needs to be understood at the local scale. Landscape, as a central concept of cultural geography for more than 80 years, provides the means to think through distinctive spatialities and temporalities of climate and its change, together with intimate relations involving people, flora, fauna, topography, environment, and weather. Climate change could not only be observed in relation to landscape but could also be felt and sensed as part of everyday life. The gradual real and imagined past, present, and future changes in land and weather become identified as the artifacts of climate change, similar to the scientific artifacts of CO_2 in the atmosphere and increases in mean temperature. The scientific community has not fully and effectively communicated the science of climate change and potential adaptation and mitigation approaches to the general lay public, which is evidenced by research showing that people are not compelled enough to change their behavior from climate change science and consensus. Furthermore, the interpretations of science are mediated by societal values, personal experience, and other factors. Alternatively, focusing on familiar local landscape is thus more productive for examining lay knowledge of climate change and finding better channels for communicating climate change science.

Geographers have highlighted that climate change must be understood, examined, and addressed in the larger context of Earth system transformation, sustainability, and the challenges of economic development, human well-being, and social justice. Current research of climate change impacts has largely focused on exposure to physical hazards such as sea-level rise, related inundation, extreme heat, and public health. In addition to standard climate change and economic impacts analysis, more in-depth studies from human geographers are needed to assess the impacts of social, socioeconomic, demographic, institutional, legal, technological, ethical, organizational, and cultural aspects. Economic geographers can carry out cost-benefit or cost-effectiveness analyses of different adaptation approaches corresponding to various climate change scenarios and thus can provide important feasibility insights for decision makers and policy makers. Human geographers focusing on behavioral research can make a significant contribution by improving our understanding of ways to encourage behavioral changes among the general public, which is needed to implement climate change mitigation and adaptation policies particularly at regional and local levels. Additionally, most current climate change impacts research at the state and national levels has been for entire key sectors (e.g., water, crop agriculture, energy). Future research from human geographers is needed to explore the impacts on such small business sectors (e.g., the winter tourism sector, the wine growing sector, organic farming, and fisheries).

Physical geographers continue to play an important role in examining and explaining the temporal and spatial signals in the longer-term historical temperature and precipitation data sets. Physical geographers have also contributed to numerical modeling for climate simulations in climate change studies. These numerical models of climate are constructed by scientists in applied mathematics, computing science, geophysics, and other disciplines to simulate global and regional responses to alterations in atmospheric composition. These models are very sophisticated in dealing with smaller-scale phenomena. Improving the prediction performance of the numerical models is the greatest challenge to all the scientists. Among them, physical geographers have been more involved in the spatial parameterizations of complex land-surface interactions combined with the role of human activities in changing near-surface mass, momentum, and energy fluxes. Exploratory research on scenarios of extreme and abrupt climate change is needed particularly from physical geographers, which will reveal hidden vulnerabilities of specific sectors or regions and allow creative thinking of mitigation and adaptation options.

Physical geographers have highlighted the importance of integrated regional assessment of climate change impacts, which attempts to examine the regional and local details and implications of global climate change in a multidisciplinary and systematic way and is involved with scientific, policy, and societal stakeholders. The notion of integrated regional assessment is mainly derived from the notion that society, policy makers, and decision makers require more socially relevant science and that impacts of climate change could be spatially uneven due to local and regional complexity and variations. Climate change mitigation can thus be designed and practiced at the regional and local levels, which is a more manageable spatial scale. One existing great challenge is that climate change scenario outputs from global circulation models (GCMs) cannot be readily used for integrated regional assessment, since the spatial resolution of those GCMs, typically hundreds of kilometers, is too coarse for regional assessment. Therefore, GCMs are often downscaled by statistical or dynamic techniques to provide regional climate scenarios. An ideal spatial unit of integrated regional assessment is river basins or watersheds. Extensive studies highlight the hydrologic (physical) and water resource (socioeconomic) impacts resulting from climate change. Based on modeling simulations of basin hydrologic systems, there exist considerable differences in estimated hydrologic effects from climate change either by different scenarios or by various basins. Some topic areas of water resource impacts include impacts on urban water supply, irrigation, flood control, reservoir operations, hydropower

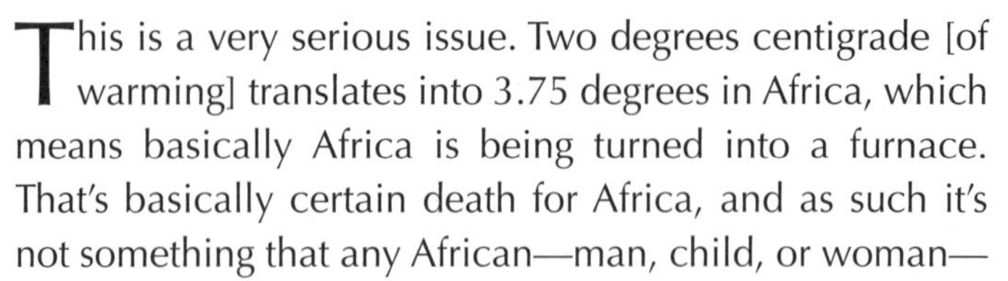

This is a very serious issue. Two degrees centigrade [of warming] translates into 3.75 degrees in Africa, which means basically Africa is being turned into a furnace. That's basically certain death for Africa, and as such it's not something that any African—man, child, or woman—will accept. I cannot imagine which president of which African state will choose to condemn his citizens to total devastation and destruction in order to keep the supremacy of Western economies.

[. . .]

The world's scientists and policy decision makers have publicly stated that this is the greatest risk humanity has ever faced. Now if that's the case, it's very strange that $10 billion is considered adequate financing even for a fast track. Compare $10 billion to $1.1 trillion that developed countries put on the table to handle and address the financial crisis. Think of the trillions that the U.S. Treasury poured into Wall Street to save the banking industry, which is legitimate. Compare that to the amount of money being used and put into the defense budget. There are trillions being spent on armies with imaginary battles to fight when the risk that we face is being dealt with in the most illogical manner.

Developed country leaders . . . need to rise to the level of leadership required. That's why we appeal to all of them, from Australia to Canada and from the U.S. to Japan, to rethink the policy decisions and the proposals put forward because they are condemning developing countries to an unimaginable sentence.

Lumumba Stanislaus Di-Aping, chief negotiator, Group of 77, Sudan, 2009

production, and environmental protection. One example case study is the Susquehanna River Basin Integrated Assessment (SRBIA). The Susquehanna River Basin spans the three states—Pennsylvania, New York, and Maryland—and covers an area of 27,500 square miles (71,227 square kilometers). The SRBIA consists of four related, collaborative, and interdisciplinary projects. The first project is about methodologies in integrated regional assessment, namely, downscaling, economic modeling, and representation of data in time and space. A second component examines decision-maker and public perceptions of vulnerability to climate change via surveys. A third study is concerned with the vulnerability of the basin's community water systems to present climate variation and climate change (e.g., floods and droughts). A fourth part is determining the impacts of climate-related waterborne diseases in the basin. The SRBIA underscores linkages between global, regional, and local scales and has a modular structure with multiple interactive components relating to the causes, impacts, and human responses of climate change at the regional scale.

A recent case study of climate change impacts on water resources is an integrated modeling research from physical geographers carried out in the Tualatin River Basin in Oregon. The researchers applied a GIS-based hydrological model, the U.S. Environmental Protection Agency's Better Assessment Science Integrating Point and Nonpoint Sources, to investigate future impacts on water resources under a range of climate change and land-use scenarios. Eight climate change scenarios were obtained from seven statistically downscaled GCMs, for example, from the National Center for Atmospheric Research in the United States, the University of Bergen in Norway, and Meteo France in France. The results are that the general pattern is increases in winter flows of 10 percent, decreases in summer flows of 37 percent, and increases in fifth-percentile flows of up to 80 percent as a result of climate change in the study area. The next generation of integrated assessment and modeling of climate change impacts is to examine climate change impacts and mitigation and adaptation options with interactions within and across sectors at various spatial levels and to further assess the additional direct and indirect impacts of mitigation and adaptation responses to climate changes.

Zhongwei Liu and William James Smith Jr.

Further Reading

Balling, R. C., Jr. "The Geographer's Niche in the Greenhouse Millennium." *Annals of the Association of American Geographers* 90 (2000): 114–122.

Brace, C., and H. Geoghegan. "Human Geographies of Climate Change: Landscape, Temporality, and Lay Knowledges." *Progress in Human Geography* 35(1) (2010): 1–19.

Clifford, N., S. French, and G. Valentine. *Key Methods in Geography.* 2nd ed. London: Sage, 2010.

Goodchild, M. F. "GIScience, Geography, Form, and Process." *Annals of the Association of American Geographers* 92(4) (2004): 709–714.

Hanson, S. "Who Are We? An Important Question for Geography's Future." *Annals of the Association of American Geographers* 94(4) (2004): 715–722.

Hulme, M. "Geographical Work at the Boundaries of Climate Change." *Transactions of the Institute of British Geographers,* new series, 33 (2008): 5–11.

Moser, S. C. "Now More Than Ever: The Need for More Societally Relevant Research on Vulnerability and Adaptation to Climate Change." *Applied Geography* 30 (2010): 464–474.

Pattison, W. D. "The Four Traditions of Geography." *Journal of Geography* 63 (1964): 211–216.

Praskievicz, S., and H. Chang. "Impacts of Climate Change and Urban Development on Water Resources in the Tualatin River Basin, Oregon." *Annals of the Association of American Geographers* 101(2) (2011): 249–271.

"What Is Geography?" Association of American Geographers, 2006, http://web.archive.org/web/20061006152742/http://www.aag.org/Careers/What_is_geog.html.

Yarnal, B. "Integrated Regional Assessment and Climate Change Impacts in River Basins." *Climate Research* 11(1998): 65–74.

Geothermal Energy

Geothermal energy is a possible energy source of vast proportions. In the United States alone, there are enough recoverable geothermal resources to meet all our energy needs for more than 1,000 years. These geothermal resources are located at depths of 2,000–10,000 meters below Earth's surface, where temperatures often exceed 200°C. This heat is produced by the radioactive decay of naturally occurring radioisotopes of potassium, uranium, and thorium that occur throughout Earth and from the slow conduction of heat away from Earth's core and mantle. To extract this geothermal energy, holes are drilled into deep rock strata to gain access to the heat. Then this heat is extracted by using superheated water that can be pumped to the surface via a production well. In a few places, water already exists in a reservoir of porous rock filled with water. But in most cases, water needs to be injected into the rock, and the rock also needs to be fractured so that it is more porous in order for the water to move through the rock. In order to produce an economically viable geothermal power plant, the production well needs to supply 15 gallons or more of water per second or more.

Iceland not only generates electricity with geothermal but also uses geothermal energy for space heating. Iceland happens to have a very good geothermal resource because there are very good thermal resources relatively close to the surface, and there is also a very good natural source of water to recharge the hot porous rock strata. In the geothermal designs used today, the existing water reservoir recharges with new water naturally as the heated water is pumped out. Enhanced geothermal energy allows greater amounts of water to be heated and brought to the surface. However, the water reservoir must be artificially recharged with injection wells and/or the rock structure must be fractured so that it is more porous for water movement. In the United States, most geothermal energy (excluding geothermal heat pumps) is used for generating electricity, but worldwide, most geothermal energy is used for space heating, hot water, and providing heat for industrial uses.

Most geothermal plants in operation today use steam that comes either directly from the production well or indirectly from superheated water that flashes to steam upon leaving the production well. This steam is then used in a steam turbine just like in a coal, natural gas, solar thermal, or nuclear power plant. These two methods require very hot rock to heat water to temperatures well over 100°C. Binary geothermal can be used where rock temperatures can only produce hot water but not boiling water. This hot water is then used to make a second working fluid boil at low temperatures, similar to the way an air conditioner uses a refrigeration fluid. The only difference is that a refrigerator uses electrical energy to power a compressor, whereas a binary geothermal system uses the working fluid to turn a turbine (that replaces the compressor) and generates electricity. After turning the turbine, the working fluid cools, is condensed, and is circulated to repeat the process. When this thermal source exists in a porous rock formation that coexists with a sufficient reservoir of water that will naturally recharge itself as the water is withdrawn, this is called a hydrothermal system.

The water in the production well needs to be continuously refilled to keep a geothermal power plant operating. Enhanced geothermal systems (EGSs) denotes a geothermal resource location that needs further geoengineering to either make the rock formation more porous to allow water to flow in quicker or an artificial method to recharge the water reservoir. One method to make the rock more porous is a drilling technology called hydraulic shearing. This is similar to hydraulic fracturing, or fracking, used by the natural gas industry for tapping unconventional natural gas shale. If the thermal rock formation does not have a sufficient water reservoir that recharges naturally, an injection well can be drilled to inject water into the reservoir. It is estimated that such hydraulic shearing and injection wells would have a limited usable lifetime of less than 10 years each. Thus, it is expected that for a given power plant, several sets of production and injection wells would need to be drilled over the years. There are currently only a few experimental enhanced geothermal facilities around the world, but this technology shows much promise.

Like solar and wind, geothermal could easily provide all the energy needed in the United States for many centuries and would mainly produce electricity. However, unlike these other renewables, geothermal has the distinct advantage of being able to produce electricity on demand anytime of day or night. On the downside, geothermal is also more geographically limited than either wind or solar. There are very good geothermal resources from Colorado to the West Coast, but there are very few resources east of the Mississippi River. Thus, for geothermal to be used as a power source for the entire country or the Northeast, a vastly expanded national grid would be needed. Such an expanded national grid would not only benefit geothermal energy but would also benefit solar and wind.

Geothermal electricity does not emit any CO_2, and since it is renewable it will never run out until Earth's center finally cools in a billion years or so. Some geothermal plants require the use of water, which in many areas is a valuable and limited resource, particularly in the western United States. This problem can be mitigated by injecting wastewater in place of freshwater and/or by having a closed water loop. In some cases, geothermal releases pollutants such as sulfur oxides that are the primary acid rain–causing pollutant. Since the sulfur comes from mineral deposits within the hot rocks, a closed loop design would send that sulfur and other minerals back into the earth, resulting in little or no emissions. Although evidence is limited, there is a concern that extracting large amounts of thermal energy could increase seismic activity, particularly since geothermal resources are generally high in regions with active seismology and near tectonic plate boundaries.

In the United States less than 1 percent of all energy comes from geothermal, and the growth of geothermal has stagnated over the past decade. Current installation costs are comparable to wind energy, which should make geothermal cost competitive with fossil fuel energy sources. However, geothermal receives only 2 percent of the federal incentives that are allocated for wind generation. The amount of federal financial subsidies for geothermal energy are very small compared to the subsidies for other renewable energy sources such as solar and biomass as well. Furthermore, the geothermal subsidies are well below those for traditional energy sources such as nuclear, coal, natural gas, and petroleum. This is unfortunate, since a comprehensive study by MIT in 2006 concludes that the cost of generating electricity from geothermal resources could drop from the current range of $0.10–$0.30 to $0.03–$0.09 per kilowatt-hour (kWh) with an investment of about $1 billion, mainly to be used to demonstrate and prove that new EGS technology works.

With that investment (or subsidy), the reduced cost of geothermal would make it economically competitive with other commercially used sources of electricity. The total current yearly federal investment in geothermal is only $14 million, compared to $720 million for wind, $1,270 for nuclear, $3,000 for coal, and $3,250 for ethanol. One reason that such a modest investment can have such a dramatic impact on geothermal energy is that the cost of geothermal is strongly governed by the cost of drilling wells into thermal rock structure 2–10 kilometers deep. Fortunately, oil- and gas-drilling technology has now been developed to allow drilling to these depths.

Geothermal can be used not only for generating electricity but also for heating systems. It is possible in some locations to take hot water or steam coming out of the ground and heat homes with this energy, but this doesn't usually happen. Most of the time when people uses the phrase "geothermal heat," they are referring, somewhat incorrectly, to geothermal heat pumps or ground-source heat pumps. For this technology, instead of extracting heat at a very high temperature (less than 100°C), heat can be extracted at a low temperature (greater than 20°C) to assist an electric-powered heat pump for space and water heating. A ground-source heat pump is similar to a regular air-source heat pump. Electricity is used to extract usable heat energy from something that would be considered cool (0–10°C), and this heat is then used to heat a home. Heat pumps can also work in reverse as an air conditioner by extracting heat energy from the warm house and dumping it either outside (air-source) or into the ground (ground-source).

This is usually the most economical way to heat a home or business when considering the operating costs only. If the large capital costs are factored into the accounting, ground-source heat pumps are often more costly overall than natural gas. In places where natural gas is unavailable, a ground-source heat pump is usually cheaper than heating oil, propane, or direct electricity for space heating. There are now well over 1 million geothermal heat pumps in use in the United States, with over 100,000 new heat pumps installed every year. These heat pumps are absolutely the most efficient methods of heating and cooling a home; however, because they are expensive to install, they may not be the cheapest method of heating or cooling.

Richard Flarend

Further Reading

Massachusetts Institute of Technology. *The Future of Geothermal Energy: Impact of Enhanced Geothermal Systems (EGS) on the United States in the 21st Century.* Idaho Falls: Idaho National Laboratory, 2006. http://geothermal.inel.gov.

U.S. Energy Information Agency. *Federal Financial Interventions and Subsidies in Energy Markets, 2007.* Washington, DC: U.S. Government Printing Office, 2008.

Glaciers and Glacial Extent

A glacier is a permanent body of ice that moves under the weight of its own mass. Glaciers are formed in places with low temperature where the rate of precipitation is greater than the melting of snow, which promotes permanent ice cover. Two dominant glacier types are the alpine glaciers associated with mountain ranges and continental glaciers that are unconfined glacial masses that flow outward in all directions. Glacial extent defines the distribution and area covered by these glaciers on Earth's surface. Understanding glaciers and changes in glacial extent is very important to understanding current climate change, because the decline of glaciers is well correlated with the increase in global mean temperature since the mid-19th century.

At present, glaciers cover 10 percent of the land surface and store about 75 percent of the world's freshwater. An overwhelming majority of this global

Hot Spot

Switzerland

In an effort to help slow global climate change, Switzerland has ratified both the United Nations Framework Convention on Climate Change (1994) and the Kyoto Protocol (1997). The phenomenon of global warming has become increasingly noticeable in Switzerland, particularly in its mountainous areas. In July 2006, a massive rock slide on Eiger Mountain dumped 20 million cubic feet of snow. Scientists attributed the crumbling to melting glaciers, which had been shrinking at a pace of nearly three feet per day. As the glaciers retreated, the rock below developed fissures, and the mountain began to lose its solidity. Global warming has both environmental and economic consequences for Switzerland, where ski resorts contribute greatly to the tourism and service industries. With these concerns in mind, the government has made a concerted effort to reduce carbon dioxide emissions and lessen its reliance on fossil fuels.

Table 1 Regional overview of the distribution of glaciers and ice caps

Macroregion	*Glaciers/Ice Caps Area (km²)*
Arctic	275,500
North America	124,000
Central Asia	114,800
Antarctica	770,00
Northern Asia	59,600
South America	25,500
Central Europe	3,785
Scandinavia	2,940
New Zealand	1,160
Africa	6
New Guinea	3

Source: Dyurgerov and Meier (2005).

ice extent is due to the Antarctic ice sheet, the largest mass of ice on Earth that covers an area of 12.3 million km^2, and the Greenland ice sheet, which covers an area of 1.7 million km^2. Glaciers and ice caps are scattered throughout the world, outside of these two ice sheets, in various mountainous regions (e.g., the European Alps, the Andes, the Brooks Range in Alaska, the Himalayas, and the Karakoram Mountains) and around the Arctic Ocean Basin in the Northern Hemisphere (table 1). These glaciers and ice caps cover an areal extent of around 785,000 km^2. Glaciers are found on every continent except Australia, and most of the glacierized area is concentrated in the Northern Hemisphere.

Over Earth's long history, glacial extent has been greater than it is currently. Two million years ago (during the Quaternary period) glaciers covered up to one-third of Earth's surface. Since then Earth has undergone several glacial and interglacial cycles that have subsequently resulted in fluctuations in global glacial extent due to glacial retreat during warmer conditions and glacial advance during colder conditions. It has been evidenced that these cycles are driven by cyclical changes in Earth's orbit around the sun that cause variations in the solar radiation received on Earth. Alpine and continental glaciers covered 32 percent of the land area at the peak of the last glacial maximum, around 20,000 years ago. The most recent glacial event, known as the Little Ice Age of the late 16th to early 20th century, was a

Hot Spot

Greenland

Ice melt in Greenland has accelerated significantly since 1990, according to a report in the *Journal of Climate* on January 15, 2008, coauthored by Konrad Steffen, director of the Cooperative Institute for Research in Environmental Sciences and professor of geography at the University of Colorado. A scientific team surveyed the rate of summer melting there between 1958 and 2006 and found that the five largest melting years all had occurred since 1995. The year 1998 was the biggest (109 cubic miles), followed by 2003, 2006, 1995, and 2002. Preliminary data suggest that the melting in 2007 may have exceeded all previous years. "Ice is moving faster into the ocean, and that will add to the sea-level rise," said Steffen.

In Ilulissat (meaning "iceberg") on the west coast of Greenland, rain fell during December 2007 and January 2008. "Twenty years ago, if I had told the people of Ilulissat that it would rain at Christmas 2007, they would have just laughed at me. Today it is a reality," said Steffen. Melting of the Greenland ice sheet increased 30 percent between 1979 and 2007. By that time, Greenland was losing 200 cubic kilometers of ice per year—from actual melting as well as from ice sliding into the ocean from outlet glaciers along its edges—which far exceeds the volume of all the ice in the European Alps. "Everything is happening faster than anticipated," Steffen said. Air temperatures on the Greenland ice sheet increased by about 7°F between 1991 and 2008.

"The amount of ice lost by Greenland over the last year [2007] is the equivalent of two times all the ice in the Alps, or a layer of water more than one-half mile deep covering Washington, D.C.," according to Steffen. The 2007 melt extent on the Greenland ice sheet broke the 2005 summer melt record by 10 percent, making it the largest ever recorded there since satellite measurements began in 1979. Ice is melting most rapidly at the edges of the ice sheet. Although Greenland has been thickening at higher elevations due to increases in snowfall (warmer air holds more moisture, thus more snow), the gain is more than offset by an accelerating mass loss, primarily from rapidly thinning and accelerating outlet glaciers.

Bruce E. Johansen

period of lower temperatures on a global scale that resulted in the advance of glaciers at high latitudes and altitudes.

Glacier Formation

Glaciers form as a result of snow accumulation that gradually turns into ice and flows in response to gravity. Glacier mass balance is the difference between accumulation (due to precipitation and deposition) and ablation (due to melting and sublimation) and is usually measured at the end of the melt season. Glaciers grow when climatic and topographic conditions allow accumulation to exceed ablation, and they recede when the losses are greater. Glaciers are more extensive at higher latitudes due to less available energy for melting and at higher altitudes where thinner air is less efficient at holding heat energy. When accumulation and ablation are affected by a change in climate, the glacier will change its extent toward a size that makes the mass balance zero. However, changes in glacier extent exhibit a lag behind climate changes that are in the timescale of few years (short steep glaciers) to several centuries (largest glaciers and ice caps).

While glacier mass balance can be understood as a direct undelayed response to annual atmospheric conditions, changes in glacier length and extent are delayed responses reflecting integrative atmospheric changes at longer timescales. Therefore, the glacial extent is determined by climate, topography, and distance from moisture sources along with the physical properties of ice.

Global Glacier Monitoring

Field methods, along with historical aerial and ground photographs, can be used to study changes in glacial extent. Although field methods provide the most reliable information of glacier mass balance and other properties, satellite remote sensing has been critical for recording a comprehensive observation of glaciers globally. In the last few decades, mapping glacial extent globally is the most important application of multispectral satellite data through sensors such as Landsat and Advanced Spaceborne Thermal Emission and Reflection Radiometer (ASTER). Glacier extent calculated by extraction from multispectral images can further be used to infer ice volume. The World Glacier Monitoring Service, established in 1986, compiles and publishes standardized information of distribution and ongoing changes in glaciers and ice caps in the World Glacier Inventory (WGI). The Global

Hot Spot

Himalayas

About 90 percent of the Tibetan Plateau's glaciers have retreated during the past century; their rate of decay has increased substantially during the past decade, according to "Global Outlook for Ice and Snow," a report released during 2007 by the United Nations Environment Programme.

The snowpack in this region could shrink a further 43–81 percent by 2100, according to projections from the Intergovernmental Panel on Climate Change (IPCC). Glaciers on the Himalayas contain 100 times as much ice as the Alps and provide more than half the drinking water for 40 percent of the world's population in seven major Asian rivers, including the Ganges and the Indus.

Temperatures in the region have risen by 1°C since the 1950s, causing thousands of glaciers to retreat by an average of 30 meters a year. Retreating glaciers may leave behind natural dams of frozen rocks, setting the stage for potential disasters. The worst recorded collapse of one of these dams occurred in 1954, when 300,000 cubic meters of water and rock poured without warning into China in a 40-meter-high flood surge from the Sangwang Dam on the Tibet-Nepal border. The city of Gyangze, 120 kilometers away, was destroyed. The dead totaled many thousands.

The Indian Space Research Organization has used satellite imaging to measure changes in 466 glaciers, finding more than a 20 percent reduction in their size between 1962 and 2001. Another study found that the Parbati Glacier, among the largest, was retreating by 170 feet a year during the 1990s. Another glacier, the Dokriani, lost an average of 55 feet a year. Temperatures in the northwestern Himalayas have risen by 2.2°C in the last two decades.

Bruce E. Johansen

Land Ice Measurements from Space (GLIMS) project was launched in 1995 as an international effort and an extension to WGI with the goals of mapping and monitoring a majority of the world's glaciers using satellite multispectral imagery, maintaining a spatio-temporal global glacier database, analyzing them for glacial extent and changes, and understanding these changes in terms of various forcings. The updated WGI currently has information on 131,000 glaciers, and GLIMS database has a repository of 62,000 glacier outlines with information on glacier extent, length, elevation, etc. Half of the glacierized area is covered by both WGI and GLIMS.

Indicators of Climate Change

Glaciers are also sensitive indicators of climate change, so the past and present glacier fluctuations provide information regarding the global climate system and also serve as an overall temperature indicator. Glacier moraines (accumulations of glacial debris) from the Little Ice Age mark a Holocene (past 10,000 years) maximum extent of glaciers in many mountain ranges. Glaciers and ice caps have shown general recession from these positions, with accelerating glacial retreat since the mid-1980s, and are close to or exceeding the lowest values since the Holocene. This overall decline of glaciers and ice caps is well correlated with the increase in global mean temperature since the mid-19th century. On a shorter timescale (decadal), glaciers in various mountain ranges have also shown intermittent readvancement, such as in the Alps in the late 1970s and in coastal Scandinavia and New Zealand in the 1990s. Regional declines in glacier extent and volume include the European Alps, the Andean glaciers, the Central Asian glaciers, the mountains in North America, and glaciers in Africa and New Guinea. Glaciers in the Antarctic Peninsula terminating in the sea indicate 87 percent of glaciers retreating over the last six decades. Similarly, Arctic glaciers and ice caps show evidence of a general retreating trend. The ongoing observed trend of worldwide glacial retreat is likely to be nonperiodic in nature and may lead to deglaciation of many mountain ranges by the end of the 21st century.

Prajjwal Panday

Further Reading

Benn, D. I., and D. J. A. Evans. *Glaciers and Glaciation.* London: Hodder Arnold, 1998.

Cogley, J. G. "A More Complete Version of the World Glacier Inventory." *Annals of Glaciology* 50 (2009): 32–38.

Dyurgerov, M., and M. F. Meier. "Glaciers and the Changing Earth System: A 2004 Snapshot." Occasional Paper 58. Institute of Arctic and Alpine Research, University of Colorado, Boulder, 2005.

Paterson, W. S. B. *The Physics of Glaciers.* Portsmouth, NH: Butterworth-Heinemann, 2000.

Raup, B., A. Kääb, J. S. Kargel, M. P. Bishop, G. Hamilton, E. Lee, F. Paul, F. Rau, D. Soltesz, and S. J. S. Khalsa. "Remote Sensing and GIS

Technology in the Global Land Ice Measurements from Space (GLIMS) Project." *Computers & Geosciences* 33 (2007): 104–125.

Solomon, S., D. Qin, M. Manning, Z. Chen, M. Marquis, K. B. Averyt, M. Tignor, and H. L. Miller. *Climate Change 2007: The Physical Science Basis; Contribution of Working Group I to the Fourth Assessment Report of the Intergovernmental Panel on Climate Change.* New York: Cambridge University Press, 2007.

World Glacier Monitoring Service. *Global Glacier Changes: Facts and Figures.* Zurich: UNEP, World Glacier Monitoring Service, 2008.

Global Atmospheric Research Program

Spurred by the launch of Sputnik I on October 4, 1957, and the meteorological satellite Explorer VIII on October 13, 1959, climate scientists realized that advances in radiometric measurements could allow the creation of an economically feasible observational system of the global climate. Toward this end, the United Nations (UN) General Assembly adopted Resolutions 1721 and 1802 on December 20, 1961, and December 14, 1962, respectively. Resolution 1721, entitled International Cooperation in the Peaceful Uses of Outer Space, called upon member states to initiate programs to (1) advance the state of atmospheric science and technology so as to provide greater knowledge of basic physical forces affecting climate and the possibility of large-scale weather modification and (2) develop existing weather forecasting capabilities and to help member states make effective use of such capabilities through regional meteorological centers. A year later, the UN General Assembly expanded the original purview of Resolution 1721 by adopting Resolution 1802, which was designed to (1) strengthen weather forecasting services and encourage their scientific communities to cooperate in the expansion of atmospheric science research; (2) develop in greater detail the plan of the World Meteorological Organization (WMO) for an expanded program to strengthen meteorological services and research, placing particular emphasis on the use of meteorological satellites and on the expansion of training and educational opportunities in these fields; and (3) grant technical and financial assistance to requests from member states to supplement their own resources for these activities, including the improvement of meteorological networks.

In accordance with these resolutions, the International Council of Scientific Unions (ICSU) and the International Union of Geodesy and Geophysics (IUGG) jointly established the Committee on Atmospheric Sciences (CAS) in June 1964. As a result of deliberations over the next two years, the CAS suggested 1972 as a 12-month period for an intensive international observational study and analysis of the global circulation of the troposphere and lower stratosphere (below 30 kilometers). These included studies in the tropics, theoretical and observational studies of the sea-air and land-air exchange processes, improved techniques for incorporating radiative transfer of energy into dynamical models of general circulation, and scientific design specifications for the global observational program. As a way of referring to all of these studies collectively, the CAS adopted the name Global Atmospheric Research Program (GARP).

While the overall idea of GARP was appealing, two logistical problems soon became evident. First, the target date of 1972 was highly optimistic. Second, the idea of a single global experiment was unrealistically restrictive. To resolve these issues, a group of meteorologists (the GARP Study Conference) from around the world met at Skepparholmen, near Stockholm, from June 28 to July 11, 1967. After much deliberation, they decided that GARP should be a long term program without a foreseeable end date and should include multiple experiments. The ICSU/IUGG CAS accepted their recommendations, and in October 1967 the WMO and the ICSU created the Joint Organizing Committee (JOC) on GARP, which became "the main scientific organ for the consideration, endorsement and recommendations on all proposals to GARP and its Sub-programmes."

The JOC, consisting of 12 climate scientists, met for the first time in April 1968 and named Swedish meteorologist Bert Bolin and Richard W. Stewart as chairman and vice chairman, respectively. Bolin (March 15, 1925–December 30, 2007) would later serve as the first chairman of the Intergovernmental Panel on Climate Change from 1988 to 1997. Stewart is presently a physical scientist at the Goddard Space Flight Center, a position he began in 1978 after a stint with the Goddard Institute for Space Studies (from 1969 to 1978).

Early on, the JOC conceived GARP as a practical endeavor to develop predictive theories regarding the behaviors of the lower atmosphere, not merely as an academic exercise. Those involved conceived of themselves as public servants with high aspirations, although they were clearly cognizant of the potential hazards. First, was it possible in principle to predict something as unpredictable and complex as the atmosphere? Second, even if it was possible to predict in principle the atmosphere, could they accomplish the task in practice?

Indeed, the problem of prediction was sobering. First, those involved with GARP had to limit their expectations to a fairly low-level spatial resolution. Predicting a local storm system was beyond their reaches. Second, the technological limitations were clear: even if enough observations could be collected to predict a local storm, the sheer amount of data would take too long to process. Third, the physico-mathematical tools available to understand global patterns were inadequate to account for local phenomena. Thus, spatial scope had to be balanced with predictive accuracy or quality (i.e., the greater the resolution, the lower the accuracy). Instrumental errors, parameterization, nonlinearity, and instability, among many other problems, made understanding how small-scale phenomena affect larger-scale atmospheric processes a significant challenge.

Despite such logistical and scientific difficulties, the climate scientists were highly successful. Projects were completed with the participation of more than 20 nations, including the United States, Great Britain, the Soviet Union, Japan, and France. Over the next 15 years, projects included the GARP Atlantic Tropical Experiment (GATE), the Barbados Oceanographic and Meteorological Experiment (BOMEX), thc First GARP Global Experiment (FGGE), the Polar Experiment (POLEX), the Complex Atmospheric Energetics Experiment (CAENEX), the Air Mass Transformation Experiment (AMTEX), the Monsoon Experiment (MONEX), and the Tropical Experiment (TROPEX).

Similar to the construction of the Mount Wilson Observatory in the early 20th century, the development of the atomic bomb in the 1940s, and the manned space program in the 1960s, GARP is an example of the emerging importance of big science for scientific progress. Based on an ardent desire to understand the deeper processes of the atmosphere, climate scientists, governments, and nations invested vast amounts of manpower and resources that would ultimately allow scientists to determine the effects of humans on the global climate. As Richard Reed presciently pointed out in 1971, GARP also helped to "accelerate the trend towards interdisciplinary programs of study" and to "affect the topics which students choose for thesis research and the manner in which the research is carried out." Potentially even more relevant, as the UN General Assembly pointed out in the early 1960s, GARP was an expression of humanity's attempt to better itself through international cooperation, science, and technology.

Gabriel Henderson

Further Reading

Global Atmospheric Research Program (GARP). Publication Series, nos. 1–27, World Meteorological Organization (1969–86).

Global Atmospheric Research Program (GARP). Special Reports, nos. 1–12, World Meteorological Organization (1969–73).

"Global Atmospheric Research Program (GARP) Records, 1966–1979." NCAR Archives, https://nldr.library.ucar.edu/archon/?p=collections/controlcard&id=24.

ICSU/IUGG CAS and WMO. *Report on the Study Conference on the Global Atmospheric Research Programme (GARP): Stockholm, 28 June–11 July 1967.* Stockholm: World Meteorological Organization, 1967.

Reed, Richard. "The Effects of GARP and Other Future Large Programs on Education and Research in the Atmospheric Sciences." *Bulletin of American Meteorology* 52(6) (June, 1971): 458–462.

United States Committee for the Global Atmospheric Research Program, National Academy of Sciences (U.S.). *Plan for U.S. Participation in the Global Atmospheric Research Program.* Washington, DC: National Academy of Sciences, 1969.

United Nations General Assembly, 16th Session. "Resolution 1721 (XVI) [International Cooperation in the Peaceful Uses of Outer Space]." December 4–20, 1961, http://daccess-dds-ny.un.org/doc/RESOLUTION/GEN/NR0/167/74/IMG/NR016774.pdf?OpenElement.

United Nations General Assembly, 17th Session. "Resolution 1802 [International Cooperation in the Peaceful Uses of Outer Space]." December 14, 1962, http://daccess-dds-ny.un.org/doc/RESOLUTION/GEN/NR0/193/10/IMG/NR019310.pdf?OpenElement.

Global Cooling in the 1970s

In the late 1960s, climate scientists noticed that atmospheric turbidity was linked with the heat balance of the earth-atmosphere system. In 1967 two researchers in Ohio, Robert McCormick and John Ludwig, suggested that in spite of CO_2 emissions, anthropogenic emissions of aerosols and fine particulates were causing a reduction in the amount of solar energy absorbed due to an increase in Earth's albedo (the fraction of radiation that is reflected). In 1968 University of Wisconsin climatologists Reid Bryson and James Peterson confirmed that atmospheric turbidity was increasing. The conclusion was that aerosol pollutants were backscattering and absorbing fractions of incident radiation with the net effect of cooling Earth's surface.

Given an observed cooling trend since the 1940s, researchers attempted to gauge the conditions in which a global cooling (or an ice age) could occur as well as the sensitivity of the climate to those changing conditions. For instance, a 1974 study by Bryson showed that small variations in solar output, atmospheric transmittance (how much energy is able to reach Earth's surface), Earth's albedo, and the greenhouse effect could cause significant changes in the climatic system. Bryson's claims were not without merit. An earlier study in 1969 by W. D. Sellers suggested that a 2–5 percent drop in the solar constant would be sufficient to initiate another ice age, while in a 1973 study Sellers lowered his estimate and suggested that only a 1 percent drop of the solar constant (a measure of the amount of radiation from the sun) would be sufficient to reduce the temperature by 5°C. Bryson's and Sellers' results were confirmed by two researchers with the National Center for Atmospheric Research, Peter Chylek and James Coakley Jr. Citing a 1969 study by the Soviet climatologist Mikhail Budyko and Sellers, they similarly argued that "a very thin layer of aerosol layer may cause a 2 to 5 percent change in the planetary albedo, which is sufficient for dramatic climatic changes."

While the idea of a global cooling concerned many within the climatological community, the relevant causes of increased turbidity were debated. On the one hand, climate scientists such as C. H. Reitan and K. K. Hirschboeck suggested that volcanic emissions explained the observed drops in temperature, while a 1974 study by Richard Kalnicky proposed that the observed cooling was due to a shift in hemispheric circulation from zonal (more or less parallel to lines of latitude) to meridional (relatively curved). Others, including Bryson, argued that man-made particulates in conjunction with volcanic emissions were leading to an overall cooling of the surface. In sum, researchers not only attempted to construct a cause-and-effect relationship between atmospheric turbidity and the climate but also examined the sensitivity of climate.

Many scientists were skeptical about the possibility of long-term global cooling. In 1975 in an attempt to put into perspective the observed cooling between 1940 and 1970, Wallace Broecker asserted that some climate scientists were complacent in discounting the warming effects of carbon dioxide. While acknowledging that the effects of carbon dioxide had been countered by natural cooling, he nonetheless argued that "this compensation cannot long continue both because of the rapid growth of the CO_2 effect and because the natural cooling will almost certainly bottom out." That same year, S. H. Schneider and Clifford Mass suggested that the cooling could be equally explained in terms of sunspot numbers or internal natural exchanges between the atmosphere, oceans, and ice masses. Indeed, while acknowledging the effects of volcanic eruptions as a cooling agent, Alan Robock's 1978 study in the *Journal of the Atmospheric Sciences* suggested that natural variability is the reigning factor in deciding the climate. Robock's reasoning was reflective of the work of Edward Lorenz, who argued that climate change is the result

Climate-History Connection

The Inuits Move South in Greenland

The present Greenland Inuit people are the descendants of the prehistoric Thule people archaeologically described by the Thule material culture. Beginning 1,000 years ago, these people colonized what is known as the Thule District of northern Greenland, and by the late 11th or early 12th century prehistoric Inuit people lived throughout the region. Primarily adapted to hunting sea mammals, including large baleen whales, they entered northern Greenland after having spread east across the Canadian Arctic archipelago during a period of climatic warming. For a century or two their material culture was virtually identical to the earlier Thule culture in eastern Alaska and Canada.

Sometime during the late 12th and early 13th centuries, the Inuit people expanded south across Melville Bay to the west coast of Greenland, where archaeological evidence indicates that they encountered the Norse. This period has been termed Inugsuk, but the degree of Norse influence on prehistoric Inuit culture as well as the nature and extent of the Inuit-Norse interaction remains uncertain. By the late 15th century, the Inuit people occupied the entire west Greenland coast as far south as Cape Farewell. Subsequent climatic changes after the 15th century, namely the Little Ice Age, have been associated with a decrease in settlement in the northerly regions, coupled with migration to more southerly regions. Similarly, during this period evidence indicates that whaling among the Inuits greatly decreased, possibly as a result of increased sea ice. There is adequate archaeological evidence from the decreasing amounts of baleen to argue for a steady decline in bowhead whale hunting after 1400. Rather, evidence indicates that the hunting strategy increasingly focused on seal, walrus, caribou, musk-ox, and other terrestrial animals.

During the late 17th or early 18th century as the climate began to change again, Inuit social organization and residential patterns also changed. By the 17th and 18th centuries, evidence indicates that the presence of Europeans had a decisive influence on the Inuit communities in Greenland. For example, archaeological data indicate that Inuit communities relocated to European landing places, mostly in the south, for the purpose of trade and exchange.

Peter N. Jones

Further Reading

Gullov, Hans Christian. *From Middle Ages to Colonial Times: Archaeological and Ethnohistorical Studies of the Thule Culture of South West Greenland, 1300–1800 AD.* Copenhagen: Meddelelser om Grønland, 1997.

Maxwell, Moreau S. *Prehistory of the Eastern Arctic*. Orlando: Academic Press, 1985.

of complex nonlinear interactions among various components of the climatic system. In essence, the cooling trend was a natural climatic variation.

While climate scientists debated, nations nonetheless proceeded to develop plans to defend against shifts in agricultural production and a potential glacial period. American and Soviet proposals, for instance, included plans to dam the Bering Strait, spread black soot on the Arctic ice cap to melt it, or break up the arctic cap with atomic bombs to warm Canada and portions of the Soviet Union. Due to droughts in the Sahel region of Africa and global food shortages during the mid-1970s, the Central Intelligence Agency (CIA) in 1974 also produced a working paper examining the potential threats of climate perturbations to the stability of governments. As the CIA warned, climate "is now a critical factor."

The belief in a global cooling or the threat of an impending ice age was never uniformly accepted by climate scientists. In addition to the earlier work of Broecker, Mass and Schneider, and Robock, Paul Damon and Steven Kunen in 1976 proposed that the measured cooling was restricted only to the Northern Hemisphere and that a marked warming existed in the Southern Hemisphere. They further remarked that the evidence for cooling was merely the result of measurements restricted to high northern latitudes and that natural causes could not be ruled out. For them, they asserted that the observed Southern Hemispheric warming was potentially a first sign of an impending global warming.

While global cooling and the threat of aerosols could not be absolutely ruled out, the issue was slowly relegated to the shadows. By April 1979 a report was released by the JASON Committee (a secret group of scientists who advised the U.S. government on science and technology since the early 1960s) titled "The Long Term Impact of Atmospheric Dioxide on Climate." The report concluded that levels of CO_2 would double by the year 2035 and would cause mean global temperatures to rise by 2–3°C. That same year, the National Research Council's Ad Hoc Study Group on Carbon Dioxide and Climate, headed by Jule Charney, released a report that confirmed the JASON report's conclusions: if carbon dioxide levels continued to rise, there was little reason to doubt subsequent increases in mean global temperature.

Within the past few years, historians and climate scientists alike have debated whether a global warming consensus had emerged by the late 1970s and early 1980s and, by the same token, when the global cooling threat was coming to an end. Thomas Peterson, William Connolley, and John Fleck suggested that "by 1978, the question of the relative role of aerosol cooling and greenhouse warming had been sorted out. Greenhouse warming . . . had become the dominant forcing." In a slightly different argument, Naomi

Oreskes, Erik Conway, and Matthew Shindell stated in a 2008 study that "in the early 1980s, a consensus emerged among climate scientists that increased atmospheric carbon dioxide from burning fossil fuels would lead to a mean global warming of 2–3°C."

However, Nicolas Nierenberg, Walter Tschinkel, and Victoria Tschinkel suggest a more complex picture: "although in 1979 and 1980 a scientific consensus was emerging that increasing CO_2 would probably cause the atmosphere to grow significantly warmer, no consensus had formed about the level of uncertainty, the seriousness of the problem, or what if anything should be done about it." While the issue of global cooling during the 1970s has become a distant memory, this period nonetheless serves as a significant historical moment when climate scientists faced the real possibility that humankind was influencing the global climate and inaugurating a period of what James Fleming refers to as an "age of vastly enhanced environmental awareness."

Gabriel Henderson

Further Reading

Broecker, Wallace. "Climatic Change: Are We on the Brink of a Pronounced Global Warming?" *Science* 189(4201) (August 8, 1975): 460–463.

Bryson, R. A. "'All Other Factors Being Constant . . .': A Reconciliation of Several Theories of Climate Change." *Weatherwise* 21 (1968): 56–61, 94.

Bryson, Reid. "A Perspective on Climatic Change." *Science* 184 (May 17, 1974): 753–760.

Budyko, Mikhail. "The Effect of Solar Radiation Variations on the Climate of the Earth." *Tellus* 21 (1969): 611–619.

Central Intelligence Agency. "A Study of Climatological Research as It Pertains to Intelligence Problems." August 1974. Reprinted in *The Weather Conspiracy: The Coming of the New Ice Age,* 161–196. New York: Ballantine Books, 1977.

Charney, Jule, et al. *Carbon Dioxide and Climate: A Scientific Assessment, National Research Council, Ad Hoc Study Group on Carbon Dioxide and Climate.* Washington, DC: National Academy Press, 1979.

Damon, Paul, and Steven Kunen. "Global Cooling?" *Science* 193(4252) (August 6, 1976): 447–453.

Fleming, James. *Historical Perspectives on Climate Change.* Oxford: Oxford University Press, 1998.

Hirschboeck, K. K. "A New Worldwide Chronology of Volcanic Eruptions." *Palaeogeography, Palaeoclimatology, and Palaeoecology* 29 (1980): 223–241.

Kalnicky, Richard. "Climate Change since 1950." *Annals of the Association of American Geographers* 64(1) (March 1974): 100–112.

MacDonald, Gordon, et al. "The Long Term Impact of Atmospheric Carbon Dioxide on Climate." JASON Technical Report JSR 78-07, prepared for the U.S. Department of Energy, 1989.

McCormick, R. A., and J. H. Ludwig. "Climate Modification by Atmospheric Aerosols." *Science* 156 (1967): 1358–1359.

Nierenberg, Nicolas, et al. "Early Climate Change Consensus at the National Academy: The Origin and Making of Changing Climate." *Historical Studies in the Natural Sciences* 40(3) (Summer 2010): 318–349.

Oreskes, Naomi, et al. "From Chicken Little to Dr. Pangloss: William Nierenberg, Global Warming, and the Social Deconstruction of Scientific Knowledge." *Historical Studies in the Natural Sciences* 38(1) (2008): 109–152.

Peterson, Thomas, et al. "The Myth of the 1970s Global Cooling Scientific Consensus." *Journal of the American Meteorological Society* 89(9) (2008): 1325–1337.

Reitan, C. H. "A Climatic Model of Solar Radiation and Temperature Change." *Quaternary Research* 4 (1975): 25–38.

Robock, Alan. "Internally and Externally Caused Climate Change." *Journal of the Atmospheric Sciences* 35 (1978): 1111–1122.

Schneider, Stephen, and Clifford Mass. "Volcanic Dust, Sunspots, and Temperature Trends." *Science* 190(741) (1975): 741–746.

Sellers, W. D. "A New Global Climatic Model." *Journal of Applied Meteorology* 12(2) (1973): 241–254.

Global Warming Literature

Writers of fantastic fiction have taken up the challenge of presenting a human face to the possibility that global warming will result in disastrous climate change for the entire planet. While issues concerning environmental degradation on planet Earth have been addressed by a long list of authors and for decades, it is only in recent years that authors have explored more specifically the effects of global warming on humanity and the planet. Using accessible literary forms, they convey threats to the current world climate, which numerous scientific studies have seen to be the result of human intervention. For example, in his book *Deep Future: The Next 100,000 Years of Life on Earth* (2011), the paleoclimatologist Curt Stager has noted that we are currently in a period of relatively warm climates between ice ages and that the end of the last major ice age corresponds with the development and stabilization of human cultures. From this perspective, threats to the climate are threats to humanity's survival.

The authors below have written literary speculations that are often based on particular climate change debates found in popular science works or in media coverage of scientists' discoveries. Their fiction has the potential to explain the findings of climate scientists to a wider public. Of course, they are also interested in how the predicted changes will affect individuals and change socioeconomic systems around the world. In resonance with the many dire predictions about our future made by climate scientists, these fictional works are sometimes similar to the older genre of literature known as disaster fiction. Many make a gesture toward the potential of global warming to cause dramatic climatic effects, including flooding, freezing, or a combination thereof, thus signaling the end either of the human race or of advanced civilization. Major categories of fiction concerning abrupt climate change can be outlined, even if some of the novels fit into more than one category. It should also be noted that while attempts are made by each of these authors to look at the lives of everymen and everywomen, some have remained true to older more heroic traditions of fantastic fiction in telling the stories of important people endowed with opportunities to change a seemingly inevitable downward arc of human society.

Many of the fictional works mentioned below address humanity's near future, within the next 100 years or so. These include John Barnes, Stephen Baxter, David Brin, Michael Crichton, Saci Lloyd, Kim Stanley Robinson, Mary Rosenblum, and Bruce Sterling. Others focus on our future hundreds of years ahead, such as Margaret Atwood, Paolo Bacigalupi, and Sheri Tepper. We have also been treated to popular films that speculate about the near future, including *2012* (2009) and *The Day after Tomorrow* (2004), and the far future, including *Waterworld* (1995), which takes place after global warming has caused abrupt climate change, accompanied by massive floods or freezing. In most cases the authors assume that abrupt climate change will be accompanied by ecological readjustment, animal extinctions, and human suffering (especially in *The Day after Tomorrow*). They also depict concomitant social change and disintegration.

A few authors, such as Michael Crichton, propose that the issue of global warming is a conspiracy on the part of environmental groups to gather financial support from national governments, nongovernmental organizations (NGOs), and private donors. Crichton dismisses the evidence gathered by generations of researchers as little more than a record of the normal climatological changes that have been going on for centuries that have little or nothing to do with human practices on the planet. At first blush, it might seem that Curt Stager's work of popular science, *Deep Future,* lends credence to Crichton's views. However, Stager supports rather than dismisses the need for human attention to a possible climate crisis. He concludes that humanity already has affected the deep future. In fact, anthropogenic global warming may have prolonged the interglacial period that has enabled virtually all of human civilization. To writers such as Crichton this might seem like a good thing, but Stager also stresses that human intervention has disrupted the planetary climate system and likely will have complex and disastrous side effects.

In works of fiction by Lloyd, Rosenblum, and Bacigalupi, humans eventually adapt somewhat successfully to the near future but only after giving up on the many advantages of advanced civilization such as clean and abundant water, energy, and, in many cases, global trade. Kim Stanley Robinson's works stand out as he charts a near future that ends on a note of hope in which political leaders are eventually elected who choose to mitigate the potential disasters and create the intellectual and financial capital to do so. Other authors posit, in the pattern of older disaster literature and film, that there may be other natural forces added to trends toward human-created climate change from the use of fossil fuels that will accelerate the disintegration of advanced technological cultures as we know them. Thus, they undercut the idea that climate change itself is a serious threat by assigning the greater threat to an as yet undiscovered nonhuman agency.

Predictions of disastrous climatic events was brought to the attention of a wider public in a debate between two fiction authors, Kim Stanley Robinson and Michael Crichton, aired by National Public Radio (NPR) on January 28, 2006. Robinson speaks for quick action on the climate crisis, but Crichton, a medical doctor famous for many previous works of fiction concerning medicine and biology, is one of the more prominent climate change deniers. An article by Colin Luckhurst embeds Crichton's *State of Fear* in the politics of the Ronald Reagan and George H. W. Bush administrations, suggesting that Crichton was an apologist for the big business interests that influenced such political figures as Newt Gingrich. One can trace a number of questionable environmental decisions as well as economic, social, and scientific ones made by the U.S. government directly to Gingrich. Crichton's work, set like Kim Stanley Robinson's in a recognizable immediate future, is a clear attempt to discredit the science of climatology as well as NGOs and related organizations whose support comes from fund-raising and secondary governmental grants. However, support for Crichton's position comes instead from big business interests surrounding mining, lumbering, oil, natural gas, and the nuclear industry who have also invested in supporting scientists who downplay the threats of global warming. Crichton follows clashes between governmental organizations investigating the claims of a private research organization about to be funded by a rich philanthropist. This fictional work attempts to credit the claim that the findings pointing to global warming are at best inconclusive and at worst falsified. In the NPR transcript, Robinson appears to be more aware of scientific fact, and Crichton's notions are best understood in the context of conspiracy theory. Crichton's sole supportable claim, as identified in such science-based sources as RealClimate: Climate Science from Climate Scientists, is that no one can exactly predict the effects of global warming. He presents scientists' unwillingness to be categorical in their predictions as a major weakness in their research findings, a disingenuous misdirection on his part.

In Crichton's opinion, the scientific understanding that no theory is ever totally proven (the very next case might disprove this) becomes a reason for discounting all otherwise logically sound predictions. Although there are many sound critiques of Crichton's work and public speaking, it also has been heavily supported by conservative elements in the U.S. government and by the major players in the oil, gas, and nuclear energy industries. Like Robinson, Crichton recognizes that global warming and climate change are political as well as scientific issues. However, they come down on opposite sides of its significance for the human race and on whether or not global warming can be safely ignored. Crichton asserts that scientists are alarmist and self-serving, while Robinson asserts that they have established the need for immediate action. Crichton uses questionable rhetorical devices to make his point. This is especially clear in an appendix of *State of Fear,* which compares the "science" of racism in the early 20th century to the plethora of research in climate studies, geology, paleogeology, paleoclimatology, and biology in the last quarter of the 20th century to the present. The appendix relies on the device of misdirection, portraying climate sciences to theories of eugenics that played into Adolf Hitler's extermination efforts directed not only at Jews but also at other "inferior races" as well as the feeble-minded (as they were then called), homosexuals, and any politically inconvenient critics. While one cannot disagree that eugenics was massively misguided, the comparison is flawed. Global warming and climate change are not about human genetics but instead are about human deformation of the planetary ecosystems. To compare this issue to attempts at proving that some humans are more worthy than others is a rhetorical sleight of hand.

Crichton's second argument, concerning T. D. Lysenko's theories to improve crop yields in the Soviet Union, is more defensible as a comparison in that it deals with soil science but not in the context of Joseph Stalin's monolithic regime. The only similarity between Lysenko's efforts and our current debate over climate science is that they both concern the land. But climate science is a worldwide research endeavor and is not limited to one nation or to a small research community such as the one that surrounded Lysenko. One can accept the need for caution when science and the political arena intersect, but this intersection is universal, pervasive, and ongoing. Modern civilization cannot work in any other way. Crichton's cautionary historical examples are valid as reminders to tread carefully, but this applies to many cases and many sciences. Perhaps a more apt comparison for climate science might be to the debates of the 1970s and 1980s over whether cigarette smoking was a public health hazard, a struggle that often pitted scientists against corporate giants. The picture that emerges from that case—of big money suppressing or deforming scientific results—arguably bears some resemblance to what is now happening in the global warming debates.

Other authors who address the breakdown of society expected from climate change either pursue familiar disaster plots or imply causes in addition to the human-generated climate change factors. John Barnes's *Mother of Storms* and Bruce Sterling's *Heavy Weather* both foreground potential climatic instabilities that produce frequent and massive storm fronts. For Barnes, by 2088 methane trapped in the North Pacific's ocean beds creates recurring hurricanes and flooding. Sterling's disaster novel is structured around the possibility of a pervasive ongoing superstorm. In each case, human civilizations have already begun to devolve under the financial burdens of coping with weather-based disasters.

The futures that Barnes and Sterling portray are similar to that found in Stephen Baxter's *Flood* and its sequel, *Ark.* Baxter addresses climate instability and massive floods caused by global warming, and his extrapolation moves humanity from the near future to the mid-21st century. By 2016, human civilization around the world begins to disintegrate under the pressure of rising water levels, while tens of thousands die from disease, hunger, and lack of infrastructure that currently ensures access to water, food, and energy (at least in the developed world). Two major rescue efforts are attempted: the first a giant survival ship and the second the exploration of other planets.

In Baxter's two novels, scientists originally interpret the disaster as having been caused by global warming, which then causes freshwater ice melts in the Arctic and Antarctic, thus desalinizing the oceans. However, one scientist notes that the floodwaters are rising too fast for that explanation to be true. Her observations eventually prove correct: a large crack in the ocean floor is admitting waters previously held in the center of the planet. By 2081 few humans can survive, and the attempts to find an alternative planet for human habitation have largely failed. Baxter's is a story of human culture ending as we know it. He even speculates that children born into a flooded landscape will become transformed physiologically to adapt to an aquatic culture, reminiscent of Tepper's story of far-future genetic manipulations. Baxter's work certainly cannot be said to conclude on a hopeful note.

The return to life in the oceans is an underlying theme in Tepper's *The Waters Rising.* Her projection of the demise of human society depends on previous and ongoing flooding of the planet's landmasses and intentional genetic manipulation of humanity by intelligent sea creatures. Her novelistic technique is to create mysterious, seemingly magic, agents who attempt to achieve great power but gradually reveal that they too are being manipulated. Earth's sea mammals have evolved intelligence and will only allow selected members of the human race to survive if they agree to the transformation to life in the sea. Others humans are not offered the option of survival. As with many others of Tepper's works, these fictions may be compelling, but they are set so far in the future that they contain little that is recognizable about current discussions of global warming. They instead focus on innate human aggression that interrupts attempted solutions of the destruction of the planet. This fiction reads almost as a fantasy, with little information on how the landmasses could have became so inundated.

Margaret Atwood's two novels *Oryx and Crake* and *The Year of the Flood,* like Tepper's fiction, are motivated more by telling fantastic stories than they are by the science of climate change itself. Both Tepper and Atwood seem to see the climate debates as a jumping-off point for imagining totally new civilizations. They create some of the most hopeful speculations about the future of humanity, but this is not a humanity that we would recognize today. Atwood especially explores avenues that human civilization might take after Earth's climate has been so violently changed that we can no longer live on it. In *Oryx,* the first of two connected novels, both highly acclaimed for their literary as well as speculative excellence, Atwood looks at the future of large corporations and of the upper classes trying to ignore both the physical and social disintegration around them, through isolation, gated communities, and protection from the suffering hoards (the sort of solution for certain small groups in Baxter's work). The second novel, *Flood,* focuses on people who are unprotected by privilege but nevertheless are struggling to maintain a semblance of humanity. Perhaps because they are so far in the future and use the threat of abrupt climate change only as a jumping-off point, both Tepper and Atwood offer some hope for humanity, though not for the majority of humans.

Young adult novelists have also contributed fiction with global warming as a topic. Here, the perspectives on flooding and massive social dislocation are usually filtered through the experiences of a teenager or young adult. The far futures of Phillip Reeve's works, including *Mortal Engines,* imply social collapse due to climate change. Other writers with such concerns include Julie Bertagna and Peter Gould. In two novels the British author Saci Lloyd relates, through the experience of a precocious musically inclined college student, a sequence of dislocations caused by the United Kingdom's attempts to reduce the release of carbon dioxide into the atmosphere. *The Carbon Diaries 2015* (2008) and *The Carbon Diaries 2017* (2009) chronicle the near future for 16-year-old Laura and her family. This series is predicated on such popularizations as Bill McKibben's website 350.org, which tracks the worldwide carbon score (now around 390 parts per million). Laura's family shows a range of reactions to large-scale government intervention. Laura's older sister and parents struggle to cope with the narrowing of their hopes as they are monitored for their carbon footprint and penalized for their inability to keep their emissions low enough. Corruption surrounds them as black market participants flaunt the regulations. In the first book, the controls are initiated with monitors on the family and training sessions on their government-assigned responsibilities. Car ownership, airplane travel, and even domestic energy use are sharply curtailed, but it is not enough. The family and many others are physically displaced as the result of increased flooding, and Laura's hopes to finish college in London, to keep her music band together, or to have any version of the life she expected as a 21st-century teenager are gradually dismissed. Again the government of Britain, and the world by implication, is not able to cope with sequential disasters, and millions of humans and animals die. Lloyd continues her articulation of the personal face of global warming with a new book, *Momentum* (2011), set in the middle of energy wars.

Both Mary Rosenblum and Paola Bacigalupi have come to the attention of fans, critics, and popular science writers. They focus on extrapolations about the effects of abrupt climate change around access to freshwater and the production of alternative fuels to run modern technology. But their works are set at virtually opposite sides of Earth. Rosenblum's first novel speculating that climate change would cause extreme drought was *Drylands,* published in 1994. In both this work and the more recent *Water Rites,* she follows the path of drought in central and western North America. Using speculations that recall the Great Depression, she mixes the disintegration of governments, the corruption of special interest groups, and the gradual dying out of human populations who try to survive in marginalized landscapes. Freshwater is controlled by what is left of the U.S. government, which is largely in the grip of a few massive corporations. These corporations are still organized around the motive of immediate profit and thus enact measures to force subsistence farmers to produce cash crops of plants that survive on saltwater. The saltwater slowly renders the soil unusable for food crops. Rosenblum's focus is both on middle-range governmental officials who are gradually educated to the injustices being perpetrated and working-class and agricultural workers trying to survive in the increasingly hostile economic climate. Although Rosenblum's work is well written, compelling even, it offers little hope that governments can rein in large interlocking corporate interests. Her descriptions bring to mind the greed of monoculture cash crop–promoting companies that have caused desertification in Asia and African and corruption and suffering in South American countries. Although set in the future, Rosenblum's descriptions of the rapid acceleration of ecological disaster may seem all too familiar.

Bacigalupi, in *Pump Six and Other Stories* (2008), *Ship Breaker* (2010), and *The Windup Girl* (2010), sets his stories in East and Southeast Asia and primarily in Thailand. In a middle-range future, flooding and the breakdown of international trade have created alternative mechanical devices and energy sources, including waterwheels and animals. Social structures also become less forgiving as immigrant populations flee the flooding and infrastructural breakdown of their own countries. Bacigalupi's lower-middle-class and working-class characters lead impossible lives caught between floods, starvation wages, and immigration woes. One protagonist is a clone doll toy abandoned to humiliation and death because it is cheaper to create another such toy than pay for her travel back to Japan from Thailand. Human crisis and degradation found in other fictional projections are here much more graphic in such settings. Those who try to maintain a semblance of normalcy are constantly frustrated. Farming communities are repeatedly embroiled in flooding disasters and are subject to raids by corrupt government officials. Bacigalupi, like Baxter, offers little hope for a global future in which social controls have broken down and the country with the most starting capital strong-arms the rest. Like Rosenblum, Bacigalupi envisions a grim future where abrupt climate change has caused massive drought, flooding, water shortages, and food riots as well as control of water access by large corporations. Her work also seems to be predicated on results of forced cash crop conversions, which have been seen in the present day, in Africa and South America.

These are only a few examples of the attempts of novelists to create cautionary fictions that communicate the threats of climate change to nonscientists. One stumbling block to motivating the population of North America, and indeed the world, to activism is the perceived remoteness of global warming's effects. People often are unwilling to support scientific and governmental efforts to reduce carbon emissions, decrease the use of fossil fuels, and lessen the waste of the modern industrialized world because they do not believe that the threats predicted by climate science are immediate. While some of the authors described above may be capitalizing on a public interest in global warming and climate change science, many have a stated commitment to rendering what they see as a serious threat to the planet and organic organisms that inhabit it. Their commitment, their hope, is that the vehicle of fiction can guide human imagination to understand our own agency, negative and positive, and tip that agency toward the positive.

Janice M. Bogstad

Further Reading

Atwood, Margaret. *Oryx and Crake.* New York: Harper, 2002.

Atwood, Margaret. *The Year of the Flood: A Novel.* New York: Harper, 2010.

Bacigalupi, Paolo. *Pump Six and Other Stories.* San Francisco: Night Shade Books, 2010.

Bacigalupi, Paolo. *Ship Breaker.* New York: Little, Brown, 2010.

Bacigalupi, Paolo. *The Windup Girl.* San Francisco: Night Shade Books, 2010.

Barnes, John. *Mother of Storms.* New York: Tor, 1994.

Baxter, Stephen. *Ark.* New York: Roc, 2010.

Baxter, Stephen. *Flood.* New York: Roc, 2009.

Bertagna, Julie. *Exodus.* New York: Walker, 2008.

Brin, David. *Earth.* New York: Bantam/Spectra, 1991.

Crichton, Michael. *State of Fear.* New York: Harper, 2004.

Gould, Peter. *Write Naked.* New York: Farrar, Straus and Giroux, 2008.

Liptak, Andrew. "ReaderCon Session on Global Warming in Science Fiction." http://www.tor.com/blogs/2010/07/readercon-panel-recap-global-warming-and-science-fiction.

Lloyd, Saci. *The Carbon Diaries, 2015.* New York: Holiday House, 2008.

Lloyd, Saci. *The Carbon Diaries, 2017.* New York: Holiday House, 2009.

Lloyd, Saci. *Momentum.* New York: Holiday House, 2010.

"Michael Crichton's State of Confusion." RealClimate, http://www.realclimate.org/index.php/archives/2004/12/michael-crichtons-state-of-confusion/comment-page-2/.

Reeve, Philip. *Mortal Engines*. New York: HarperCollins, 2003.

Rosenblum, Mary. *The Drylands*. New York: Del Rey, 1994.

Rosemblum, Mary. *Water Rites*. Auburn, WA: Fairwoods, 2007.

"Science Fiction Writers Target Global Warming." National Public Radio, transcript of show from January 28, 2006. Debbie Elliot, host, Michael Crichton and Kim Stanley Robinson, guests. http://www.npr.org/templates/story/story.php?storyId=5176592.

Sterling, Bruce. *Heavy Weather*. New York: Bantam/Spectra, 1994.

Strieber, Will. *The Day after Tomorrow*. New York: Pocket, 2004.

Tepper, Sheri. *The Waters Rising*. New York: Harper, 2010.

350.org. http://www.350.org/.

Gore, Albert Arnold, Jr. (1948–)

When future historians look back on the introduction of the concept of climate change, it is very likely that no single figure will stand out as prominently as Albert Arnold "Al" Gore Jr. He became committed to the issue as a college student, and his life in politics presented him with an extraordinary opportunity to lead the American reaction to climate change. Much less likely, though, a convergence of environmental interest and international cooperation brought Gore international recognition on a scale never before seen.

Gore's life in politics was remarkable in its own right. The son of a well-known congressman, Gore served as an elected official for 24 years, representing Tennessee in the U.S. House of Representatives (1977–1985) and later in the U.S. Senate (1985–1993) and finally becoming vice president in 1993. He served in this last position for two terms to President Bill Clinton. Finally, Gore ran against George W. Bush in the 2000 U.S. presidential election, one of the most contested elections in American history. Gore won the popular vote by a margin of 500,000 votes; however, he ultimately lost the electoral college election when the U.S. Supreme Court decided the Florida vote recount in favor of Bush.

Following this difficult loss, Gore left politics and dedicated himself to the issue about which he was most passionate: climate change. He had been involved with environmental issues since 1976, when as a freshman congressman he held the first congressional hearings on climate change and cosponsored hearings on toxic waste and global warming. In addition to sponsoring bills or speaking out on environmental issues, Gore authored the best-selling *Earth in the Balance* in 2006 and wrote the introduction to the contemporary version of Rachel Carson's *Silent Spring,* the environmental classic from 1961. These efforts proved to be the prelude to how he would spend his life after politics.

Returning specifically to the issue of climate change, Gore became an outspoken leader on the issue he had heard about during his college days at Duke University. He began to speak out publicly about the scientific findings behind the claims of climate change and gradually mastered the art of simplifying these complex ideas for the general public. In 2006 director Davis Guggenheim, working with Laurie David, used Gore's lecture—with PowerPoint slides—as the organizing device for a near-documentary film called *An Inconvenient Truth.* Similar to *Silent Spring* and other seminal events in modern environmentalism, the most remarkable aspect of the film was the public response that it stirred.

The film catapulted Gore to global stature. He penned a book to accompany the film as well as other guides about how to take action on climate change. Even more important, he sparked a network of knowledge through the film: interested people all over the world could request training to speak publicly on climate change, and Gore's organization would provide them with materials. This grassroots dissemination of climate change data and understanding was seminal for global consideration of the issue. The culmination, however, was certainly the 2007 decision to award the Nobel Peace Prize to Gore and the Intergovernmental Panel on Climate Change.

Undoubtedly, the film caused a dramatic turn for the issue of climate change in general. In addition to bringing this environmental issue to an immense audience, the film made Gore into the 21st century's first environmental celebrity. As he helped to accept the Academy Award for best documentary for the film, Gore became the darling for a generation searching for ways to combat this problem. Of course, he also moved into the crosshairs of the interest groups arguing that global warming was a great hoax.

Following the Oscar ceremony, Gore made his first visit to Congress since appearing there to grudgingly accept the results of the 2000 presidential election. He came to Capitol Hill to educate Congress about the implications of global climate change. During his appearance, Gore appeared intermittently as a nerdy science teacher teaching the older generation about scientific theories and cutting-edge findings and at other times as a preacher attempting to convert his listeners. He forecasted what lawmakers would hear from their grandchildren in a few decades as they asked, "What in God's name were

Former vice president Al Gore (left) talks with Ted Turner (right), chairman of the United Nations Foundation, at the 2005 Institutional Investor Summit on Climate Risk at UN headquarters in New York on May 10, 2005. Gore attended the meeting in his capacity as chair of Generation Investment Management LLP, a global investment management firm. (UN Photo/Eskinder Debebe)

they doing? Didn't they see the evidence? Didn't they realize that four times in 15 years the entire scientific community of this world issued unanimous reports calling upon them to act?"

At other times, Gore appeared as a historian attempting to put the climate crisis in perspective. He reminded lawmakers of the resolve it took to fight Nazism and Communism. He told them that climate change requires the same kind of commitment: "What we're facing now is a crisis that is by far the most serious we've ever faced."

This was hardly the first time Gore tried to motivate Congress to respond to global warming. He first held a hearing on the topic in 1984, not long after he started serving in the House of Representatives. At that time, he served with another young idealist, Representative Ed Markey (D-MA). Markey told Gore that he was ahead of his time on climate change and other issues. "What you were saying about information technologies, what you were saying about environmental issues back then, now retrospectively really do make you look like a prophet," Markey said. "And I think that it would be wise for the Congress to listen to your warnings, because I think that history now has borne you out."

Today, Gore has become a businessman with roots in green causes and green consumerism. He is the founder and current chair of the Alliance for Climate Protection, the cofounder and chair of Generation Investment Management, the cofounder and chair of Current TV, a member of the Board of Directors of Apple Inc., and a senior adviser to Google. Through the issue of climate change, Gore has clearly achieved a global stature nearly unmatched.

Brian C. Black

Further Reading

Gore, Al. *Earth in the Balance.* New York: Rodale, 2006.

Gore, Al. *An Inconvenient Truth.* New York: Rodale, 2007.

Government Concerns for Human Health

The advance of modern efforts to use technology to solve problems clearly overwhelmed American life in the late 1800s. As historian Alan Trachtenberg and others have written, American culture tolerated social and environmental injustices if they were accompanied by a commensurate economic benefit. In short, environmental or health impacts were of little concern in the 19th century.

Reforming this ethic required generations of basic cultural changes in American society. The calls for these reforms came from many different types of Americans. Although these demands became much more specific by the end of the 20th century, 100 years prior one of the first issues to galvanize reformist interest was urban squalor. Whereas some observers argued that the priority of cities needed to be facilitating the industries that grew the economy, reformers factored in new information about influences on the health of urban residents to demand change.

Progressive reform in order to ensure Americans' health initiated the long tradition of relying on government regulation to temper economic development. Ultimately this sensibility would use the federal government to ensure environmental health and even to attempt to mitigate the influence of climate change. Such federal action, however, only came about following a century of work by activists and politicians to blend new scientific knowledge into laws and regulations that made citizens' health part of the federal government's responsibilities.

Demanding Change in New York City

One of the most important new understandings of the late 19th century was the origin of diseases. Concentrated in cities, diseases such as cholera became epidemics during the early 1800s because of a lack of understanding. For instance, when New York City was struck by epidemic cholera in 1832, ordinary citizens pointed to the prevalence of immoral behavior as the cause of this latest malady. In such instances, cholera was seen as God's judgment. Historian Charles Rosenberg writes that "Cholera was a scourge not of mankind but of the sinner. . . . Most Americans did not doubt that cholera was a divine imposition."

When the disease returned to New York in 1849, medically trained observes focused on specific parts of the city such as the tenements inhabited by poor immigrants. Some nativists used the moral argument to call for controls on immigration. By the mid-1860s, public health workers and sanitarians observed a connection between dirt and pollution and the outbreak of disease. Soon, city inspectors used this logic on New York tenements to argue that cleaner rear-yard areas indicated healthy occupants.

The concept of germs was not yet understood, but discerning viewers were growing more able to pick out indicators of good or ill health. That said, most ordinary Americans showed little concern for the implications of how they did basic things such as cooking or storing food, going to the bathroom, or acquiring their drinking water. Thus, many of the first efforts to reform urban areas occurred from the top down. Cleaning streets, emptying privies, disinfecting tenement buildings, and inspecting food and beverage manufacturing all helped to mitigate the dangers posed by unsanitary conditions.

One of the nation's first organizations devoted to urban reform was the Citizen's Association of New York Council on Hygiene and Public Health, which was founded in 1865. Under the supervision of John Griscom, the association set out to document the living conditions of the working class and the poor. The association consisted of wealthy merchants and city leaders, and it was in the association's best interest to improve urban squalor. Many urban elite feared that disease that began in tenements would wind up in their neighborhoods as well.

Widely distributed in a variety of forms, the 1865 Citizen's Association report laid bare the links between poverty, unsanitary living conditions, and ill health and was used to compel city and state governments to create a permanent health department. The Metropolitan Board of Health used the reports to exert new controls over the urban environment. With new authority to quarantine and disinfect unsanitary houses and rear yards, the Metropolitan

"Over the coming years and decades, climate change is likely to have a significant impact on health in the United States and globally. The United States and other developed countries with well-developed health infrastructure and the involvement of government and non-governmental agencies in disaster planning and response will be better able to address the health effects from climate change than will be countries in the developing world. Nevertheless, Americans may experience difficult challenges, and different regions of the Country may experience these challenges at varying degrees.

The anticipated health impacts of climate change have been well-reviewed and articulated by the Intergovernmental Panel on Climate Change and by the United States Climate Change Science Program through their Synthesis and Assessment Products. While knowledge of the potential public health impacts of climate change will advance in the coming years and decades, these entities have identified the following, which are current best estimates of major anticipated health outcomes . . . :

- Direct effects of heat,
- Health effects related to extreme weather events,
- Air pollution-related health effects,
- Water- and food-borne infectious diseases,
- Vector-borne and zoonotic diseases,
- Emerging pathogens susceptible to weather conditions,
- Allergies, and
- Mental health problems.

Howard Frumkin, director, National Center for Environmental Health, Centers for Disease Control and Prevention, congressional testimony, 2009

Board of Health cleared more than 160,000 tons of manure from vacant lots, cleaned and disinfected more than 4,000 yards, emptied 771 cisterns, and cleaned more than 6,418 privies.

By the late 1800s, many reformers were willing to listen to the findings of scientific professionals. Collections of data such as John Shaw Billings's "Vital Statistics of New York City and Brooklyn" became important tools for determining strategies for reforming urban life.

The Organic City

Amid the squalor of the organic city, cities such as Chicago, Illinois, were often considered to be state of the art for using waterways to take wastes away from the city and into larger bodies of water. Few people comprehended that such practices simply spread the problem over a broader area.

Waste management became a reality first in the cities, where large-scale changes were more easily carried out. Historian Martin Melosi traces the start of a waste management effort to the work of George Waring. After 5,000 residents died from yellow fever, Memphis, Tennessee, hired Waring in 1880 to developed a system for disposing of its sewage. His design became state of the art, and he defined the new role of the sanitation engineer. Melosi writes that Waring was committed "to the positive implications of environmental santitation." Waring clarified the important connection between disease and mismanaged urban waste. Melosi writes that "Despite his advocacy of an outdated theory of disease, Waring instinctively recognized many of the potential dangers of unsanitary surroundings."

Waring became an advisor to the federal government and worked with the National Board of Health. In 1894, political shifts in New York pushed out the Tammany Hall machine, and the new mayor brought in Waring, the world's leading authority on waste management. Under Waring, the Department of Street Cleaning unleashed thousands of military-like workers to clean up New York City and make it a national leader in sanitation. A corps of 1,450 sweepers cleaned each of the city's 433 miles of paved road one to five times per day.

By the 1910s, approximately 50 percent of the nation's cities operated municipally owned collection systems for solid waste. The Waring system was adopted in most cities; however, many cities remained a patchwork of collection practices. In terms of water, by 1910, 7 out of 10 cities of over 30,000 residents had constructed municipally owned wastewater disposal infrasturures.

The influence of Waring helped to diminish the "out of sight, out of mind" mentality that allowed many communities to dump waste into nearby bodies of water. Most often, dumps were sited in less-desirable areas of the city. Crematories and incinerators, a practice imported from Britain, became popular in the 1900s. Landfills became more popular by 1920. Melosi argues that as cities grew after 1920, the organization of waste management systems significantly increased. Although industrialization created more waste with which to deal, it also helped to created an engineering mind-set in which solutions were sought through planning and technology.

Living with Technology

As industrialization increased in the late 1800s, radically new ways of doing things changed many very basic parts of American life. Historian Thomas Hughes refers to the emergence of these new technologies as "the American genesis." Inventors, he writes, "persuaded us that we were involved in a second creation of the world." For instance, the number of patents issued annually more than doubled between 1866 and 1896. Humans began to interact closely with technology both at work and at home. Food, travel, lighting, and heating are just a few of the essential parts of life that changed in the 1890s. Those who seized these new opportunities also had the capability to make some of the greatest fortunes the world had ever seen in nonroyalties.

Although the ethics of many of these wealthy Americans earned them the name "robber barons," the era of big business additionally helped to shape a middle class of managers and businesspeople. This younger generation also began to ask serious questions of the American model of progress. Overall, though, this model of progress continued to prioritize economic expansion in the late 19th century. Industrialists of the era viewed mechanization as a boon because they could use cheaper labor. This social organization, of course, created a distinct working class. The ethics with which the few successful tycoons managed their workforces are considered extremely dubious by today's standards.

During the Gilded Age, many capitalists dismissed considerations of workplace safety and human welfare in order to create the greatest possible profits. For instance, child labor became a norm in factories, mines, and other extremely dangerous environments, largely because children required the least pay. All industrial growth surged after the American Civil War. This growth and a general faith in economic development allowed a few corporations to gain control of entire commodities and their production. Called trusts, these conglomerates were near monopolies during an era when government and society had not yet defined such an entity as evil. With such power concentrated in a single entity, any efforts for reform faced a mighty challenge.

American cities had by and large borne the brunt of rapid industrial development. In the late 1800s, cities were more often viewed as entities for producing economic development than as places to live. Reformers called for change beginning in the early 1900s.

Today, environmental policies regulate many different aspects of the world around us. This is particularly true of industries and factories that produce air, water, and other types of pollution. Today, we even know that there are different types of pollution: point, which affects the area immediately adjacent to the pollution's creation, and nonpoint, which affects a wider region possibly even distant portions of Earth such as Antarctica. Scientists have given us a much clearer idea about how such pollution damages the surrounding environment as well as the human body.

These ideas, of course, are radically different from those of the 19th century, when few questions were asked of economic development. Personal injuries or health problems were considered by many to be the price of having a job. This expectation began to change during the Progressive Era of the early 20th century. Muckraking journalists alerted the public and politicians to examples in which industry exploited resources, including the natural environment and the human worker. Unions called for more attention to be paid to workers' rights. In each case, the focus of concern became the massive factories that could be found concentrated in many American cities.

Reforming the City

Against a culture in which these ethics were the norm, a younger generation came of age in the 1890s and began to demand reform. Initially, one of the most common outlets for voicing discontent was journalism. The impassioned pleas of concerned journalists, then, found receptive ears among the elite women of the era. Activists such as Jacob Riis, who wrote *How the Other Half Lives* in 1890 to describe life in New York City slums, and Jane Addams, who started Hull House to aid immigrant acclimation to American culture, led a movement for progressive reform. Ironically, the wealth of some robber barons would contribute to the evolution of a public consciousness on issues such as ghettos, environmental degradation, and unfair labor practices. Riis observed that in New York City, "three-fourths of [the city's residents] live in the tenements, and the nineteenth-century drift of the population to the cities is sending ever-increasing multitudes to crowd them. . . . We know now that there is no way out; that the 'system' that was the evil offspring of public neglect and private greed has come to stay, a storm-centre forever of our civilization."

Riis used his writing and drawings to paint a picture of urban life that was largely unknown—or at least unacknowledged—by wealthy Americans. In *How the Other Half Lives,* he sets out to overcome this oversight:

> The name of the pile is not down in the City Directory, but in the public records it holds an unenviable place. It was here the mortality rose during the last great cholera epidemic to the unprecedented rate of 195 in 1,000 inhabitants. In its worst days a full thousand could not be packed

into the court, though the number did probably not fall far short of it. Even now, under the management of men of conscience, and an agent, a King's Daughter, whose practical energy, kindliness and good sense have done much to redeem its foul reputation, the swarms it shelters would make more than one fair-sized country village. . . .

. . . It is curious to find that this notorious block, whose name was so long synonymous with all that was desperately bad, was originally built (in 1851) by a benevolent Quaker for the express purpose of rescuing the poor people from the dreadful rookeries they were then living in. How long it continued a model tenement is not on record. It could not have been very long, for already in 1862, 10 years after it was finished, a sanitary official counted 146 cases of sickness in the court, including "all kinds of infectious disease," from small-pox down, and reported that of 138 children born in it in less than three years 61 had died, mostly before they were one year old. Seven years later the inspector of the district reported to the Board of Health that "nearly ten per cent of the population is sent to the public hospitals each year." When the alley was finally taken in hand by the authorities, and, as a first step toward its reclamation, the entire population was driven out by the police, experience dictated, as one of the first improvements to be made, the putting in of a kind of sewer-grating, so constructed, as the official report patiently puts it, "as to prevent the ingress of persons disposed to make a hiding-place" of the sewer and the cellars into which they opened. The fact was that the big vaulted sewers had long been a runway for thieves—the Swamp Angels—who through them easily escaped when chased by the police, as well as a storehouse for their plunder. The sewers are there to-day; in fact the two alleys are nothing but the roofs of these enormous tunnels in which a man may walk upright the full distance of the block and into the Cherry Street sewer—if he likes the fun and is not afraid of rats. Could their grimy walls speak, the big canals might tell many a startling tale. But they are silent enough, and 80 are most of those whose secrets they might betray. The flood-gates connecting with the Cherry Street main are closed now, except when the water is drained off. Then there were no gates, and it is on record that the sewers were chosen as a short cut habitually by residents of the court whose business lay on the line of them, near a manhole, perhaps, in Cherry Street, or at the river mouth of the big pipe when it was clear at low tide. "Me Jimmy," said one wrinkled old dame, who looked in while we were nosing about under Double Alley, "he used to go to his work along down Cherry

Street that way every morning and come back at night." The associations must have been congenial. Probably "Jimmy" himself fitted into the landscape.

Combining his writing and drawing talents with care and concern, Riis brought the plight of the poor to many wealthy Americans. He made it more difficult for Americans to look the other way and, as a result, reformers began to take effective action. Soon city, state, and federal government had little choice but to listen to their calls for reform and to take action.

Sewage and Water Technology

The technology for managing water and wastewater in urban areas changed steadily from the decentralized privy vault-cesspool system of the early 1800s. By 1850, centralized water-carriage sewer systems had been placed in some urban areas. By the end of the 19th century, centralized management systems had become the preferred urban wastewater management method, and most communities installed them by the end of the 20th century.

The primary reason for improving water treatment was a new understanding about the role that mistreated water played in breeding disease. The new technologies did not necessarily consistently prevent contamination of nearby surface water or groundwater. By the mid-19th century, engineers, public health officials, and the general public were searching for alternative wastewater management options.

Imported from Europe, the new concept of centralized water-carriage waste removal entailed planning a coordinated system of conduits and channels that used water to convey the wastes away from the sources to a central disposal location. The primary ingredient was piped-in water, which allowed each home's water closet to send out its waste into the system. The water closet marked a significant shift in sewage by increasing both the quantity that could be processed and the remaining fecal matter in discharges. In an odd twist, this increase in discharge actually made the risk of disease transfer greater.

The transition slowed at this point when the increased wastewater levels overwhelmed the privy vault-cesspool system. Few municipalities were able to plan for the new needs. Left to handle the problem on their own, many residents either continued to utilize the existing privy vault or cesspool or created illegal connections to storm sewers or gutters along streets. Each of these illegal options, of course, created additional problems.

The population in the United States surged more than fourfold from 1850 to 1920. This population increase was accompanied by an increase in the number of cities with populations greater than 50,000 (from 392 to 2,722). During the same time period, total U.S. population in urban areas increased from 12.5 percent to 51 percent. By the end of the 19th century, sewage technology began to have a noticeable effect on inner-city areas. By that time, most major U.S. cities had also constructed some form of a sewer system. In 1909, cities with populations over 30,000 had approximately 24,972 miles of sewers, of which 18,361 miles were combined sewers, 5,258 miles were separate sanitary sewers, and 1,352 miles were storm sewers. In larger cities (populations over 100,000), there were 17,068 miles of sewers, of which 14,240 miles were combined sewers, 2,194 miles were separate sanitary sewers, and 634 miles were storm sewers.

By the early 1900s, the most common technologies for treating wastewater were dilution, land application and irrigation of farmlands, filtration, and chemical precipitation. Each of these practices was carried out with untreated wastewater. By 1905, more than 95 percent of the urban population discharged their wastewater untreated to waterways. Little changed over the first quarter of the 20th century, and in 1924 more than 88 percent of the population in cities of more than 100,000 people continued to dispose of their wastewater directly to waterways.

Federal regulations demanded change beginning with the Water Pollution Control Act of 1948, which demanded and provided funds for planning, technical services, research, financial assistance, and enforcement. The renewal of this act through 1965 set water quality standards for every community. While plans of the late 1960s continued to prioritize the protection of public health, they also began to stress the need for preserving the aesthetics of water resources and protecting aquatic life. In 1972, the Water Pollution Act set the unprecedented goal of eliminating all water pollution by 1985 and authorized expenditures of $24.6 billion in research and construction grants. A new environmental era had descended on America's water and wastewater management program.

Taming Trash

Ideas for organizing and properly disposing of waste continued to evolve in the 20th century. A frequent sight in the 1950s, burning dumps were employed by many urban areas to condense waste. The premise, of course, was that burning reduced the volume of refuse received at the dump and therefore extended the life of the site. However, the burning dumps' impact on local air quality was a primary reason that early efforts after World War II to address the problems of dumps were directed toward putting out the fires. On a smaller scale, many homeowners also had burning pits in their backyards. Overall, open burning of refuse stopped in the 1950s.

Discontinuation of the fires created additional problems at dumps throughout the United States. Size increased more rapidly, and often the trash was more likely to breed disease and threaten human health. These needs stimulated a national movement in solid waste management during World War II toward the ideal that became known as the sanitary landfill.

The model for such landfills derived from the military. Due to the tremendous growth of new military bases during the Cold War, studies determined that the sanitary landfill was adaptable to changing conditions and would accommodate varying quantities of refuse with little significant change in equipment need or operating procedures. The essential element was a heavy piece of equipment called a bullclam. Resembling a bulldozer, the bullclam carried a movable flap or blade that could form a bucket or basket to hold significant quantities of refuse or push material around the landfill in order to cover the surface. Almost constantly, the bullclam moved material and compacted it more and more. Finally, it moved earth-cover material over the surface of the fill. As landfills became commonplace at military bases, they also influenced civilian refuse operations. By the end of 1945, almost 100 cities in the United States were using sanitary landfills, and by 1960 some 1,400 cities were using sanitary landfills.

Clearly, the management of human waste has been one of the most persistent environmental issues confronting American policy makers. The following is a timeline of some of the efforts to control human waste first in England and then extending to the United States.

Trash Timeline

1842 A report in England links disease to filthy environmental conditions.

1874 In Nottingham, England, the "destructor" burns garbage and produces electricity. Eleven years later, the first American incinerator opens in New York.

1898 The first energy recovery from garbage incineration in the United States started in New York City.

1900s Pigs are used to help get rid of garbage in several cities. One expert said that 75 pigs could consume one ton of garbage a day.

1904 The first major aluminum recycling plants open in the United States.

1920s Landfilling becomes the most popular way to get rid of garbage.

1959 The first guide to sanitary landfilling is published.
1965 Congress passes the first set of solid waste management laws.
1987 A garbage barge circles Long Island with no place to unload its cargo. Americans perceive a new garbage crisis.
1989 An Environmental Protection Agency report titled "The Solid Waste Dilemma: An Agenda for Action" advocates recycling as a waste management tool.
1997 The first American Recycles Day occurs.

The Russell Sage Foundation Studies Urban Problems in Pittsburgh

It was no easy task to convince Americans that industrial cities—with the economic opportunity that they offered—were generating social difficulties for those living in them. By the early 1900s, sociologists began to actively study cities in order to quantify the claims being made by activists and reformers. It is not surprising that one of the first sources of study was Pittsburgh, Pennsylvania, one of the most active industrial cities on Earth.

Rivers were the corridor that first initiated settlement at Pittsburgh. The confluence of three mighty rivers, Pittsburgh hosts the Monongahela, Ohio, and Allegheny Rivers. Travel westward on the rivers also initiated a great deal of industry in the city. Building and outfitting boats became Pittsburgh's first big business in the early 1800s. This industry was concentrated primarily on the shores of the Monongahela River near Pittsburgh's Point Park. During this era, inexpensive flatboats were powered solely by river current and steered with a 30–40-foot oar at the back. This traffic traveled only in one direction. Once they arrived at points in the West, settlers would break up the boats and use the wood as building material.

In 1852, though, new railroads ensured efficient connections to the East. Pittsburgh became the leading city not only in western Pennsylvania but also in many nearby states as well. Even so, shipping by water remained much cheaper than sending raw materials via rail. Pittsburgh offered the confluence of these transportation technologies. Pittsburgh became a city in which trains and rivers worked together to lay the foundation for the city's industrial future. Often, trains shifted their loads to barges in order to make the trip downriver. Then barges might empty their loads onto trains to send finished goods into the countryside.

The heavy industries of the 1800s also centered on Pittsburgh's transportation connection. When the coke-burning blast furnace was developed, ironmakers moved their operations close to the rivers. This way, the coke could be delivered by barge or rail to a riverside furnace near rolling mills and other ironworking operations whose engines and processes demanded water.

By 1850, Pittsburgh possessed a remarkably diversified economy. With railroads to deliver materials to Pittsburgh's factories and carry off finished products to markets in other cities, other industries began to flourish as well. For instance, by 1857 five large textile mills employed more than 3,000 workers. In addition, during the 1860s Pittsburgh also became the world's greatest petroleum-refining center. When western Pennsylvania brought the world its first supply of oil, the boats full of crude oil headed down the Allegheny River toward Pittsburgh. The refinery period was brief, though, since John D. Rockefeller's Standard Oil Company attracted the petroleum shippers to Cleveland.

By the 1870s, however, Pittsburgh's involvement in industry had just begun. The growth of the steel and iron industry changed the city's population, economy, and environment forever. Thanks to its transportation infrastructure and its access to raw materials, Pittsburgh became one of the world's greatest symbols of the new industrial era.

When sociologists working for the Russell Sage Foundation looked for a city in which to study the human effects of industrialization and pollution, they immediately found Pittsburgh to be the best case study. Here is how Paul U. Kellogg described the process:

> The main work was set under way in September, 1907, when a company of men and women of established reputation as students of social and industrial problems spent the month in Pittsburgh. On the basis of their diagnosis, a series of specialized investigations was projected along a few of the lines which promised significant results. The staff has included not only trained investigators but also representatives of the different races who make up so large a share of the working population dealt with. Limitations of time and money set definite bounds to the work, which will become clear as the findings are presented. The experimental nature of the undertaking, and the unfavorable trade conditions which during the past year have reacted upon economic life in all its phases, have set other limits. Our inquiries have dealt with the wage-earners of Pittsburgh (a) in their relation to the community as a whole, and (b) in their relation to industry. Under the former we have studied the genesis and racial make-up of the population; its physical setting and its social institutions; under the latter we have studied the general labor situation; hours, wages, and labor control in the steel industry; child labor, industrial education, women in industry, the cost of living, and industrial accidents.

The findings of the Pittsburgh Survey remain one of the best representations of the ills of unregulated industrialization. This study was an important step in the efforts of reformers to help make everyday life in American cities safer from industrial hazards. Such an attitude for reform, though, directly contradicted many Americans' view of what was responsible corporate behavior.

Alice Hamilton Connects Social Reform with Human Health

The critiques of social reformers in the early 1900s often contained much more passion than substance. Their criticism of environmental degradation most often concerned the appearance or other physical manifestations of industrial pollution. By the 1910s, physicians and scientists began to connect such critiques with verifiable medical findings. Very quickly it became clear to many Americans that industrial development claimed a serious impact on human health. It is remarkable to learn that such a dramatic alteration to American ideas emanated from the actions and efforts of very few individuals. In fact, our expectations of health protection can be traced to one individual: Alice Hamilton.

Although President Theodore Roosevelt and his progressive allies worked to change laws to help Americans in general, other progressive reformers used medical skills at the grassroots level to help urban Americans. Hamilton is one of the best known. She established the field of occupational medicine, served as the first woman professor at Harvard Medical School, and was the first woman to receive the Lasker Award in public health.

After taking her first academic appointment in 1897, Hamilton was appointed professor of pathology at the Women's Medical School of Northwestern University in Evanston, Illinois, and in 1902 she accepted a position as a bacteriologist at the Memorial Institute for Infectious Diseases in Chicago, Illinois. Dr. Hamilton became familiar with Jane Addams's Hull House, where Hamilton began to apply her medical knowledge to the needs of the urban poor. During her stay at the Hull House, she established medical education classes and a well-baby clinic.

During the typhoid fever epidemic in Chicago in 1902, it was Hamilton who connected improper sewage disposal and the role of flies in transmitting the disease. Based on her findings, the Chicago Health Department was entirely reorganized. She then noted that the health problems of many of the immigrant poor were due to unsafe conditions and noxious chemicals, especially lead dust, to which they were being exposed in the course of their employment. In 1910 the State of Illinois created the world's first Commission on Occupational Disease, with Hamilton as its director. The commission led to new laws and regulations for Illinois and contributed to the workers' rights movement in the United States. The commission also introduced a new notion that workers were entitled to compensation for health impairment and injuries sustained on the job.

Because of her work in Illinois, Hamilton was asked by the U.S. commissioner of labor to replicate her research on a national level. She noted hazards from exposure to lead, arsenic, mercury, and organic solvents as well as to radium, which was used in the manufacture of watch dials. Although many reformers argued that the industrial environment had become unsafe, Hamilton was one of the first scientists to prove it. Industrialists could do little but agree with her scientific findings.

Federal Efforts to Regulate Human Health Begin with Diet

It was one thing for Americans to slowly decide that their federal government needed to be more involved in stimulating development by financing certain large-scale transportation projects; it was entirely another case for the federal government to actively oversee public health. As new scientific understanding connected human health to outside stimulants, including diet, information and action was slow to come to the public.

The connection between science and the health of the American public evolved late in the 19th century. The first federal efforts at monitoring health focused on one of the most important sectors of the nation's population in the late 1700s: seamen. From 1798 to 1902 when national security and global trade relied on the stability of this segment of the population, the Marine Hospital Service (MHS)—and from 1902 to 1912 the Public Health and Marine Hospital Service—made sure sailors had the best health care the nation could muster. Members of the maritime industries, of course, are not government employees; however, their well-being had a direct impact on the nation's economy. Seamen traveled widely and often became sick at sea or in foreign nations. Therefore, their health care became a national problem.

In 1798, Congress established a network of marine hospitals in port cities around the world. Here, doctors cared for these sick and disabled seamen in facilities financed by taxing American seamen 20 cents per month. This payment was one of the nation's first direct taxes as well as its first medical insurance program.

The Progressive Era of the late 1800s and early 1900s brought new calls from the public for the federal government to help improve the American standard of living. Of primary concern were well-known epidemics of contagious diseases, such as smallpox, yellow fever, and cholera that had caused many deaths worldwide. Congress's interest in enacting laws to stop the importation and spread of such diseases resulted in a significant expansion of the responsibilities of the MHS.

The increase in passenger travel by steamship, for instance, meant that the MHS was responsible for supervising national quarantine, including ship inspection and disinfection, the medical inspection of immigrants, the prevention of interstate spread of disease, and general investigations in the field of public health, including yellow fever epidemics.

The effort to diagnose and treat infectious disease required the application of new science. To help inspect and diagnose passengers of incoming ships, the MHS established a small bacteriology laboratory in 1887. The Hygienic Laboratory was first located at the marine hospital on Staten Island, New York, and then later moved to Washington, D.C., where it ultimately became the National Institutes of Health.

In 1902 Congress passed an act to expand the scientific research work at the Hygienic Laboratory and provide it with reliable funding. In an effort to spread the impact of good health practices, the bill also required the surgeon general to organize annual conferences of local and national health officials. To reflect these new responsibilities, the name of the MHS became the Public Health and Marine Hospital Services. Finally, the Public Health Service was established in 1912, just in time to confront one of the nation's most serious health issues—a flu pandemic.

The Food and Drug Administration as the Focus of Federal Efforts

The Food and Drug Administration (FDA) is an agency of the Public Health Service of the Department of Health and Human Services. The agency is charged with ensuring that food is safe and pure, drugs and medical devices are safe and effective, cosmetics are safe, and products are labeled truthfully and informatively. Six bureaus within the FDA carry out these responsibilities: Foods, Drugs, Medical Devices, Veterinary Medicine, Radiological Health, and Biologics. Enforcement includes inspecting factories, testing food and drugs for purity and potency, approving premarket clinical trials on the safety and efficacy of drugs, informing industries of legislation and working to establish procedures that will prevent violations, publicizing irresponsible industry activities, and taking action against violators that includes seizure of substandard products, injunction suits, and criminal prosecutions. The requirement of FDA approval before new products can be used by the public has put the agency at the center of controversial issues such as birth control and AIDS treatment. At times, critics charge that the agency's caution can cost American lives; of course, other observers point out that the FDA is the last protection for American consumers.

Public concern over consumer protection issues remained strong throughout the 20th century, reflected in legislation ranging from the landmark Pure Food and Drug Act of 1906 to regulations governing food labeling in the 1990s. The concern over food safety is attributed to the muckraker Upton Sinclair. In 1906 Sinclair published *The Jungle,* which aimed, as he later said, at people's hearts but hit their stomachs instead. His few pages describing filthy conditions in Chicago's packing plants, widely reported and confirmed by governmental inquiry, cut meat sales in half, angered President Roosevelt, and pushed a meat inspection bill aimed at protecting the domestic market through Congress.

Working with Dr. Harvey W. Wiley, Roosevelt helped press a much-debated bill into becoming law. The 1906 law forbade interstate and foreign commerce in adulterated and misbranded food and drugs. Offending products could be seized and condemned; offending persons could be fined and jailed. Drugs had either to abide by standards of purity and quality set forth in guidebooks prepared by committees of physicians and pharmacists or meet individual standards chosen by their manufacturers and stated on their labels. An effort failed to place in the law food standards as defined by the agricultural chemists, but the law still prohibited the adulteration of food by the removal of valuable constituents, the substitution of ingredients so as to reduce quality, the addition of deleterious ingredients, and the use of spoiled animal and vegetable products. Making false or misleading label statements regarding a food or a drug constituted misbranding. The presence and quantity of alcohol or certain narcotic drugs had to be stated on proprietary labels. Although some fuzziness existed in the administrative provisions, the law gave Wiley's Bureau of Chemistry the task of spotting violations and preparing cases for the courts.

The Influenza Pandemic Mobilizes Federal Action

Increasing interactions because of global trade increased the occurrence and awareness of disease. Although World War I was a global tragedy, it also contributed to one of the most significant pandemics in world history. The influenza (flu) pandemic of 1918–1919 killed between 20 million and 40 million people. It has been cited as the most devastating epidemic in recorded world history. More people died of influenza in a single year than in four years of the Black Death bubonic plague pandemic from 1347 to

1351. Known as Spanish flu or La Grippe, the influenza of 1918–1919 was a global health disaster.

Ultimately influenza in the autumn of 1918 infected approximately one-fifth of the world's population. The flu usually most affects the elderly and the young, but this strand hit hardest on people aged 20 to 40. It infected 28 percent of all Americans, killing an estimated 675,000. The flu's connection to World War I was clear: of the U.S. soldiers who died in Europe, half of them fell to the influenza virus—an estimated 43,000 servicemen. The *Journal of the American Medical Association* wrote in 1918 that "The effect of the influenza epidemic was so severe that the average life span in the U.S. was depressed by 10 years. The death rate for 15- to 34-year-olds of influenza and pneumonia were 20 times higher in 1918 than in previous years."

As the influenza pandemic circled the globe, there were very few regions that did not feel its effects. Primarily, the spread of the flu followed the path of its human carriers along trade routes and shipping lines. Outbreaks swept through North America, Europe, Asia, Africa, Brazil, and the South Pacific in 1919. In India the mortality rate was extremely high, at around 50 deaths from influenza per 1,000 people.

Ports such as Boston, where materials were shipped out to the battlefront, were the most heavily affected American cities. The flu first arrived in Boston in September 1918. Then, however, the disease became more diffuse when soldiers brought the virus with them to those they contacted. In October 1918 alone, the virus killed almost 200,000 people. As people celebrated the end of the war on Armistice Day with parades and large parties, many U.S. cities suffered from public health emergencies. Throughout the winter, millions became infected and thousands died.

In one effort to stall the spread of the disease, public health departments distributed gauze masks to be worn in public. Basic parts of everyday life changed: places of business closed, funerals were limited to 15 minutes in order to fit in more services, and some towns and railroads required a signed certificate to enter. Bodies pilled up throughout the nation. Besides the lack of health care workers and medical supplies, there was a shortage of coffins, morticians, and gravediggers. The public emergency very closely resembled the Black Death of the Middle Ages.

Philadelphia, the hardest hit of all U.S. cities, was struck in October 1918. By the end of the first week, 700 residents were dead; 2,600 died by October 12, and the death toll continued to rise. Although no single group or neighborhood was entirely spared, immigrant neighborhoods—where basic sanitation and overall health were poorest—were the hardest hit. By November 2, the death toll in Philadelphia from the flu reached a staggering 12,162 people.

Through this public health disaster, Americans learned a valuable lesson: the federal government needed to be used proactively to assist in preventing such outbreaks and helping to ensure better health practices and in educating American about these practices.

Learning the Limits of Federal Action on Health

The McNary-Maples Amendment of 1930 clarified the agency's role by authorizing FDA standards of quality and fill-of-container for canned food, excluding meat and milk products. In the 1960s, the Fair Packaging and Labeling Act required all consumer products in interstate commerce to be honestly and informatively labeled, with the FDA enforcing provisions on foods, drugs, cosmetics, and medical devices. The FDA rapidly became the trusted authority in testing new items and assuring consumers of the ingredients in packaged materials.

In the 1970s, however, free-market advocates initiated a strong countermovement against consumer protection forces. Whereas consumer groups favored more stringent regulation and stressed safety considerations, conservative proponents of the free market pointed to increased research and development costs and a slower pace of product innovation as liabilities associated with federal drug regulations. A stalemate developed. In 1974 and 1976 the Senate passed bills that would have increased FDA authority, but the bills died without House action. On the other hand, in 1976 a House investigation subcommittee concluded that nine major U.S. government agencies, including the FDA, were biased in favor of regulated industry. Also in 1976, the FDA was granted greater authority to regulate medical devices, such as heart valves and kidney dialysis machines.

By the early 1980s the FDA was taking a less aggressive stance toward the drug industry. President Ronald Reagan's administration promoted voluntary compliance with federal regulations and in 1984 relaxed approval for generic versions of drugs in an effort to increase price competition. The result was a decrease in inspections, seizures, and legal actions. Several FDA officials were convicted of taking bribes from generic drug makers. In 1989 FDA commissioner Frank E. Young was demoted amid criticism of laxity toward the drug industry. David A. Kessler was named as his replacement and continued as head of the FDA under the Bill Clinton administration. In 1988 the Food and Drug Administration Act established the FDA as an agency of the Department of Health and Human Services, with a commissioner of food and drugs appointed by the president. In this position, Kessler pursued a more active role for the FDA, broadened by a passage of the omnibus health package in

1988 that included the first rules for coping with the AIDS epidemic. In 1990 further legislation provided for consolidation of the 23 Washington offices of the FDA, an automated system for processing drug applications, and higher salaries for biomedical scientists. In 1994 Commissioner Kessler took his most controversial action by declaring that tobacco should be declared an addictive substance and regulated by the federal government.

The FDA has found itself in other highly politicized controversies. In the early 1970s these included the use of diethylstilbestrol (DES) as a growth stimulant in livestock and in a morning-after contraceptive pill. In the early 1990s controversy erupted over approval of genetically engineered recombinant bovine somatotropin (BST) to stimulate milk production in dairy cows. The debates were complicated by concerns among environmental and religious groups over genetically altered products (as in the case of BST) and the growing intensity of the abortion debate. In the 1980s women's health and family-planning organizations, along with several drug companies, became interested in the possibility of introducing the morning-after pill, RU-486, into the United States. The drug was widely used in France as a safe and effective alternative to early abortions. Religious and antiabortion groups sprang into opposition, because RU-486 would make abortion clinics and traditional abortion procedures obsolete. In 1993 President Bill Clinton authorized research in the United States on the French abortion pill, making its availability to U.S. women possible before the end of the 20th century.

One of the major controversies of the 1980s and 1990s was over the government's response to the AIDS crisis. When AIDS emerged in the 1980s, gays and lesbians were already a well-organized political force. Their experiences had often led them to be skeptical of the medical and scientific communities. AIDS activist groups, such as the New York–based ACT UP (AIDS Coalition to Unleash Power) and the San Francisco–based Project Inform, pressured the FDA to provide promising but still-experimental drugs to people with AIDS on a parallel track with standard clinical trials required before FDA approval of drugs. Despite resistance, the FDA moved toward such a policy and established accelerated approval procedures in 1992. Accelerated approval is intended to get promising but still-unproven drugs for life-threatening diseases to patients as quickly as possible. The drugs must still be shown to be safe, but the usual standards of efficacy are relaxed. While drug companies have long resisted increasing regulations, this particular case did not meet with resistance. Unlike other consumer groups, AIDS activists had an interest in less stringent regulations, and by the late 1980s prominent researchers became convinced that there was a moral obligation to provide promising therapies as early as possible. Another FDA concern has been the reform of food labeling. Responding to reports of nutritional deficiencies, the FDA in 1973 adopted voluntary labeling that emphasized vitamins, minerals, and proteins. While nutritional deficiencies are now uncommon, problems with labeling became apparent. As consumers became concerned about the link between diet and disease, the food industry began adding terms such as "light," "healthy," and "low fat" to labels. What these phrases meant was unclear. The Nutrition Labeling and Education Act of 1990 required food labels, and the FDA set standard serving sizes. In 1994 requirements for a standardized food label took effect, making it easier for consumers to check the nutritional content of packaged foods.

Today, the FDA is one of our nation's oldest consumer protection agencies. Its approximately 9,000 employees monitor the manufacture, import, transport, storage, and sale of about $1 trillion worth of products each year. It does that at a cost to the taxpayer of about $3 per person. The FDA has more than 1,100 investigators and inspectors who cover the country's almost 95,000 FDA-regulated businesses. These employees are located in district and local offices in 157 cities across the country

Obama Presses for Federal Health Care

In its efforts to federalize health care in 2009, President Barack Obama's administration may have pressed to the limits Americans' interest in making citizens' health part of the responsibilities of the federal government. He explained the problem to Americans in this fashion:

> [T]he problem that plagues the health care system is not just a problem for the uninsured. Those who do have insurance have never had less security and stability than they do today. More and more Americans worry that if you move, lose your job, or change your job, you'll lose your health insurance too. More and more Americans pay their premiums, only to discover that their insurance company has dropped their coverage when they get sick, or won't pay the full cost of care.

To rectify this problem, the Affordable Care Act, passed by Congress and signed into law by the president in March 2010, established a plan to provide comprehensive health insurance for many more Americans, to hold insurance companies accountable, lower health care costs, guarantee more choice, and enhance the quality of care for all Americans.

In this system, the federal government was no longer simply regulating threats to Americans' health or providing nutritional information about our food. Instead, this health care reform placed the federal government as a primary actor in and regulator of a very large segment of the nation's economy. As the nation faced additional financial difficulties and as powerful lobbying forces fought back, by 2011 much of the new reform—dubbed "Obamacare" by critics—grew threatened. It appeared that possibly Americans were reluctant to give the federal government such a dramatic role in maintaining citizens' health. Even after the law's approval by the U.S. Supreme Court, federalizing health care continues to pit many Americans against one another.

Conclusion: Minimizing the Human Footprint

Humans affect the environment around them. It then makes sense that where humans are most concentrated, in urban areas the impact is most acute. During the 1900s, Americans came to realize that they impacted the environment around them. In addition, these impacts created health implications for each of us. Therefore, over the course of the 20th century, Americans took positive action to minimize in many ways their impact on the world in which they lived. Trash and waste are very likely the most undeniable parts of the human life cycle. However, air pollution of all sorts clearly joined with these problems to demand action during the 20th century. What began with small-scale examples had become problems of Earth-wide proportions by the 21st century.

Brian C. Black

Further Reading

Anderson, Oscar E., Jr. *The Health of a Nation: Harvey W. Wiley and the Fight for Pure Food.* Chicago: University of Chicago Press, 1958.

Burian, Steven J., Stephan J. Nix, Robert E. Pitt, and S. Rocky Durrans. "Urban Wastewater Management in the United States: Past, Present, and Future." *Journal of Urban Technology* 7(3) (2000): 33–62.

Crosby, Alfred. *America's Forgotten Pandemic: The Influenza of 1918.* New York: Cambridge, 1990.

Hughes, Thomas. *American Genesis.* New York: Penguin, 1989.

Kellogg, Paul U. "What Was the Pittsburgh Survey?" http://www.clpgh.org/exhibit/stell30.html.

Kraut, Alan M. *Silent Travelers: Germs, Genes, and the Immigrant Menace.* New York: Basic Books, 1994.

Melosi, Martin. *Coping with Abundance.* New York: Knopf, 1985.

Melosi, Martin. *Sanitary City.* Baltimore: Johns Hopkins University Press, 1999.

Obama, Barack. "Remarks by the President to a Joint Session of Congress on Health Care." September 9, 2009, http://www.whitehouse.gov/the_press_office/Remarks-by-the-President-to-a-Joint-Session-of-Congress-on-Health-Care/.

Opie, John. *Nature's Nation.* New York: Harcourt Brace, 1998.

Riis, Jacob. *How the Other Half Lives: Studies among the Tenements of New York.* New York: Scribner, 1914.

Rosenberg, Charles. *The Cholera Years: The United States in 1832, 1849, and 1866.* Chicago: University of Chicago Press, 1962.

Rosner, David. *Hives of Sickness: Public Health and Epidemics in New York City.* New Brunswick, NJ: Rutgers University Press, 1995.

Rosner, David. "The Living City: Engineering Social and Urban Change in New York City, 1865 to 1920." *Bulletin of the History of Medicine* 73(1) (Spring 1999): 124–129.

Tarr, Joel. *The Search for the Ultimate Sink.* Akron, OH: University of Akron Press, 1996.

Tarr, Joel, ed. *Devastation and Renewal.* Pittsburgh: University of Pittsburgh Press, 2003.

Tenement Museum. http://www.tenement.org/encyclopedia/diseases_cholera.htm.

Tomes, Nancy. *The Gospel of Germs: Men, Women, and the Microbe in American Life.* Cambridge, MA: Harvard University Press, 1998.

Trachtenberg, Alan. *Incorporation of America.* New York: Hill and Wang, 1982.

Young, James Harvey. *The Medical Messiahs: A Social History of Health Quackery in Twentieth-Century America.* Princeton, NJ: Princeton University Press, 1967.

Young, James Harvey. *Pure Food: Securing the Federal Food and Drugs Act of 1906.* Princeton, NJ: Princeton University Press, 1989.

Green Culture and American Environmentalism

As policy makers reacted to grassroots demand for environmental reform in the 1970s and 1980s, many Americans sought out ways to integrate their newfound environmental ethics into their everyday lives. More than at any other time in American history, then, the living patterns of everyday American life in the 1980s included a thought or awareness of humans' impact on the world around them.

Once this environmental awareness made it into basic patterns of American mass culture, it often held little identifiable connection to its roots in the ecological principles of Aldo Leopold and Rachel Carson. However, many Americans had clearly added impact on the environment to their list of considerations when they made choices about which products to buy, where to eat, and what to do in their free time. This cultural foundation would serve as a critical tool as humans began to wrestle with large-scale environmental problems such as climate change.

When these choices reflect a bit of environmental conscience or reflection, it can be grouped with a cultural pattern termed "green culture." Often, this change is marked by alterations to tradition and practices already ingrained in American life, including residential patterns, leisure culture, and film preferences. Many scientists and active environmentalists decried such efforts as depthless efforts to exploit environmental greenness without understanding the real issues. They described green culture as a consumer America's example of "green washing" seen in corporate America.

Regardless of the debate over its true meaning, green culture represents the fashion in which the late 1970s and the 1980s mark a period of applied cultural change in American ideas about nature. Of course, once one begins to study green culture, one finds, of course, that there are examples of this form throughout the 20th century. In certain cases, green culture can derive from forms that seem incongruous with the environment. However, there are also plenty of clear patterns within American mass culture that suggest a growing interest in the environment, possibly even a more widespread environmental ethic.

Imaging Conservation: Smokey and Woodsy

Smokey Bear, the guardian of our forests, has been a part of the American popular culture for over 60 years. Although Smokey's story is interesting for a variety of reasons, it especially symbolizes an era when resource conservation issues could be taken directly to an interested and educated American public. Today, Smokey Bear is one of the most famous advertising symbols in the world and is protected by federal law. He has his own private zip code, his own legal council, and his own private committee to ensure that his

Illinois EPA Green Schools Checklist
☐ Conduct an energy audit. Contact your energy utility, an energy services company, the Illinois Department of Commerce and Community Affairs' Energy Bureau, or Illinois EPA's Office of Pollution Prevention to arrange a site visit. Consider involving students in the audit as a learning project.
☐ Make sure your building systems (e.g., boilers, fans and pumps) are operating efficiently. Optimize efficiency through regular inspections and preventative maintenance.
☐ Use compact fluorescent bulbs instead of standard incandescent "screw-in" bulbs.
☐ Clean lights and fixtures every two years to keep light output high.
☐ Convert to higher efficiency fluorescent lamps and electronic ballasts for most general lighting applications.
☐ Take advantage of natural light or daylighting, particularly when a school undergoes significant remodeling or when new structures are added.
☐ Consider high intensity discharge lights (e.g., high pressure sodium) instead of standard fluorescent lights for outdoor areas.
☐ Replace incandescent bulbs in exit signs with a light-emitting diode (LED) or compact fluorescent replacement kit.
☐ Install double pane windows and/or windows with a low-emission coating.
☐ Plug holes and caulk windows to stop heat loss.

Awareness of climate change has caused some localities to take action in schools. Checklist developed by the Illinois Environmental Protection Agency to help schools go "green." (Illinois Environmental Protection Agency)

name is used properly. Smokey Bear is much more than a make-believe paper image; he exists as an actual symbol of forest fire prevention.

Smokey Bear and the interest in bringing fire prevention to the public actually relates directly to events of World War II. As the official Smokey Bear website explains:

> In the Spring of 1942, a Japanese submarine surfaced near the coast of Southern California and fired a salvo of shells that exploded on an oil field near Santa Barbara, very close to the Los Padres National Forest. Americans throughout the country were shocked by the news that the

Green culture brings environmental concerns into the public sphere. In one of the first examples, Smokey the Bear, the memorable mascot of the U.S. Forest Service, led a campaign to prevent forest fires. The character was created by artist Harry Rossoll in the 1930s. (U.S. Forest Service)

war had now been brought directly to the American mainland. There was concern that further attacks could bring a disastrous loss of life and destruction of property. One of the areas that seemed ripe for destruction were valuable forests on the Pacific Coast. With experienced firefighters and other able-bodied men engaged in the armed forces, the home communities had to deal with the forest fires as best they could. Protection of these forests became a matter of national importance, and a new idea was born. If people could be urged to be more careful, perhaps some of the fires could be prevented.

This very genuine security concern eventually combined with the emerging American interest in the environment.

During World War II, the Forest Service worked with the Wartime Advertising Council to create a marketing campaign with such posters and slogans as "Forest Fires Aid the Enemy" and "Our Carelessness, Their Secret Weapon." Throughout the campaign, the suggestion was clear: more careful practices by humans could prevent many forest fires. Indirectly, fewer fires could help the war effort by maintaining timber supplies and minimizing necessary manpower.

Disney also got into the act when it released the motion picture *Bambi* in 1944. The company allowed the character to be used in the forest fire prevention campaign for one year. The "Bambi" poster was a success. Although the Forest Service was contractually bound not to use a fawn following Bambi's one-year run, the Forest Service did feel certain that an animal mascot had helped to raise public interest in fire prevention. After internal debates, the Forest Service settled on using a bear as its new mascot.

The first Smokey Bear poster was released later in August 9, 1944. The poster, which depicted a bear pouring a bucket of water on a campfire, was a hit. Within a decade, Smokey Bear was a moneymaker for the federal government. His character was officially moved from the public domain and placed under the secretary of agriculture's control. Whether Smokey was a stuffed toy or a poster, the federal government could now collect fees and royalties while also publicizing.

Smokey was joined a few decades later by Woodsy Owl. Created by the U.S. Department of Agriculture's National Forest Service in celebration of Earth Day 1970, Woodsy Owl told Americans: "Give a Hoot! . . . Don't Pollute!" These ad campaigns were especially effective with schoolchildren, and Woodsy was used on many informational programs that were developed to educate American children about these environmental issues.

Interpreting Green Culture

Green culture did not only emanate from conservation agencies. The dissemination of greener ethics has also greatly impacted the popular culture created by mass media. This development seems to have occurred at a pace with the growing interest of Americans in the environment; therefore, one can argue that the popular images fed the evolving desire of many Americans to be environmentally aware.

Most prevalent might be the genre of culture that seeks to give viewers access to the natural world, which lay quite distant from the professional worlds of most viewers. Mutual of Omaha's *Wild Kingdom* began this tradition in the 1970s. In the tradition of National Geographic, Marlon Perkins created adventure from far-off locations that was based in the unknown secrets of the natural world. Breeding an entire genre of television—even an entire network—*Wild Kingdom* continues production but has spawned a great many programs, particularly for young viewers. Finally, Perkins's search for showing animals in their natural surroundings contributed to the interest in ecotourism, in which the very wealthy now travel to various portions of the world not to shoot big game but only to view it.

Zoos and wildlife parks have also seized on this interest and attempted to manufacture similar experiences for visitors. Possibly the most well-known cultural manifestation of environmental themes is Sea World, the marine theme park that first opened in 1964 and now includes parks in Florida, California, and Ohio. Unlike Disneyland and other amusement parks, Sea World carries a full-blown theme: the effort to bring visitors into closer contact with the marine world. As this agenda has become more routine since 1980, performing mammals have taken center stage. The most famous of these performers is Shamu the killer whale. In the highly competitive amusement industry, Sea World has exploited its niche by focusing since 1990 on environmental themes deriving from threats to marine life.

Such cultural interest in natural history and science is most clearly evident for children. While entire school curricula have been altered to include environmental perspectives, juvenile popular culture has guided such interest. From Disney's *Bambi* to Dr. Seuss's *The Lorax,* artists have clearly identified a sensitivity in juveniles. Each of these tales stresses overuse, mismanagement, or cruelty toward the natural world. The typical use of easily recognizable examples of good and evil that support children's media have been radically expanded.

Mixing science with action, environmentalism proved to be excellent fodder for American educators. More importantly, though, the philosophy of fairness and living within limits merged with cultural forms to become mainstays in entertainment for young people, including feature films such as *The Lion King, Free Willie,* and *Fern Gully;* environmental music; and even clothing styles. Contemporary films such as *The Lion King* bring complex ecological principles of balance between species to the child's level.

Many parents find children acting as environmental regulators within a household. Shaped by green culture, a child's mind-set is often entirely utopian, whereas parents possess more real-world stress and knowledge. Even so, many adults long for such simplicity and idealism, and scholars say that children awaken these convictions in many adults. In fact, a growing number of adults hold jobs that are involved with the environment. Environmental regulation and green culture created a mandate for a new segment of the workforce: technically trained individuals to carry out new ways of managing waste and consumption.

Imaging Environmental Cataclysm: *The Late Great Planet Earth*

Throughout the history of mass and popular culture, creators of popular reading material and film especially science fiction have sought to strike American anxieties by choosing topics about which the public has great trepidation. During the 1970s, one of the most popular paradigms revolved around human exploitation of natural resources that then resulted in widespread human suffering. This form of tragic environmental drama continues to be popular in the 21st century.

One of the first examples of this line of thought was Hal Lindsey's *The Late Great Planet Earth,* originally published in 1970. In this text, Lindsey offered readers a guide to finding the future in the text of the Bible. With 15 million copies in print, this best seller obviously struck a nerve in the modern world. Specifically, Lindsey offered order to the chaotic close of the 20th century by arguing that many of the predictions of the Old and New Testament have come true. Such a connection offered hope to many Judeo-Christians that the Bible and the morality that it imposes had resonance in contemporary life. Lindsey's book also made many readers turn to the Bible in order to prophesy future events. In this fashion, Lindsey spurred contemporary readers to read the Bible with care and helped to reenergize Christianity. Many critics, however, suggest that few of his predictions for the 1980s came true and that he preyed on readers' hopes and fears. Regardless, the prophetic rhetoric of *The Late Great Planet Earth* made it one of the most popular books of the 1970s.

Although there have been many science fiction films designed to depict cataclysm and the end of the world, one of the most recent blockbusters was more specific than all of the others. *The Day after Tomorrow* (2003) marked a new approach and a new awareness of the potential impact of human living patterns.

The Whole Earth Catalog and Green Consumption

Aspects of the early environmental movement's roots in the 1960s counterculture persisted in alternate forms of consumption during the late 20th century. The roots of this movement can be found in Stewart Brand's *Whole Earth Catalog,* which first introduced Americans to green consumerism in

Stewart Brand, founder of the *Whole Earth Catalog*. The catalog became a staple among its countercultural readers for its tools and products that advocated self-sufficiency. (Tom Graves)

1968. A national best seller, the *Whole Earth Catalog* soon became the unofficial handbook of the period's counterculture and won the National Book Award. The book contained listings of products as well as related philosophical ideas based in science, holistic living, and metaphysics that resonated with many Americans who sought to turn their backs on the nation's prevailing culture of consumption. The *Whole Earth Catalog* offered an alternative paradigm based in counterculture values, and even those who did not strictly adhere to its message gained a valuable lesson in discerning consumption.

Brand's book became an instruction manual for participants in the counterculture who entered communes or sought out other ways of returning to the land. The first page declared that "the establishment" had failed and that the *Whole Earth Catalog*'s goal was to supply tools to help an individual "conduct his own education, find his own inspiration, shape his own environment, and share his adventure with whoever is interested." The text offered readers advice about organic gardening, massage, meditation, and even do-it-yourself burial: "Human bodies are an organic part of the whole earth and at death must return to the ongoing stream of life." Within the pages of the *Whole Earth Catalog,* many Americans found the rationale and resources to live as rebels against the American "establishment." Interestingly, however, Brand did not urge readers to reject consumption altogether, and the *Whole Earth Catalog*'s enlightened philosophy had a significant impact on consumption patterns in American mass culture as a whole.

In particular, the *Whole Earth Catalog* helped to create what is known today as green consumerism, which rejects the use of synthetic raw materials, applications of intrusive technologies, or products that contribute to or derive from polluting the environment. The *Whole Earth Catalog* sought to appeal to this consumer niche by offering green products, such as recycled paper, as well as the rationale for their use. The *Whole Earth Catalog* is viewed by many Americans as the trend-setting publication that started the movement toward purchasing whole grains, environmentally friendly products, and other hallmarks of healthy living. Today, these green products constitute a significant portion of all consumer goods, and national retail chains have been based on the sale of such goods.

The original catalog combined the best qualities of the venerable *Farmer's Almanac* with the ubiquitous Sears catalog, merging the values of wisdom and consumption with those of environmental awareness. Even though today's green culture has infiltrated mass society, the *Whole Earth Catalog* continues to serve as a network of experts gathering information and other tools to live a better life and even, for some, to construct "practical utopias." For example, the *Millennium Whole Earth Catalog* promotes itself as integrating the best ideas of the past 25 years with the best for the next 25 years, all based on such *Whole Earth Catalog* standards as environmental restoration, community building, holistic thinking, and self-administered medical care. Despite the increased environmental awareness of the American public, the *Whole Earth Catalog* continues to find a niche for its unique ideas about soft living and careful consuming. It has also led to the growth of organic products and markets.

Nature Company Empowers Green Consumption

As with many aspects of green culture, some of the forms appear to be false attempts to exploit consumer interest in the environment. A revealing example of this green culture occurred during the 1980s when the American shopping mall—the quintessential example of artifice—played host to green consumerism. The Nature Company, which sold scientific and naturalistic gadgets as well as holistic and Third World crafts, had originally been founded in Berkeley, California, in 1973 by Tom and Priscilla Wrubel, who had been members of the Peace Corps in the 1960s.

By 1994, the Wrubels had sold their interest to a corporation that specialized in newer models of consumption targeting yuppie consumers. Now there were 124 stores in the United States and approximately 20 more worldwide. Since then, Nature Company's fortunes have declined. However, the interest in this type of green consumption has not diminished. The niche that the Nature Company identified has now become part of mainstream consumption and can be found in many different types of stores. In this niche, Americans unapologetically mix consumptive practices with symbols and forms of the natural environment.

The Slow Food Movement and Making Locovores

The Slow Food movement began in Italy when in 1986 Carlo Petrini protested the opening of a McDonald's in Rome. The movement was first known as Arcigola and was designed to resist fast food. Today the object has expanded to not only combat fast food but also to help preserve traditional plants, seeds, animals, and farming practices, in addition to the cultural cuisine associated with these traditions. The movement includes more than 80,000 members and is active in more than 50 countries.

As it stands today, each slow-food chapter is responsible for promoting local farmers, flavors, and culinary artisans through programs such as farmers markets and taste workshops. These programs are designed to meet the objectives of the Slow Food movement: preserving and celebrating local culinary traditions, forming and maintaining seed banks to preserve local heirloom varieties, supporting small-scale production and processing, and educating people about the qualities of good food and the risks of fast food and agribusiness. The Slow Food movement also lobbies to include organic farming concerns within agricultural policies and to limit funding of genetically modified organisms and agrochemical use. In addition, the movement works with primary and secondary school students creating school gardens, educating the students on nutrition, and perpetuating the skills of farming.

Beyond the local programs, the Slow Food movement also takes some international actions. The movement opened the University of Gastronomic Sciences in Emilia-Romagna, Italy, in 2004. The goals are to work with students from many locations to not only preserve the artistic elements of traditional cooking but also to promote awareness of good food and nutrition.

Within the United States, the Slow Food movement membership totals 12,000 and includes more than 140 local chapters. Slow Food U.S.A. is a nonprofit organization dedicated to supporting, celebrating, and educating people about the food traditions in North America. Programs focus on everything from the purity of the organic movement to animal breeds and heirloom varieties of plant species to handcrafted beer and farmhouse cheeses. The goal is to find pleasure and quality in everyday life by slowing down, respecting traditions of the table, and celebrating the diversity of Earth's bounty.

Critics of the Slow Food movement claim that it is elitist. They also suggest that the movement is disparaging cheaper alternative methods of food processing and preparing methods that are important to lower-income people and families. The Slow Food movement counters these critiques by suggesting that they support local production and preparation because ultimately these processes are less expensive, as they do not necessitate long-haul transportation or energy and chemical use. According to the Slow Food movement, most food travels long distances, utilizing both substantial amounts of fuel and preservatives and other additives. The movement notes that the so-called cheaper alternatives do not accurately reflect their true costs due to government subsidies, which keep transportation and other costs artificially low.

Overall, the Slow Food movement believes that it is protecting not only cultural heritage but also regional diversity and biodiversity. By supporting local small farms and organics, the movement works to preserve each regional ecosystem and the culture that has developed over generations of living within each ecosystem.

Conclusion: What Is the Environmental Ethic of the Plastic Pink Flamingo?

How willing are we to accept elements of green culture into the environmental movement? Another way of asking this question is to ask this question: Can a pink artificial plastic decoration actually connect the everyday life of Americans to the natural forms around them? Some scholars argue that the pink flamingo placed in many American lawns does just that.

The pink flamingo was first sold to the public in 1957. By the 1970s, the flamingos became a prevalent part of the landscape of middle-class suburbs. For suburbanites, the plastic bird signaled a hint of disconformity in the homogeneous world of suburbia. Jenny Price, however, writes that the flamingo swiftly came to also signal definitions of nature.

The baby boomers did not invent the bird. But as with television, we were born with it and grew up with it. And we appropriated it for ourselves. Through the 1970s, we used the pink flamingo as a ubiquitous signpost for crossing the various overlapping boundaries of class, taste, propriety, art, sexuality, and nature.

During the 1980s and 1990s, the flamingo became more popular than ever before. During this ever more fluid era of boundaries and definition, the flamingo became a symbol of the connection between nature and art or artifice.

Whether it be through lawn ornaments or through the environmental themes contained in many feature films or even the remarkable popularity of *An Inconvenient Truth,* green culture became one of the most noticeable portions of environmentalism to mass-culture Americans. In fact, green culture might be the only environmentalism many of them ever get to know.

Brian C. Black

Further Reading

Cronon, William, ed. *Uncommon Ground: Rethinking the Human Place in Nature.* New York: Norton, 1996.

Hopkins, J. "'Slow Food' Movement Gathers Momentum." *USA Today,* November 25, 2003.

Kirk, Andrew. *Counterculture Green.* Lawrence: University Press of Kansas, 2007.

Nabhan, G. P. *Coming Home to Eat: The Pleasures and Politics of Local Foods.* New York: Norton, 2002.

Nash, Roderick. *Wilderness and the American Mind.* New Haven, CT: Yale University Press, 1982.

Opie, J. *Nature's Nation.* New York: Harcourt Brace, 1998.

Petrini, C., and G. Padovani. *Slow Food Revolution: A New Culture for Eating and Living.* New York: Rizzoli, 2006.

Price, J. *Flight Maps.* New York: Basic Books, 2000.

Rothman, H. K. *The Greening of a Nation.* New York: Harcourt, 1998.

Rothman, H. K. *Saving the Planet: The American Response to the Environment in the 20th Century.* Chicago: Ivan R. Dee, 2000.

Slow Food International. http://www.slowfood.com/.

Slow Food U.S.A. http://www.slowfoodusa.org/.

Steinberg, T. *Down to Earth.* New York: Oxford University Press, 2002.

Greenhouse Effect

The greenhouse effect is a natural phenomenon that has existed since the formation of Earth's atmosphere. We could not live without it, but human activity has recently accelerated the rate at which the process occurs by increasing the amount of greenhouse gases in the atmosphere, especially the amount of carbon dioxide (CO_2). There is no one person responsible for discovering and explaining the greenhouse effect. The theory developed slowly over two centuries as scientists devised methods to analyze the composition of the atmosphere and explain Earth's average temperature.

The temperature of Earth is regulated by balancing the input of solar radiation and the reflection of some of this energy back into space. Solar radiation arrives in the form of short-wavelength radiation (visible and ultraviolet light). Approximately a third of the radiation that reaches Earth's surface is reflected back into space. Some of the remaining radiation is absorbed by the atmosphere, but most is absorbed by land and ocean surfaces. The surface of Earth warms and emits long-wavelength (heat) radiation, which warms the atmosphere further. The atmosphere contains a number of gases, called greenhouse gases, that are critical to balancing air temperature. These include water vapor, carbon dioxide, ozone, methane, and nitrous oxide. These gases contain the radiated heat somewhat like a warm coat or a blanket. Without greenhouse warming, Earth's average surface temperature would be unable to support life. An increase in the concentration of greenhouse gases in the atmosphere caused by human activities (including the deforestation and the burning of fossil fuels and vegetation) increases the amount of heat contained in the atmosphere and thereby raises the global surface temperature.

During the 1820s, the French physicist Jean Baptiste Joseph Fourier asked, what determines the average surface temperature of Earth? When radiation from the sun warms the planet, why doesn't it keep getting hotter and hotter? Fourier calculated that an object the size of Earth, at its distance from the sun, heated only by solar radiation, should be much colder than it actually is. He examined a variety of possible explanations and sources for the additional heat. Fourier recognized that Earth's atmosphere was somehow acting as an insulator. He compared the process to a box covered with a pane of glass. The interior of the box warms in the sun, but the glass contains the heat so

In this image released by NOAA, Chris Carparelli, adjusts one of the glass flasks that line the walls of an air sample processing room at NOAA's Earth System Research Laboratory in Boulder, Colorado, Wednesday, May 30, 2012. Researchers at the lab measure the levels of carbon dioxide and other greenhouse gases in air sent in weekly from sites that are part of an international cooperative air sampling network. (AP Photo/NOAA, Will von Dauster)

that it cannot escape. Fourier is credited with being the first to propose the greenhouse effect theory. Although his explanation was not fully developed, his work gave the central idea for others to follow: that the atmosphere had an influence on the amount of heat radiating from Earth's surface.

The correct explanation came a few decades later. In 1861 John Tyndall, an Irish physicist, speculated on the existence of an "atmospheric envelope" and suggested that water vapor and various gases in the atmosphere are responsible for retaining heat from the sun. He developed experiments that demonstrated the radiative and absorptive properties of gases (especially water vapor and carbon dioxide) and used this knowledge to propose that the climate might warm or cool based on the quantity of these gases in the atmosphere. Tyndall found that simple water vapor blocks infrared radiation as readily as a solid wall. However, he did not imagine that carbon dioxide could be a strong source of warming, since the percentage in the atmosphere seemed too small (about a 20th of a percent) to have such a huge effect.

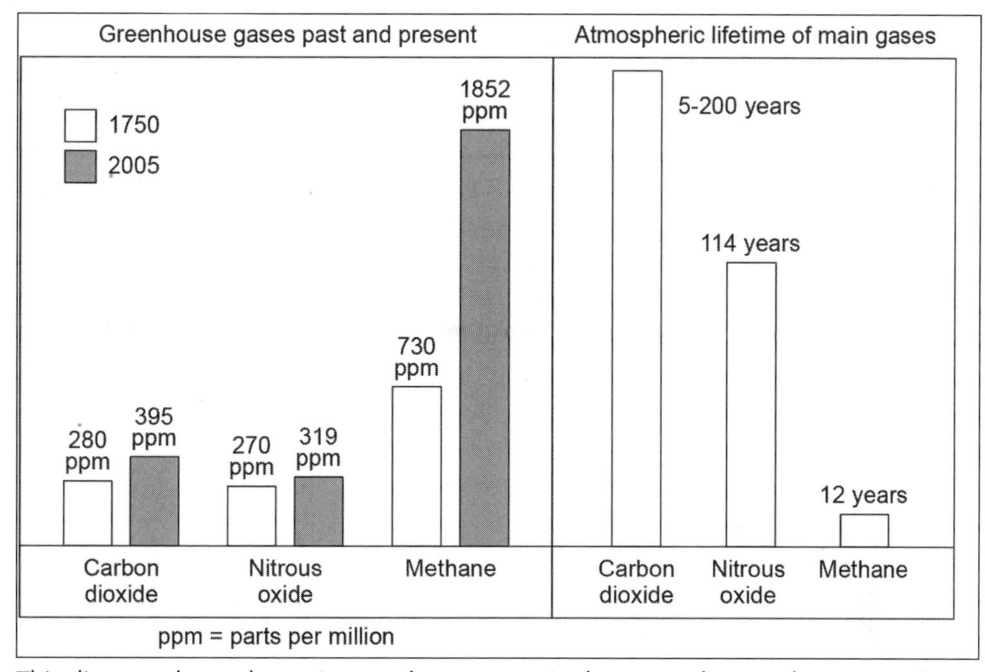

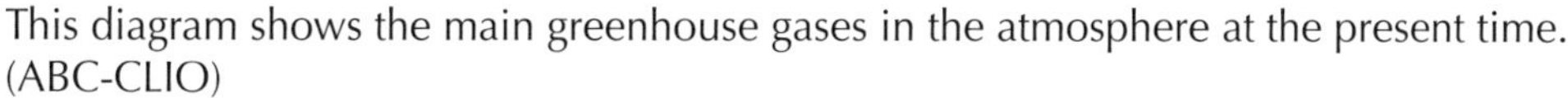
This diagram shows the main greenhouse gases in the atmosphere at the present time. (ABC-CLIO)

Many of the scientists who studied the greenhouse effect were not interested in the warming of our world but instead were looking for the mechanism responsible for causing past ice ages. It was clear that ice once covered vast areas that now have moderate climate. What process could cause the world to freeze and then make the process reverse? More importantly, could it happen again?

In the 1870s, the British geologist James Croll considered the first example of what we would call a positive feedback loop. Croll reasoned that if something caused Earth to cool, then snow and ice would cover much of Earth's surface. This would lead to more sunlight being reflected back into space, thus contributing to Earth's cooling trend. On the other hand, if something caused Earth to warm, then the warming would expose areas of dark soil, vegetation, and bare rock. This would lead to more sunlight being absorbed, which would contribute to Earth's warming trend. Croll considered the possibility that tiny changes in Earth's orbit around the sun were the original cause of the cooling and warming trends, but he was not able to confirm this in detail.

Influenced by his predecessors, the Swedish chemist Savante August Arrhenius focused on the idea that a rise in the concentration of atmospheric carbon dioxide could raise Earth's temperature. He developed his theory in

1896 by using equations to calculate that doubling the concentration of carbon dioxide in the atmosphere would raise global surface temperatures about 10°F. He also believed that it would take 3,000 years for carbon dioxide levels generated by human activity to double, which we now know is an underestimation. Although his estimate was incorrect, Arrhenius was the first person to predict that carbon dioxide emissions from human industry would impact the temperature of the planet. Nevertheless, little attention was paid to Arrhenius's greenhouse effect theory during his lifetime. It might come as a surprise that Arrhenius applauded his result that Earth was warming; this way, future generations would have an increasingly pleasant environment and ample resources to feed an increasing population.

While Arrhenius's mathematical model of Earth's greenhouse effect was not widely discussed, the idea did not go away at the start of the 20th century. Alfred Lotka, an American physicist, warned the public in 1924 that people are radically altering the carbon cycle by releasing far more carbon dioxide into the air. His calculations of fossil fuel use at the very beginning of the automotive age resulted in an estimate of doubling the carbon dioxide in the atmosphere within 500 years as a direct result of human activity, one-sixth of the time forecast by Arrhenius.

By the late 1930s, the concept of global warming caught the attention of Guy D. Callendar, a British meteorologist. He gathered temperature records from more than 200 weather stations around the world and used the data to demonstrate that Earth had warmed 0.4°C between the 1880s and the 1930s as a direct result of carbon dioxide emissions from industrialization. Scientists were skeptical, but Callendar established the groundwork for modern efforts on temperature trends and their possible link to changes in carbon dioxide levels.

The development of the greenhouse theory took on added dimensions in the 1950s. Scientists Roger Revelle and Hans Suess warned in a 1957 paper that vast amounts of carbon that took millions of years to be deposited in sedimentary rock were being returned to the atmosphere and oceans. Estimates of temperature increase were argued fiercely among researchers. To make matters worse, until the end of the decade there were not enough reliable records of greenhouse gas levels in the atmosphere to establish a consensus.

At about the same time that Revelle and Suess issued their warning, a researcher named Charles David Keeling (working with Revelle on a project associated with the International Geophysical Year of late 1957 and 1958) constructed equipment that collected high-precision systematic data on the carbon dioxide content of the atmosphere (measured in parts per million). He found that the carbon dioxide readings were very similar, no matter where he used the machine. This fact contradicted the general assumption of the time that levels would vary depending on geographic location. He continued to track the readings and demonstrated that carbon dioxide levels were rising steadily every year. The graph charting this data became known as the Keeling Curve. Keeling also discovered that the readings showed seasonal variation, especially in the Northern Hemisphere. The levels were lower in the spring and summer, when vegetation processed carbon dioxide, and were higher in the autumn and winter, when plants were dormant and decaying vegetation released additional carbon. The data gathered by Keeling and his associates illustrated clearly that human activity was raising levels of carbon dioxide in the atmosphere faster than the land and ocean surfaces could absorb it.

Debate over the greenhouse effect and global warming became a political issue in 1979. A variety of agencies and scientific teams studying greenhouse gas concentrations, global temperatures, and related concerns all concurred on the main point. The world appeared to be getting warmer. Nine years later, global warming became known to the general public thanks to wide media coverage. During the hot summer of 1988, James Hansen, director of the Goddard Institute for Space Studies, testified to a congressional committee that the unusually warm temperatures of the 1980s were a warning sign of impending global warming. He continued to push for recognition of the issue throughout the Ronald Reagan and George H. W. Bush administrations. Around the world a scientific consensus had finally been reached that human activity has already altered Earth's atmosphere. Studies also showed that daily minimum temperatures were increasing faster than daily maximums, another change that was forecast in climate warming models.

The major concern of climate scientists today is that the warming being experienced currently is due to the burning of fossil fuels during the last 50 or so years. The climate system is never in a state of perfect equilibrium. Daily and seasonal variations in temperature reflect the ground and oceans catching up to changes in the balance of energy in and out of Earth's surface. Surface changes lag behind changes to trace gases in the atmosphere, and oceans and land absorb radiation at different rates. Considering the carbon dioxide already dumped into Earth's atmosphere, we are already committed to a global temperature rise of 1.5–5°C in the coming decades. Any changes made to emissions right now will not be reflected in our climate for many years. To address such difficulties, it is likely that human society must move from short-term planning to a more long-term and global approach.

Jill M. Church

With rocks and bushes sticking through the snow pack, a pair of skiers ride a lift at Homewood Ski Resort in Homewood, California in January 2007. Scientists say climate change in the Sierra Nevada could lead to shorter winters and less snow, threatening Lake Tahoe's ski and winter tourist industry. (Rich Pedroncelli/AP/Wide World Photos)

Further Reading

Christianson, Gale E. *Greenhouse: The 200-Year Story of Global Warming.* New York: Walker, 1999.

Fleming, James R. "Climate Change and Anthropogenic Greenhouse Warming: A Selection of Key Articles, 1824–1995, with Interpretive essays." http://wiki.nsdl.org/index.php/PALE:ClassicArticles/GlobalWarming.

Fleming, James R. *Historical Perspectives on Climate Change.* New York: Oxford University Press, 2005.

Jones, M. D. H., and A. Henderson-Sellers. "History of the Greenhouse Effect." *Progress in Physical Geography* 14 (1990): 1–18.

Ruddiman, William F. *Plows, Plagues and Petroleum: How Humans Took Control of Climate.* Princeton, NJ: Princeton University Press, 2005.

Stevens, William K. *The Change in the Weather: People, Weather, and the Science of Climate.* New York: Delacorte, 1999.

Weart, Spencer R. *The Discovery of Global Warming.* Cambridge, MA: Harvard University Press, 2008.

Green Party in the United States

The political impact of environmentalism became very great by the end of the 20th century. Most often, members of each major American political party might have an environmental commitment. By the 1980s though, a growing number of Americans continued an international trend and initiated a party organized around environmental and other humanitarian principles: the Green Party. The party continues to have some influence in American politics, which makes many observers argue whether or not such a third party helps the causes it supports.

The U.S. Green Party traces itself to the European Greens, who first organized as an antinuclear propeace movement at the height of the Cold War. It was the German Greens, organized by Petra Kelly, who were most specifically influenced by the U.S. environmental movement of the early 1970s.

The U.S. Green Party convened in 1984 in meetings that focused on local elections. By organizing itself on the local level, the Greens first influenced policy and government in towns, cities, and villages. The first Green Party candidate appeared on the ballot in 1986. In 1990, Alaska became the first state to grant the Greens ballot status. California followed in 1992, and then many other state parties followed.

During the 1990s, the Green Party went through internal changes. Some Greens grew impatient with the strategy of radical, slow, long-term organizing. These Greens called for the immediate creation of state Green parties, and some called for a national Green party. Their calls created internal division, and two distinct factions took shape within the party. One faction believed that the Greens were a social movement that should work up from the grassroots, while the other faction urged the party to pursue national power and to challenge the American two-party system. In 1991 the groups changed their names, respectively, to the Greens/Green Party USA (G/GPUSA). While many members remained focused on activism, the Green Party USA sought a national stature.

The emergence to the national scene came quickly when in November 1995, the well-known activist Ralph Nader initiated the Green Party's first presidential campaign by officially entering the California Green primary. Nader's unconventional campaign aroused Green Party activity in states that up this point had little activity. By election eve, the Greens had placed Nader on 22 ballots nationwide, with another 23 states qualifying him as a write-in candidate. In August 1996, state Green parties held their first national nominating convention in Los Angeles, California. The party chose a ticket of Nader and Native American Winona LaDuke, who is a well-known activist for causes related to environmental and indigenous women's issues.

The Nader-LaDuke campaign challenged the candidates and platforms of the Democratic and Republican Parties and forced the parties to discuss issues that were important to the Green Party. The success of the campaign spurred many additional Green political efforts throughout the United States. In the end, the Nader-LaDuke campaign came in fourth place after that of Ross Perot. The campaign earned more than 700,000 votes, which accounted for approximately 1 percent of the vote nationwide. The single best state performance came in Oregon, where the Green Party earned 4 percent of the vote.

Although the Greens focused on local and state politics, some Greens continued to believe in a presidential run. Ronald Reagan and Jimmy Carter each demonstrated the ways that they each possessed the ability to alter the intensity with which federal regulation of environmental factors is carried out. The chief executive's position came to look most attractive for a Green candidate. In the 2000 presidential election, Nader ran for president again as a member of the Green Party with a platform against development and large corporations and in support of environmental causes.

Even though 70 percent of Americans call themselves environmentalists, Nader's campaign garnered less than 5 percent of the national vote. However, genuine pockets of such sentiments were seen in states such as Wisconsin,

Florida, Oregon, and California. It seems that a growing percentage of Americans are willing to entertain radical political change in the best interests of environmental causes.

Today the Green Party is said to be organized around these key values:

- Ecological wisdom
- Grassroots democracy
- Decentralization
- Community-based economics
- Feminism
- Respect for diversity
- Personal and global responsibility
- Future focus/sustainability

By forcing politicians to consider their stands on these issues, members of the Green Party have altered the national landscape of American politics. Throughout the nation, though, the Green Party members elected to state and local office have initiated basic changes to the planning and environmental issues that concern Greens.

Donna Lybecker and Brian C. Black

Green Revolution

The Green Revolution, a term coined in 1968, was an agricultural and technological revolution that increased agricultural production between the 1940s and the 1980s. The primary objective of the Green Revolution was to increase crop yields and augment aggregate food supplies to alleviate hunger and malnutrition. During the Green Revolution, agricultural programs of research and extension in addition to infrastructural development brought major social and ecological impacts to particularly developing nations. The results of the Green Revolution have drawn both praise and criticism.

Clearly, though, the Green Revolution joined with new communication technologies and the end of the Cold War to create new inroads for nations to work together. Over the last half of the 20th century, common issues became a new conduit through which nations could interact and cooperate. Although some scientists and environmentalists hypothesized that the human species was doomed because of its inevitable need to act in individual self-interest, a culture of international cooperation slowly took shape. While this required political agencies such as the United Nations, the movement for international cooperation also drew from evolving science that assisted in establishing international issues that confronted all humans.

The Green Revolution Ushers in a New Cooperative Era

Most scholars date the interest in globalization to the world wars and the Cold War. In addition, the world's worst recorded food disaster occurred in 1943 in British-ruled India. Known as the Bengal Famine, an estimated 4 million people died of hunger that year alone in eastern India (which included today's Bangladesh). The international community blamed poor food production for the shortage and began to consider ways of helping India and other less developed nations. The gap separating developed and less developed nations had become most pronounced by 1950, when modern conveniences and technologies made the standard of living in the United States and many other developed countries leap forward, while other nations lagged further behind.

When British occupiers left India in 1947, the Indian government set about to try to close this gap in terms of food production. However, two decades later, India realized that its efforts at achieving food self-sufficiency were not entirely successful. This awareness led to the importation of new agricultural technology from around the world. Referred to as the Green Revolution, the effort to bring new agricultural know-how to developing nations radically altered the possibility of famine while also attacking the primary issue dividing developed nations, such as the United States, and less developed nations, such as India and much of Africa.

Judged as successful by the major funding agencies, including the Rockefeller Foundation and the Ford Foundation, there was a feeling that these programs could be important to the future of other nations. Thus began the movement to spread the techniques to other countries. Beginning in Mexico in the 1940s, collaboration between the Rockefeller Foundation and Mexico's President Manuel Avila Camacho's administration brought the expectation that Mexico's developing agriculture would aid in the nation's industrial development and economic growth.

The Green Revolution then moved to India, bringing programs of plant breeding, irrigation, and agrochemicals, including the use of pesticides and synthetic fertilizers that created substantial increases in grain production.

Headed by Norman Borlaug, a plant breeder from the University of Minnesota, American agricultural technology was given to many Third World nations. Although there were many successful examples, Borlaug was awarded the Nobel Prize for his work on a high-yielding wheat plant. Work on this wheat had begun in the mid-1940s in Mexico as Borlaug and others developed broadly adapted short-stemmed disease-resistant wheats that excelled at converting fertilizer and water into high yields.

The winter wheat, as it was called, could be grown much more easily in areas such as Mexico and India. The impact of this plant was monumental. While in 1944 Mexico imported half its wheat, by 1956 it was self-sufficient in wheat production and by 1964 was growing enough of a surplus that it exported approximately half a million tons of wheat. In India, wheat production increased fourfold in 20 years (from 12 million tons in 1966 to 47 million tons in 1986). This success inspired rice experiments in the Philippines and elsewhere. By the 1960s, high-yielding (so-called miracle seed) varieties of wheat, corn, and rice were growing in many non–Soviet bloc countries in Asia, Latin America, and North Africa. As a part of the expanding programs, the U.S. Agency for International Development helped subsidize rural infrastructure development and agrochemical shipments. By 1992 the system included 18 centers, mostly in developing countries, staffed by scientists from around the world and supported by a consortium of foundations, national governments, and international agencies.

Supporters of the Green Revolution believe that it has helped to avoid widespread famine by feeding billions of people. Most of these supporters accept the Malthusian principle of population, believing that in many countries, population would outgrow food production, causing vast famine and malnutrition. The increased yield crops within developing countries have provided for the growing population. Supporters also note that many of the new varieties of crops are fortified with vitamins and minerals that improve the health of the people who consume them, such as golden rice grown in areas where there is a shortage of dietary vitamin A. The rice was developed as a humanitarian tool to help combat the irreversible health conditions related to a vitamin A deficiency. Finally, supporters suggest that by improving nutrition, one result would be a healthier workforce who could then work to improve the industrial and economic growth of their country.

Some Criticisms of Globalization and the Green Revolution

Critics dispute a number of the supporters' claims and suggest that although the numbers show that grain production steadily increased from the 1960s to the 1990s due to the Green Revolution and its programs, this does not translate to an increase in overall food production. This claim is backed with the notion that Green Revolution agriculture follows the idea of monoculture, while traditional agriculture often incorporates other editable plant species. Thus, despite the fact that more grains are being produced with the techniques from the Green Revolution, there is a simultaneous reduction in other food varieties. Furthermore, some traditional agriculture practices displaced by the Green Revolution are highly productive and would likely compete with the Green Revolution production. For example, systems such as *chinampas,* growing crops on the fertile arable land in shallow lake beds, have been replaced by the agricultural practices taught during the Green Revolution. Critics suggest that through these changes, the Green Revolution may have decreased food security for some of the poorest people—depleting the traditional foods of many peasants though monoculture and the use of pesticides that eliminate other plants and animals within the ecosystem.

Critics also point out the dependence of modern agriculture on petroleum products. As petroleum supplies shrink, programs from the Green Revolution may well become prohibitively expensive for many of the developing nations that they were intended to help. Furthermore, the pesticides used as a part of the Green Revolution were necessary to deal with the large amount of pest damage that inevitably occurs in monocultures. These chemicals managed pests; however, they do not easily break down in the environment, and thus many environmentalists believe that they accumulate in the food chain and spread throughout ecosystems. These outcomes can lead to water contamination and the evolution of resistance in pest organism populations.

Another area of criticism is that the Green Revolution plant varieties were developed as hybrid seeds. The developers of these seeds hold the intellectual property rights for the seeds and thus require the farmers to purchase new seeds each season rather than saving seeds from the last harvest. Although this process pays the companies for their research and development, it also increases the farmers' production costs. Critics suggest that this process goes against the original goals of the Green Revolution, finding a solution to hunger, poverty, and underdevelopment. Additionally, studies of the impact of the Green Revolution show that incomes for the larger farms in rural society have increased with the introduction of technology, while incomes for the smaller farms and the poorest strata have tended to fall. Critics suggest that this reveals that the purchasing of the seeds and necessary agrochemicals is financially difficult for smaller farmers, often pushing them into debt, while the larger farms are benefiting.

Finally, there are also accusations that the Green Revolution is a political program more than a humanitarian program. For example, the name itself—"Green Revolution"—is said to be a contrast to the Soviet Red

Revolution and the Iranian White Revolution. Additionally, U.S. journalist Mark Dowie, a major critic of the Green Revolution, suggests that the primary objective behind the programs were increasing social stability in non–Soviet bloc developing countries, thus creating beneficial relationships between these countries and the United States. These connections would not only create positive relationships for the United States but would also function as an alternative to socialist policies of expanding agrarian reform initiatives.

The Green Revolution has made changes to countries within the developing world. Due to both the praise and the criticism, institutions involved in Green Revolution programs now attempt to take a more holistic view of agriculture. The scientists are working to better understand the problems faced by farmers and are now involving the farmers in the development process. Likewise, international organizations involved in the Green Revolution, such as the United Nations (UN) Food and Agriculture Organization, are working to formulate a more equitable and sustainable Green Revolution, aiming in particular at improving the standard of living for the people involved in producing, providing, and managing food supplies within the poorest rural households—women farmers. To date, overall results have been both positive and negative, guaranteeing continued controversy surrounding the Green Revolution.

Clearly, though, the spread of agricultural technology has greatly assisted less developed nations in feeding their populations. The exchange of ideas and technology also initiated an opening that came to define the American worldview once the Cold War bipolarism had passed.

Garrett Hardin and the "Tragedy of the Commons"

After 1960 Americans' worldview changed considerably, whether influenced by photos from the moon or scientific concepts. Our lives, it became clear, were tied to many other creatures and systems. Therefore, our choices and actions had broad impacts.

Following Rachel Carson, in 1968 Garrett Hardin wrote an article titled "Tragedy of the Commons" that developed the ecological idea of the commons. This concept and his argument of its tragic (undeniable) outcome in depletion gave humans new rationale with which to view common resources such as the air and the ocean:

> The tragedy of the commons develops in this way. Picture a pasture open to all. It is to be expected that each herdsman will try to keep as

many cattle as possible on the commons. Such an arrangement may work reasonably satisfactorily for centuries because tribal wars, poaching, and disease keep the numbers of both man and beast well below the carrying capacity of the land. Finally, however, comes the day of reckoning, that is, the day when the long-desired goal of social stability becomes a reality. At this point, the inherent logic of the commons remorselessly generates tragedy.

As a rational being, each herdsman seeks to maximize his gain. Explicitly or implicitly, more or less consciously, he asks, "What is the utility to me of adding one more animal to my herd?" This utility has one negative and one positive component. . . .

Adding together the components . . . the rational herdsman concludes that the only sensible course for him to pursue is to add another animal to his herd. And another. . . . But this is the conclusion reached by each and every rational herdsman sharing a commons. Therein is the tragedy. Each man is locked into a system that compels him to increase his herd without limit—in a world that is limited. Ruin is the destination toward which all men rush, each pursuing his own best interest in a society that believes in the freedom of the commons. Freedom in a commons brings ruin to all.

Fostering Mechanisms for Cooperation

While a global perspective seemed inherent in the web of life put forward by Rachel Carson and others, it would take global issues such as the Chernobyl nuclear accident in 1986 and shared problems such as greenhouse gases and global warming to bind the world into a common perspective. Organizations, including Greenpeace and the UN, assisted members from many nations to shape a common stand on issues.

The UN presented the leading tool for facilitating global environmental efforts. With its first meeting on the environment in 1972, the global organization created the United Nations Environment Programme (UNEP). This organization would sponsor the historic Rio Conference on the Environment in 1992 and the conference on global warming in 1997. UNEP would also be the primary institution moving the UN's environmental agenda into the areas that were most in need of assistance, primarily developing nations. In response to such activities, the U.S. government declared the environment a genuine diplomatic risk in global affairs by creating a State Department undersecretary for the environment in 1996.

What began as an intellectual philosophy in the early 1800s had so impacted the human worldview that it would now influence global relations. Of course, a primary portion of this environmental worldview was scientific understandings that were communicated to the public after the 1960s. During the 1990s, a global agenda for action took shape that was organized around environmental improvement. Such an agenda, though, was not without its difficulties.

The most difficult portion of this global debate may be the fairness of each nation to be allowed to develop economically. Many less developed nations resent the environmental efforts of nations that have already industrialized, including many European nations and the United States. Such less developed nations believe they are being denied their own opportunity to develop economically simply because of problems created by industrial nations. This sentiment has been part of major demonstrations in recent meetings of global organizations, such as the World Trade Organization (WTO). Any global agreements will need to balance the basic differences of these constituencies.

Forming a Discourse of International Cooperation

During the 1980s and 1990s, the international discourse on environmental issues took a more organized and systematic form. On the 20th anniversary of the first UNEP meeting in Stockholm, the UN hosted the 1992 United Nations Conference on Environment and Development (UNCED), known as the Earth Summit and held in Rio de Janeiro, in which world leaders agreed on Agenda 21 and the Rio Declaration.

The summit brought environmental and developmental issues firmly into the public arena. Along with the Rio Declaration and Agenda 21, the summit led to agreement on two legally binding conventions: Biological Diversity and the Framework Convention on Climate Change (FCCC).

Agenda 21, in particular, functioned to place an important new idea into mainstream international environmentalism: sustainable development. Agenda 21 was a 300-page plan for achieving sustainable development in the 21st century. The United Nations Commission on Sustainable Development (CSD) was created in December 1992 to ensure effective follow-up of UNCED and to monitor and report on implementation of the UNCED agreements at the local, national, regional, and international levels. The CSD is a functional commission of the United Nations Economic and Social Council (ECOSOC), with 53 members. A five-year review of Earth Summit progress took place in 1997 by the UN General Assembly meeting in special session, followed in 2002 by a 10-year review by the World Summit on Sustainable Development.

The CSD was established as a functional commission of the Economic and Social Council by council decision 1993/207. Its functions are set out in General Assembly resolution 47/191 of December 22, 1992. The commission is composed of 53 members elected for terms of office of three years, meets annually for a period of two to three weeks, and receives substantive and technical services from the Department of Economic and Social Affairs, Division for Sustainable Development. The commission reports to the Economic and Social Council and through it to the Second Committee of the General Assembly.

The CSD now focuses on assessment and education. Most important, it must contain representatives from the nations that most need assistance with development strategies. In 2005, the membership of the CSD included 13 members elected from Africa, 11 from Asia, 10 from Latin America and the Caribbean, 6 from Eastern Europe, and 13 from Western Europe and other areas. Such a balanced membership is a priority of most UN efforts; however, with CSD it may be even more important because representation can ensure that nations receive the help they need most.

Global environmental initiatives remain just that. There remains no authority that can enforce environmental policies and regulations across political lines everywhere in the world. However, great strides have been made in creating regional cooperative initiatives.

NAFTA and Regional Agreements

Outside of the UN, the new international connections that were inherent in globalization altered relations between nations. For the United States, global free trade remained a priority, but regional development required new agreements with neighbors. The 1993 North American Free Trade Agreement (NAFTA) defined a vast new region for free trade. Organizations such as the WTO also helped to prioritize free trade over environmental protection.

Opposed to such agreements, public interest activist Ralph Nader argued that such deals would erode environmental and social legislation. The Bill Clinton administration negotiated at least two environmental side agreements. One of these created the North American Commission for Environmental Cooperation, which emphasized cross-border initiatives. The second agreement specifically focused on the Mexican border.

The success of efforts to limit the environmental impact of these types of initiatives and agreements has been limited by such factors as difficulty in regulating users and accessing documents that would allow investigators to bring suit. Most observers believe that the environmental problems along the U.S.-Mexican border have intensified under NAFTA. Green nongovernmental organizations have joined with American labor organizations to call for the United States to renegotiate or pull out altogether from the 1993 agreement.

Protesting the WTO and Globalization

A policy infrastructure for globalization had taken shape by the mid-1990s. The nature of the American worldview expanded to consider the interests of other nations as well as other ecological entities. However, just as many proponents began to think that they had created the agencies and initiatives to close the gap separating developed and less developed nations, activists—many of them associated with environmental causes—altered their view of initiatives such as the WTO and the World Bank.

Whereas instruments of change such as the World Bank, the WTO, and even the UN had been created to promote global peace and stability, critics began to argue that such organizations exerted the will of powerful nations on those of the less powerful and less developed nations. Many American environmentalists argued that not only were the agencies' actions heavy-handed, but they also fueled less developed nations to follow the less sustainable paths to progress used and favored by Western nations.

By the end of the 1990s, every meeting of the WTO became a gathering for activists denouncing its activities. In only a few short years, the attitude toward international assistance had undergone a radical shift. Although there was a clear history of cooperation on some global environmental issues, consensus seemed to shatter at the start of the 21st century.

Conclusion: Global Issues of Climate Change

In the late 20th century, the environmental issue referred to as global warming seemed to be the next great issue in the global unification of efforts to diminish the impact of humans on Earth. However, by the beginning of the 21st century, it was obvious that instead of bringing the world together, global warming was going to cause even more division. Unlike the issues behind the Green Revolution, the science behind global warming was difficult to verify. And, most important, acting on global warming could potentially hurt the U.S. economic dominance of the globe.

Brian C. Black and Donna L. Lybecker

Further Reading

Conway, G. *The Doubly Green Revolution.* Ithaca NY: Cornell University Press, 1988.

Dowie, M. *American Foundations: An Investigative History.* Cambridge, MA: MIT Press, 2001.

Food and Agriculture Organization of the United Nations. "Women and the Green Revolution." 2007, http://www.fao.org/FOCUS/E/Women/green-e.htm.

Hughes, Thomas. *American Genesis.* New York: Penguin, 1989.

McNeil, John R. *Something New under the Sun: An Environmental History of the Twentieth-Century World.* New York: Norton, 2001.

Perkins, J. H. *Geopolitics and the Green Revolution: Wheat, Genes, and the Cold War.* New York: Oxford University Press, 1997.

Pollan, Michael. *Second Nature.* New York: Delta, 1992.

Shiva, V. *The Violence of the Green Revolution: Ecological Degradation and Political Conflict in Punjab.* New Delhi, India: Zed, 1992.

UN Department of Economic and Social Affairs. "Mandate of the Commission on Sustainable Development." United Nations, 2009, http://www.un.org/esa/dsd/csd/csd_mandate.shtml.

Wright, A. *The Death of Ramon Gonzalez: The Modern Agricultural Dilemma.* Austin: University of Texas Press, 1992.

Greenwashing

While many corporations have attempted to implement environmentally friendly measures, others sought to appeal to green culture while doing business as usual. The term "greenwashing" is used for corporations who try to present an image of being environmentally friendly without necessarily making any changes in their actual business practices. In such instances, a company, an industry, a government, a politician, or even a nongovernmental organization creates a proenvironmental image in order to sell a product or a policy or try to rehabilitate its standing with the public and decision makers after being embroiled in controversy.

The use of greenwashing demonstrates how prevalent environmental perspectives have become. However, it also leads to suspicion and doubt about the true interests or capabilities of companies to alter their ethics.

Businesses Go Green, Or So It Seems

Joshua Karliner of CorpWatch traces this trend back to the late 1960s and the rise of environmentalism. Initially, a few ad executives, including Jerry Mander, who observed the trend referred to it as "ecopornography." According to Karliner,

> It seemed that everyone was jumping on the bandwagon. It was a time when the anti-nuclear movement was coming into its own. In response, notes Mander, the nuclear power division of Westinghouse ran four-color advertisements "everywhere, extolling the anti-polluting virtues of atomic power" as "'reliable, low-cost . . . neat, clean, safe.'" . . . Meanwhile, in the year 1969 alone, public utilities spent more than $300 million on advertising—more than eight times what they spent on the anti-pollution research they were touting in their ads. Overall, Mander estimated that oil, chemical, and automobile corporations, along with industrial associations and utilities, were spending nearly $1 billion a year on "ecopornography" and in the process were "destroying the word 'ecology' and perhaps all understanding of the concept."

The prevalence of greenwashing in companies peaked each year on Earth Day as many companies felt responsible for noting the holiday in some fashion. By the 21st century, though, efforts to placate environmentally minded consumers occurred year-round. Karliner is careful to point out that greenwashing occurs hand in glove with globalization. Often, green-speak seemed to be the primary transborder language. For instance, one of the most notable efforts in this arena has been that of the giant oil company BP. Formally known as British Petroleum, in 2000 BP Amoco officially made its initials stand for "Beyond Petroleum."

Wise Use and Greenwashing Politics

One of the long-term implications of the rebellion was a national movement organized around the name and idea of wise use, which was borrowed from conservationists such as Gifford Pinchot. This is a well-financed right-wing movement that blossomed in the 1980s, and most observers agree that the use of the conservation terminology derived more from greenwashing than from any genuine ethic. Wise users sought to use public frustration with government interference to attract the support of middle-class voters.

Wise use became one of the first organized responses to environmentalism. Prioritizing use, particularly of federally owned resources, wise users fought against federal regulation and environmental limitations. Some support came from property rights advocates. Other supporters of wise use argued that environmentalists were antihumans. Wise use continues to have an active influence on efforts to mitigate or abolish environmental regulations.

With environmentalism increasingly a matter of individual definitions, many Americans move into the 21st century unclear about what it means to be an environmentalist. In the presidential election of 2004 when a questioner asked President George W. Bush to assess his view on the environment, the president responded simply that "I'd say that I am a good steward of the land." Many environmental organizations would never say this about the Bush administration. In short, though, history will most certainly show that Bush policies introduced an entirely new phase in the American use of policy to administer and enforce adherence of environmental legislation. Part of this new phase in legislation is a lack of clarity in presenting this agenda to the American people. Critics also claim that greenwashing has infected American politics.

During the first decade of the 21st century, the Bush administration stressed an effort to achieve quantifiable results when discussing its no-nonsense approach to overseeing America's natural environment. A brief list of the accomplishments claimed by the administration include the Clear Skies Initiative, cuts in mercury emissions, a growth-oriented approach to climate change, tax incentives for renewable energy and hybrid and fuel-cell vehicles, and the development of domestic energy sources, including hydrogen. In each of the cases, however, initiatives and even the names of the policies demonstrate little about the ethical intent of the policy and are more an example of greenwashing. Were Americans being misled by the administration? Many critics say so.

One of the Bush administration's most vocal critics was attorney and business consultant Robert F. Kennedy Jr. In his book *Crimes against Nature,* Kennedy writes that "You simply can't talk honestly about the environment today without criticizing this president. George W. Bush will go down as the worst environmental president in our nation's history." Clearly, though, Americans continue to be able to agree that nature played a tremendously important role in the development of the nation during the 20th century.

Senator Patrick Leahy, a Democrat from Vermont, made the following statement on the floor of the Senate on April 26, 2004:

> Just recently the *New York Times* reported on the creative White House fact-spinning of the Administration's proposed retreat from strong mercury controls on power plants.

> Of course, we all recognize that [the Bush administration's] favorite tactic is just giving one of their environmental assaults a green name and hoping the American public believes it. "Clear Skies" and "Healthy Forests" are just about as accurate as "No Child Left Behind."
>
> The Administration has used all of these tactics when it comes to misleading the public about wetlands protections. Last January, on a Friday, the Administration announced one of its most sweeping rollbacks to take away protections under the Clean Water Act for 20 million acres of wetlands.
>
> The policy created such a groundswell of opposition from hunters, anglers, environmental groups and others that the President finally withdrew the proposed rulemaking last December. Unfortunately, what the Administration did not tell the public is that they were not revoking the underlying instructions to federal agencies to follow the same policy that leaves 20 million acres of wetlands at risk.
>
> That is why I found it so interesting that the President would start his election year attempts to greenwash his Administration's anti-environmental record by talking about wetlands. . . .
>
> While the President was touting his plan to restore 1 million acres of wetlands, he made no mention of his policy to revoke protection for 20 million acres. . . .
>
> The Administration's retreat from aggressive mercury controls on power plants is just the most recent missile in his all out environmental assault.
>
> Again, the President did get some nice photo ops in Maine and Florida, but his record on the environment is too mired in reversals and rollbacks for any greenwash to last for long.
>
> Greenwash—like whitewash—doesn't stick.
>
> Despite all their public relations maneuvering, the public recognizes the enormous and long-term affect of these Bush policies on our environment and our health.
>
> They will mean more pollution in our rivers and streams, more toxics in our air and less natural resources to pass on to the next generation.

Conclusion: Separating True Environmental Ethics

In 1999, "greenwash" was added to the *Oxford English Dictionary,* where it is defined as "Disinformation disseminated by an organization so as to present an environmentally responsible public image." Although many observers have become skilled at discerning authentic environmental statements from greenwashing, the mass of American consumers are assumed to accept greenwashing as reality.

At times, environmental organizations seem to complicate matters. For instance, to combat what it viewed as misleading corporate practices, Greenpeace USA began publishing the *Book of Greenwash* during the 1990s. Some critics even argue that community recycling programs also should be grouped under the heading of greenwash because they prevent calls for reducing consumption and economic growth.

Therefore, the admission by nonenvironmental political and corporate actors that greenness appeals to the general public has contributed to a confusing era in environmental politics and marketing. By obscuring the connection between corporations or policies and environmental degradation, the public often is misled about the primary agenda of such practices. The Internet has become an important resource for savvy consumers who have sought out outlets to clarify green initiatives by Wal-Mart and others, particularly in relation to the use of the term "organic."

Overall, though, these variations in thought demonstrate how the nature of environmentalism had changed dramatically by the end of the 20th century. Regardless of one's perspective, environmentalism had clearly gained a new connection to everyday patterns of living instead of being restricted to the annual dues sent to conservation organizations. Although false or misleading advertising of any type is unfair, greenwashing represents an awareness by companies, politicians, and others that environmental concerns must be addressed or at least recognized.

Brian C. Black

Further Reading

Andrews, R. N. L. *Managing the Environment, Managing Ourselves.* New Haven, CT: Yale University Press, 1999.

Cronon, William, ed. *Uncommon Ground: Rethinking the Human Place in Nature.* New York: Norton, 1996.

Gottleib, R. *Forcing the Spring: The Transformation of the American Environmental Movement.* Washington, DC: Island Press, 1993.

Karliner, Joshua. "A Brief History of Greenwash." CorpWatch, March 22, 2001, http://www.corpwatch.org/article.php?id=243.

Kennedy, Robert F., Jr. *Crimes against Nature: How George W. Bush and His Corporate Pals Are Plundering the Country and Hijacking Our Democracy.* New York: HarperCollins, 2004.

Kirk, Andrew. *Counterculture Green.* Lawrence: University Press of Kansas, 2007.

Nabhan, G. P. *Coming Home to Eat: The Pleasures and Politics of Local Foods.* New York: Norton, 2002.

Nash, Roderick. *Wilderness and the American Mind.* New Haven, CT: Yale University Press, 1982.

Opie, J. *Nature's Nation.* New York: Harcourt Brace, 1998.

Price, J. *Flight Maps.* New York: Basic Books, 2000.

Rothman, H. K. *The Greening of a Nation.* New York: Harcourt, 1998.

Rothman, H. K. *Saving the Planet: The American Response to the Environment in the 20th Century.* Chicago: Ivan R. Dee, 2000.